URBAN MOBILITY

How the iPhone, COVID, and Climate Changed Everything

Edited by Shauna Brail and Betsy Donald

Urban Mobility sheds light on mobility in twenty-first-century Canadian cities. The book explores the profound changes associated with technological innovation, pandemic-induced impacts on travel behaviour, and the urgent need for mobility to respond meaningfully to the climate crisis.

Featuring contributions from leading Canadian and American scholars and researchers, this edited collection traverses disciplines including geography, engineering, management, policy studies, political science, and urban planning. Chapters illuminate novel research findings related to a variety of modes of mobility, including public transit, e-scooters, bike-sharing, ride-hailing, and autonomous vehicles. Contributors draw out the connections between urban challenges, technological change, societal need, and governance mechanisms. The collection demonstrates why the smart phone, COVID-19, and climate present a crucial lens through which we can understand the present and future of urban mobility. The way we move in cities has been disrupted and altered because of technological innovation, the lingering impacts of COVID-19, and efforts to reduce transport-related emissions.

Urban Mobility concludes that the path forward requires good public policy from all levels of government, working in partnership with the private sector and non-profits to direct and address the best urban mobility framework for Canadian cities.

SHAUNA BRAIL is an associate professor at the Institute for Management and Innovation, cross-appointed to the Munk School of Global Affairs and Public Policy at the University of Toronto.

BETSY DONALD is a professor in the Department of Geography and Planning at Queen's University.

Urban Mobility: How the iPhone, COVID, and Climate Changed Everything

EDITED BY SHAUNA BRAIL AND BETSY DONALD

UNIVERSITY OF TORONTO PRESS
Toronto Buffalo London

Toronto Buffalo London
utorontopress.com

ISBN 978-1-4875-5059-2 (cloth)
ISBN 978-1-4875-5185-8 (paper)
ISBN 978-1-4875-5408-8 (EPUB)
ISBN 978-1-4875-5367-8 (PDF)

Library and Archives Canada Cataloguing in Publication

Title: Urban mobility : how the iPhone, COVID, and climate changed everything / edited by Shauna Brail and Betsy Donald.
Other titles: Urban mobility (Toronto, Ont.)
Names: Brail, Shauna, editor. | Donald, Betsy, editor.
Description: Includes bibliographical references and index.
Identifiers: Canadiana (print) 20240366883 | Canadiana (ebook) 20240366921 | ISBN 9781487550592 (cloth) | ISBN 9781487551858 (paper) | ISBN 9781487554088 (EPUB) | ISBN 9781487553678 (PDF)
Subjects: LCSH: Urban transportation – Technological innovations – Canada. | LCSH: Local transit – Technological innovations – Canada. | LCSH: City and town life – Canada. | LCSH: COVID-19 Pandemic, 2020– – Canada. | LCSH: Climatic changes – Canada. | LCSH: iPhone (Smartphone)
Classification: LCC HE311.C3 U73 2024 | DDC 388.40971–dc23

Cover design: Liz Harasymczuk
Cover image: Mooi Design/Shutterstock.com

We wish to acknowledge the land on which the University of Toronto Press operates. This land is the traditional territory of the Wendat, the Anishnaabeg, the Haudenosaunee, the Métis, and the Mississaugas of the Credit First Nation.

University of Toronto Press acknowledges the financial support of the Government of Canada, the Canada Council for the Arts, and the Ontario Arts Council, an agency of the Government of Ontario, for its publishing activities.

Canada Council for the Arts
Conseil des Arts du Canada

Funded by the Government of Canada
Financé par le gouvernement du Canada
Canada

Contents

Section III: Places, Patterns, and Challenges

Section IV: Governing Mobility

Section V: Moving Mobility Studies Forward

Figures

Tables

Acknowledgments

We have many people to thank. This manuscript was compiled with the excellent support of Susan Bazely (Queen's University), Dr. Emilia Bruck (University of Toronto), Marisa Klassen (University of Toronto), and Dr. Gaurav Mittal (University of Toronto). Becky Vogan provided thoughtful and clear editing guidance. Many thanks also to University of Toronto Press and especially our editor, Jodi Lewchuk, for her encouraging and sage advice. Jodi shepherded us expertly through the publication process. We are also grateful to Professor Anita McGahan for providing additional publication funding from Social Science and Humanities Research Council of Canada Insight Grant 435-2021-0064.

Much of the research highlighted in this book was supported by Social Sciences and Humanities Research Council of Canada Insight Grant 435-2019-0022, which funded a study entitled "Taking Canada for a Ride? Digital Ride-hailing and Its Impact on Canadian Cities," for which we were co-Principal Investigators. This grant allowed us to come together as a team, support emerging research talent, and share our research and ideas. This research collaboration formed the basis for this book and will continue to inform future work on mobility in Canadian cities.

We first met in 1995 as PhD students in the University of Toronto's Department of Geography & Planning. Over the course of nearly three decades, we continue to maintain a close and supportive friendship as we share the joys and challenges of academic life. The foundations for this book were laid not only in the writing and editing of these chapters, but also over almost thirty years of conferences, collaboration, and cheering for one another.

Shauna Brail and Betsy Donald

Dedicated to our families, who have been a great source of support, strength, and fun. Thank you.

Ayal, Dalia, Jacob, Naomi, and Leo
Shauna Brail

Doug, Lucy, Lydia, and Cedar
Betsy Donald

SECTION I: SETTING THE STAGE

1 Urban Mobility: How the iPhone, COVID, and Climate Changed Everything

SHAUNA BRAIL AND BETSY DONALD

Introduction

This book is about mobility in twenty-first-century Canadian cities. Mobility plays a crucial role in urbanization and urban life with respect to economic, cultural, environmental, and social progress. *Urban Mobility Changes: How the iPhone, COVID, and Climate Changed Everything* explores the extent to which the trilogy of technology, infectious disease, and climate change are transforming how we move, live, and work in cities.

The book focuses on the Canadian urban context, where more than 82 per cent of the country's 40 million residents live in an urban area and nearly two-thirds of the population lives in one of fifteen metropolitan areas (Statistics Canada, 2022a). As Canada's cities and metropolitan areas continue to grow and expand, it is increasingly important to understand how mobility affects the country's urban dwellers. Specifically, we need to know how mobility affords opportunities to work, live, and play; how it supports local, regional, and national equity and sustainability goals; and how mobility enables innovation, progress, and prosperity.

Although abundant evidence demonstrates that COVID-19 significantly disrupted mobility and urban life, this period represents just one example of a mobility disruption since the start of the twenty-first century. Well before the COVID-19 pandemic, technological change began to disrupt long-standing transportation norms, particularly related to technological advances in the prospects for automated, shared, and electric vehicles (Sperling, 2018). Furthermore, attention to the looming devastation associated with the prospects of climate change and transportation emissions presents a compelling case for refocusing urban mobility on shorter, less polluting trips alongside denser urban form (Dodman et al., 2022). Yet, progress has been slow and uneven.

Before the early 2000s, very little changed for decades in the way people moved around cities (Ranchordas, 2015). Since the widespread manufacture and adoption of automobiles, especially following the Second World War, mobility in Canada has mainly taken place via private automobile. But as Canada's cities mature and grow, as our economy builds strength in knowledge-based industries, and as societal priorities turn towards sustainability and equity, mobility is also changing (see Hutton, chapter 3). Undoubtedly, mobility transitions in Canadian cities are tied to a broad range of shifts. This includes the country's urban economic strengths in advanced manufacturing, technology-based industries, and a vibrant public sector, which ought to serve the country's efforts to continue to accommodate population growth alongside improvements in the quality of urban life. Concurrently, attention has necessarily begun to focus on the interrelationship between mobility, economic development, climate change, and social equity. Furthermore, if we hope to understand and improve upon mobility connections, shifts, and opportunities, our need for high-quality, accessible, and timely data becomes more urgent (see Vinodrai, chapter 4).

Since the early 2000s, mobility innovation has been both dramatic and disruptive. With the widespread adoption of smartphones, the increased capacity in cloud computing, and the provision of GPS, digital platform economy firms gained the ability to massively disrupt many industries (Kenney & Zysman, 2016), including ground transportation services such as taxis and public transit. The seemingly sudden emergence of ride-hailing across Canadian cities beginning in 2012 pushed almost every major city in the country to address governance challenges and develop regulations separate from those governing the taxi sector (Brail, 2018; Zwick, Young, and Spicer, chapter 13). While the roll-out of ride-hailing took cities by surprise, governments are paying more proactive attention to the presumed launch of autonomous vehicles (Council of Canadian Academies, 2021).

Although some mobility innovations emphasize the continued use of private automobiles, others focus on addressing sustainability, emphasizing micromobility devices including e-bikes and e-scooters (Rojas and Miller, chapter 5; Kong and Leszczynski, chapter 7; McNee and Miller, chapter 8). As we acknowledge the need to reduce greenhouse gas emissions through transportation activities, and we recognize that many trips in cities are less than five kilometres, shared micromobility becomes a promising option.

In March 2020, cities across Canada experienced a mobility disruption unlike any other. The declaration of the COVID-19 pandemic and subsequent measures to prevent disease spread and to maintain health care systems led to a sudden and protracted change. While all modes of mobility declined initially, it became apparent as the pandemic lingered that the switch to hybrid work would have a long-lasting impact (Statistics Canada, 2022b). In 2021, 2.8 million fewer people were commuting to work than in 2016, a direct result of the increase in working from home. This trend, which accelerated at the start of the pandemic emergency, is slowly recovering (Statistics Canada, 2023). However, a higher percentage of people continue to work from home in 2023 (20.1%) than 2016 (7.1%). Related to these shifts in work patterns, public transit use by commuters suffered dramatic declines in the lockdown period of the pandemic, reversing the gains in ridership experienced since transit commuting was first measured in the 1996 census (Statistics Canada, 2022b). In May 2023, 10.1 per cent of commuters in Canada mainly travelled to work via public transit, up from 8.5 per cent in May 2022 and 7.8 per cent in May 2021, but below pre-pandemic levels recorded in May 2016 at 12.6 per cent (Statistics Canada, 2023). Active transportation such as walking and cycling as a means of commuting to work also declined by more than 25 per cent between 2016 and 2021 (Statistics Canada, 2022b). Behavioural changes precipitated by pandemic restrictions, including an enduring increase in remote work and the concomitant reductions in workplace activity, have altered long-standing mobility patterns and may have implications for urban infrastructure investment decisions as well as the nature of urban form (Brail, 2022). It remains too soon to know how the pandemic will impact mobility over the long term (Cavalli and McGahan, chapter 2), however, it is unlikely that pre-pandemic commuting patterns will fully return. This is especially the case given rapid technological changes in mobility options and an increasingly urgent climate emergency that continues to command attention and necessitate transformation. At the core of this book lies the hypothesis that the iPhone-climate-pandemic narrative obliges us to redirect mobility in twenty-first-century Canadian cities.

The shock of COVID-19, in combination with disruptive innovations and the urgent climate crisis, has helped people to understand how mobility contributes to vibrant, equitable, sustainable, and prosperous cities. Furthermore, technological change, the climate crisis, and COVID-19 highlight challenges and opportunities for addressing equity concerns, such as who has access to adequate mobility (Palm and Farber, chapter 6), as well as labour concerns, including how platform business models rely on precarious work (Hashemi, Motaghi, and Tremblay, chapter 9; Brail and Donald, chapter 10). Deeper examination of mobility opportunities and challenges enables decision-makers to make informed choices amongst a range of options, privileging societal priorities such as those related to improving equity and reducing greenhouse gas emissions, for instance. Several contributors also emphasize the potential for policy and governance mechanisms to direct change, supporting the public good (Gorachinova, Huh, and Wolfe, chapter 11; Aoyama and Alvarez León, chapter 12; Lorinc, chapter 14). Together, these analyses draw connections amongst diverse priorities related to advancing economic opportunity, improving social equity, and addressing climate adaptation – in part through an approach that values policy innovation.

As we demonstrate throughout this book, mobility matters well beyond helping us to understand movement. Mobility is a lens through which we can examine and interpret larger societal questions and debates. *Urban Mobility: How the iPhone, COVID, and Climate Changed Everything* illuminates the role of mobility in cities, shaping access to all facets of urban life. Collectively, the book suggests that mobility underscores the future of Canadian cities. With regard to considerations that influence and contribute to innovation, disruption, and climate adaptation, we have a window of opportunity to steer Canadian cities towards mobility choices that will strengthen economic opportunities, improve social equity, build resilience against future shocks, and address the pressing need to reduce greenhouse gas emissions.

Defining Mobility

How do we define mobility? Mobility is the ability for people to move from one place to another. The quality of mobility contributes to people's ability to access a range of opportunities for work, food, education, and leisure. As such, mobility is essential to urban life (Brail, 2019).

Furthermore, mobility is distinct from, yet connected to, two other terms associated with the movement of people: accessibility and transportation. Access is a measure of which places are reachable, while accessibility refers to the "ease of reaching valued destinations" as well as a way to measure which places are reachable (El-Geneidy et al., 2015, p. 175). Transportation is the act of moving people or goods by a specific mode (e.g., bicycle, car, train). Transportation systems are the sum of transport modes (e.g., wheelchair, walking, cycling, transit, private automobile) supported by physical infrastructure (e.g., sidewalks, bike lanes, streets, and subways) in a given area (El-Geneidy et al., 2015). Transportation systems do not exist in isolation; they are both integrated with and integral to land use systems and plans (Higgins et al., 2021). It is also clear that mobility relies on public infrastructure such as streets and sidewalks. However, twenty-first-century mobilities raise new questions and debates because public and private mobility operations all rely on public infrastructure.

Multidisciplinary Perspectives

Traditionally, transportation experts in fields such as engineering and geography have dominated mobility scholarship. Until recently, there has been limited interaction between these mobility scholars, who are based in science, technology, engineering, management fields, and other social scientists. Yet understanding mobility in the twenty-first century demands multidisciplinary approaches that consider many dimensions, including distribution, culture, economics, labour, place, policy, systems, and technology. While the more abstract field of mobility studies examines the movement of bodies, ideas, and objects across space (Urry, 2007), the study of mobility in the context of urban life is less examined.

Urban Mobility brings together researchers, theories, and ideas from a range of disciplines: engineering, geography, management, policy studies, political science, and urban planning. While each contributor investigates challenges and opportunities at the intersection of cities and emerging mobility technologies, our research and contributions collectively offer greater impact.

Twin Transitions: Digital and Green

As societies transform globally, cities have an opportunity to address the twin transitions associated with work related to digital possibilities and sustainability efforts (European Commission et al., 2022; Santoalha & Boschma, 2021; Wolfe, 2019).

Digital Transitions

For urban dwellers, technological adaptations – such as digital networks that enable shared mobility services as well as innovations in autonomous and electric vehicles – offer the possibility of improved quality of life and enhanced economic opportunity. Connections between economic development and mobility continue to matter, while the relationship between urban form, urban density, and new mobility modes requires investigation.

In *Three Revolutions*, Sperling (2018) details the impending technological shifts that will likely impact transportation into the foreseeable future. Highlighting the transition to electric, shared, and autonomous vehicles, *Three Revolutions* suggests that policy, regulation, and government oversight will play a significant role in determining whether new technologies advance goals related to sustainability, equity, and efficiency.

Green Transitions

The twenty-first century has also seen that transitions focused on sustainability are influencing mobility. Efforts to promote more sustainable forms of transportation, including walking and cycling, now dominate debates about reducing greenhouse gas emissions and congestion.

Transportation is one of the leading contributors of greenhouse gas emissions and air pollution worldwide (Brail, 2019). Policymakers can intentionally direct change by applying digital technologies, in combination with other innovations, plans, and measures. Developing expertise in sustainable mobility, through investment in new "green" technologies, presents an economic opportunity, one with the potential to position regions and firms as leaders in what is estimated to be a trillion-dollar market (Docherty et al., 2018).

COVID-19 and Recovery

In March 2020, as COVID-19 was declared a pandemic across Canada, city after city witnessed dramatic shifts in mobility patterns, practically overnight. Cars disappeared from city streets as people stayed home if they could. Transit ridership declined, costing municipalities an estimated $400 million a month after March 2020 (Sevunts, 2020). Cycling and walking increased in popularity as people looked for new ways to physically distance while remaining social and active.

As contributors detail throughout this book, the impact of COVID-19 on mobility has been one of the starkest indicators of urban change during the pandemic. From a mobility perspective, recovery from COVID-19 is proving to be both complex and prolonged.

The widespread use of digital technology has facilitated many new mobility options such as on-demand ride-hailing and the convenient delivery of goods and services, supported by digital platforms. While these services helped residents reduce their exposure during COVID-19, the impact has created new challenges for cities, mobility infrastructures, workers, and small businesses. These challenges and opportunities will only continue to accelerate with technology-driven change and system shocks, such as the ongoing struggles associated with the pandemic, climate crisis, and municipal financial exigencies.

Mobility and the Twenty-First-Century City

This book argues that technological changes since the invention and widespread adoption of smartphones – beginning with the release of

Apple's iPhone in 2007 – have facilitated new mobility options that are both disruptive and opportunity-generating. These shifts are especially visible in cities, as places characterized by dense concentrations of people and activity. New mobility options, such as on-demand ride-hailing and dockless e-scooter services, reshape the challenges associated with urban life, including congestion, intracity travel, conflicts regarding the use of streets, public transit service, and transport equity.

Subsequent chapters delve into questions such as these: How do digitization and technological change influence mobility in cities? What extended impacts, if any, will COVID-19 have on urban mobility? How can collaborative action mitigate the effects of mobility on climate over the long term? Who is responsible for ensuring that emerging mobility technologies bring value to cities and people?

Structure of the Book

Urban Mobility: How the iPhone, COVID, and Climate Changed Everything was conceived to convene scholars engaged in a Social Sciences and Humanities Research Council grant that we led as Co-Principal Investigators. During pandemic lockdowns, our research collaborators from across the country were unable to continue our planned in-person meetings. This gave us pause to consider how else we might bring our research together to draw and share insights from research focused on digital technologies, ride-hailing, and mobility. In addition, some authors were invited to elaborate on specific mobility advancements (e.g., MaaS); mobility policies (e.g., regulating autonomous vehicles); and key mobility themes missing from the original research group's canon (e.g., mobility equity). As co-editors, we believe that a unique strength of this collection lies in its Canadian-focused, multidisciplinary, urban studies approach to mobility.

The book is organized thematically into five sections. This introductory chapter appears under *Section I: Setting the Stage*. Following it, *Section II: The Big Picture – Framing the Questions* highlights shifts in thinking about mobility in the context of broader changes and challenges facing Canadian cities in the twenty-first century. *Section III: Places, Patterns, and Challenges* includes five case studies, revealing the need for cities to consider unintended consequences and societal priorities when they make decisions about implementing mobility technologies. *Section IV: Governing Mobility* highlights features of the digital transformation that characterize new mobility options and the challenges and opportunities that this poses for governance. Finally, *Section V: Moving Mobility Studies Forward* addresses the need to focus on mobility as a crucial component of cities. It calls for further research into mobility studies as a way of understanding the future of cities, especially their long-term health and sustainability.

Section II: The Big Picture – Framing the Questions

Section II includes four chapters that investigate the broader environment within which mobility changes are occurring. This section considers the implications of socio-economic and political shifts in Canada as well as ongoing questions about building urban resilience. Crucially, it examines how to improve data collection and access that can help to inform policy, decision-making, and future directions.

Gabriel Cavalli and Anita McGahan, in chapter 2, address the profound impacts of COVID-19 policies, restrictions, and behavioural shifts on urban Canadians. The chapter describes how cities have transformed since the onset of the pandemic, especially in the way people move around, with a focus on the concepts of mobility and proximity. Cavalli and McGahan's analysis addresses uncertainties, exacerbated by COVID-19, related to international, national, sectoral, and individual change. The chapter concludes by discussing the broad implications for all Canadian cities of reshaping, for example, city identities, priorities, and leadership demands.

In chapter 3, Tom Hutton tells the story of the evolving meta-level debates about urban structure and circulation in the Canadian city. Tracing the connections between transit planning and urban growth, Hutton emphasizes the relationship between mobility and urban and regional change. Furthermore, the chapter highlights the mix of factors that shape transportation-related innovation and priorities, including efforts to transition to sustainable transportation, changes in communications technologies, attempts to reduce the costs of travel, and shifting demographic preferences.

Tara Vinodrai writes in chapter 4 about the difficulties of capturing data that reflect on both the characteristics of – and the changes taking place in – urban mobility across Canada's cities. Vinodrai presents a nuanced discussion related to urban mobility research and policy interest, laying out, in detail, challenges connected to measuring urban mobility in Canadian cities. Using mobility-oriented examples, the chapter illuminates the problems that researchers face accessing timely, comparable, and high-quality data for Canadian cities. Vinodrai concludes with the recommendation that governments advance the capabilities of public data agencies to supply quality data that can help us better understand Canadian cities.

In chapter 5, Lisa Losada Rojas and Eric Miller survey the state of emerging mobility technologies and transportation systems in Canadian cities. Building on a taxonomy of mobility services developed by Calderón and Miller (2020, 2022), Rojas and Miller review the state of ride-hailing, pooled ride-hailing, car-sharing, bike-sharing, e-bike-sharing, e-scooters, and demand-responsive transit. The chapter considers the impacts of COVID-19 on mobility services and examines the potential for these services to support the transition to lower carbon emissions. Rojas and Miller also address questions related to externalities that arise because of increasing attention to emerging mobilities, data transparency (or lack thereof), and the role of consumer attraction in influencing transportation modes.

Section III: Places, Patterns, and Challenges

This section of the book includes five chapters that showcase, through case studies, how emerging mobility technologies both support and evade expectations and priorities.

Matt Palm and Steve Farber begin Section 3 with chapter 6, which concerns public transit equity in the Greater Toronto Area and the impacts of COVID-19. Using a survey tool meant to identify who used public transit during the pandemic and why, the chapter examines the role of public transit in a time of crisis. The authors find that those with low socio-economic status, including recent immigrants, older adults, and people with low incomes, were far less likely to be able to work from home during the pandemic, resulting in their continued use of public transit. Palm and Farber also argue that transit is far less accessible to disadvantaged groups of people who do not live in city cores. The chapter concludes with a call for investment in on-demand transit to help address equity concerns.

In chapter 7, through a study of a shared e-scooter pilot program in Calgary, Vivian Kong and Agnieszka Leszczynski evaluate the potential of dockless micromobility to address equity considerations. Noting that previous research demonstrates a lack of equity in the distribution of bike-sharing, Kong and Leszczynzki aim to discover whether usage patterns of e-scooters are equitable. In particular, they investigate whether shared e-scooters have the potential to be equally accessible in affluent communities as well as in socio-economically deprived neighbourhoods. Significantly, they find this suggestion aspirational. By deepening understanding of the way that e-scooter programs operate across the city, Kong and Leszczynski highlight the potential for stakeholders to remedy inequities while at the same time addressing climate concerns.

In chapter 8, Spencer McNee and Eric Miller examine docked bike-sharing systems. The chapter begins with a history of four distinct waves of bike-sharing systems and identifies the City of Toronto's system, featuring docked bikes, as a third-wave example. McNee and Miller review the development and growth of Bike Share Toronto since its launch in 2017, examining system changes and features with respect to demand, the impact of COVID-19, and accessibility. They suggest that geographic and socio-economic factors specific to Toronto are critical to understanding issues such as rebalancing bike supply, positioning stations to meet ridership patterns, and identifying network effects that enable the system's operation. The authors stress the essential role of place in the planning, development, and integration of bike-sharing systems.

Chapter 9, by Moe Hashemi, Hamed Motaghi, and Diane-Gabrielle Tremblay, reviews ride-hailing in Montreal with a focus on home-grown innovative solutions and benefits to workers and communities. Relying on a mixture of interviews with business managers, officials of government entities, and members of the unions/associations in the digital mobility sector, the authors ask: Who contributes to Quebec's digital mobility ecosystem? What is their contribution? And how has the evolution towards digitization impacted the ecosystem's stakeholders? Suggesting that stakeholders have been impacted in four key ways – economically, environmentally, socially, and technologically – the chapter outlines both positive and negative outcomes associated with mobility changes.

In chapter 10, Shauna Brail and Betsy Donald document rapid changes in ride-hailing, facilitated by technology and precipitated by the pandemic, charting the declining use of ride-hailing during the pandemic and a concomitant shift towards food-hailing. The authors note that by early 2020, the global spread of COVID-19 and reductions in personal mobility raised questions about the future of ride-hailing and the role of digital platforms more broadly. During this period, some ride-hailing firms pivoted their focus from moving people to moving food and other goods. Similarly, digital platform firms focused exclusively on food delivery adapted to sudden and dramatic growth demands. Brail and Donald examine the implications of pandemic-related shifts to on-demand food delivery with respect to platform governance, platform pivots, and platform hacks. The chapter challenges understanding of the implications related to technological change, innovation, and disruption in cities.

Section IV: Governing Mobility

The penultimate section of the book emphasizes the role of governance in directing mobility goals and priorities.

Elena Gorachinova, Lisa Huh, and David Wolfe's chapter 11 synthesizes research about digital infrastructure, connected and autonomous vehicles (C/AVs), and mobility as a service (MaaS). Examining case studies from Canada, the US, and Finland, they seek to understand how these countries have developed policy, regulation, and implementation activities while adapting to integrated mobility and emerging technologies. The authors also distinguish between national-level legislation and practices and regional or local practices. Through this review, they find that despite some interconnections, economic development goals do not always align with transportation goals because the motives differ. The authors conclude that there is no consensus on the best way to format regulations surrounding the management of new transportation technologies. No single approach seems more effective than any other.

Chapter 12, by Yuko Aoyama and Luis Alvarez León, does not address the Canadian context but rather provides analysis regarding the governance of autonomous vehicles from an international perspective. As an emerging and critical subject of enquiry that Canadian cities will need to address, this chapter provides insight into the governance of emerging mobility technologies through a discussion of autonomous vehicle (AV) policies in the US. The chapter first outlines various federal and state-level legislative and promotional efforts and responsibilities regarding AVs. The authors suggest that while nation- and state-wide standards help grow the AV sector, cities are the most critical actors in this growth. The authors describe four ways that cities, as policymakers, can engage with AVs. First, cities are regulators when they manage AV testing and promoters when they allow for testing sites. Cities also mediate testing efforts between private and non-profit sectors, as well as across multiple levels of government. Finally, cities collect data about AVs, manage data distribution, and consolidate data from multiple sources across the public and private sectors.

In chapter 13, Austin Zwick, Mischa Young, and Zachary Spicer explore the dimensions of change associated with ride-hailing regulation by studying the evolution of regulation across Canada's thirty largest municipalities. The chapter identifies three distinct phases of ride-hailing regulation and details how policy learning occurred across cities in Canada over time. The authors suggest that during COVID-19, municipalities were overwhelmed with other pandemic-related work and did not implement many changes to ride-hailing policies. The chapter finds that regulatory activity has reached a state of policy convergence and is shifting away from issues such as passenger safety and permits towards broader societal issues, including employment and working conditions.

Chapter 14, by John Lorinc, discusses the concept of "mobility as a service" (MaaS). MaaS operates as a digital platform that allows customers to book and pay for trips from multiple services (e.g., bus, taxi, bike-share) all in one place. MaaS apps are designed to remove the inconvenience associated with navigating between multiple platforms for a single trip. The chapter discusses the rise of MaaS prior to COVID-19 and the challenges it has faced throughout the pandemic as public transportation use has declined. Although these platforms offer promise as aggregators of mobility services, Lorinc addresses challenges associated with getting transport operators on board. The chapter concludes by recommending that MaaS platforms be facilitated by local governments rather than private vendors, in part to maximize public investments in transit infrastructure.

Section V: Moving Mobility Studies Forward

In the concluding chapter, Betsy Donald and Shauna Brail summarize the book's cross-cutting themes. The chapter ties together the grand challenges associated with mobility shifts in relation to technological advancement, public health crises, and climate change while examining the ongoing relationship between these shifts and the future of cities. Additionally, the authors amplify questions raised throughout the book regarding the role of innovation in mobility, efforts to promote and support equity and social justice through mobility-related policy, planning, and decision-making, and the significance of place-based considerations with respect to mobility. Finally, the conclusion discusses the essential role of collective government and industry action and urban policy in fostering sustainable mobility in the twenty-first-century city.

REFERENCES

Brail, S. (2018). From renegade to regulated: The digital platform economy, ride-hailing and the case of Toronto. *Canadian Journal of Urban Research, 27(2)*, 51–64. https://cjur.uwinnipeg.ca/index.php/cjur/article/view/132/67.

Brail, S. (2019, 24 November). *Cities need to innovate to improve transportation and reduce emissions. The Conversation*. http://theconversation.com/cities-need-to-innovate-to-improve-transportation-and-reduce-emissions-125778.

Brail, S. (2022, November). COVID-19 and the future of urban policy and planning. *Current History, 121*(838), 298–303. https://doi.org/10.1525/curh.2022.121.838.298.

Calderón, F., & Miller, E.J. (2020). A literature review of mobility services: Definitions, modelling state-of-the-art, and key considerations for a conceptual modelling framework. *Transport Reviews*, *40*(3), 312–32. https://doi.org/10.1080/01441647.2019.1704916.

Calderón, F., & Miller, E.J. (2022). A conceptual framework for modeling the supply side of mobility services within large-scale agent-based travel demand models. *Transportation Letters*, *14*(6), 600–9. https://doi.org/10.1080/19427867.2021.1913303.

Council of Canadian Academies. (2021). *Choosing Canada's Automotive Future* (The Expert Panel on Connected and Autonomous Vehicles and Shared Mobility, Council of Canadian Academies, p. 233).

Docherty, I., Marsden, G., & Anable, J. (2018). The governance of smart mobility. *Transportation Research Part A: Policy and Practice*, *115*, 114–25. https://doi.org/10.1016/j.tra.2017.09.012.

Dodman, D., Hayward, B., Pelling, M., Broto, V. C., Chow, W., Chu, E., Dawson, R., Khirfan, L., McPhearson, T., Prakash, A., Zheng, Y., & Ziervogel, G. (2022). Cities, settlements and key infrastructure. In H.-O. Pörtner, D.C. Roberts, M.M.B. Tignor, E.S. Poloczanska, K. Mintenbeck, A. Alegría, M. Craig, S. Langsdorf, S. Löschke, V. Möller, A. Okem, & B. Rama (Eds.), *Climate change 2022: Impacts, adaptation and vulnerability. Contribution of working group II to the sixth assessment report of the Intergovernmental Panel on Climate Change* (pp. 907–1040). Cambridge University Press.

El-Geneidy, A., Patterson, Z., & St.-Louis, E. (2015). Transport and land-use interactions in cities: Getting closer to opportunities. In P. Filion, M. Moos, T. Vinodrai, & R. Walker (Eds.), *Canadian cities in transition: Perspectives for an urban age* (5th ed., pp. 174–93). Oxford University Press.

Higgins, C., Farber, S., Shalaby, A., Khandker, N.H., Miller, E., Brail, S., Widener, M., Diamond, S., Paez, A., Zhang, B., Zhang, Y., Palm, M., & Tiznado-Aitken, I. (2021). *An integrated approach to transit system evolution*. University of Toronto Transportation Research Institute. https://uttri.utoronto.ca/files/2022/01/An-Integrated-Approach-to-Transit-System-Evolution.pdf.

Kenney, M., & Zysman, J. (2016, Spring). The rise of the platform economy. *Issues in Science and Technology*, *32*(3). https://issues.org/the-rise-of-the-platform-economy/.

Muench, S., Stoermer, E., Jensen, K., Asikainen, T., Salvi, M., & Scapolo, F. (2022). *Towards a green & digital future: Key requirements for successful twin transitions in the European Union*. Publications Office of the European Union. European Commission, Joint Research Centre. https://doi.org/10.2760/977331.

Ranchordas, S. (2015). Does sharing mean caring: Regulating innovation in the sharing economy. *Minnesota Journal of Law, Science and Technology*, *16*(1), 413–76. https://scholarship.law.umn.edu/mjlst/vol16/iss1/9.

Santoalha, A., & Boschma, R. (2021). Diversifying in green technologies in European regions: Does political support matter? *Regional Studies*, *55*(2), 182–95. https://doi.org/10.1080/00343404.2020.1744122.

Sevunts, L. (2020, 23 April). *Canadian municipalities seek emergency federal funding*. Radio Canada International. https://www.rcinet.ca/en/2020/04/23/canadian-municipalities-seek-emergency-federal-funding/.
Sperling, D. (2018). *Three revolutions: Steering automated, shared and electric vehicles to a better future*. Island Press.
Statistics Canada. (2022a, 9 February). *Canada's large urban centres continue to grow and spread*. The Daily. http://www150.statcan.gc.ca/n1/daily-quotidien/220209/dq220209b-eng.htm.
Statistics Canada. (2022b, 30 November). *Has the COVID-19 pandemic changed commuting patterns for good?* The Daily. http://www150.statcan.gc.ca/n1/daily-quotidien/221130/dq221130c-eng.htm.
Statistics Canada. (2023, 22 August). *Commuting to work by car and public transit grows in 2023*. The Daily. https://www150.statcan.gc.ca/n1/daily-quotidien/230822/dq230822b-eng.htm.
Urry, J. (2007). *Mobilities*. Polity.
Wolfe, D. (2019). *A digital strategy for Canada: The current challenge. IRPP Insight No. 25*. https://doi.org/10.13140/RG.2.2.22510.46402.

SECTION II: THE BIG PICTURE – FRAMING THE QUESTIONS

2 The Impacts of COVID Policies, Restrictions, and Behavioural Shifts on Urban Canadians

GABRIEL CAVALLI AND ANITA M. McGAHAN

Introduction

The purpose of this chapter is to consider the enduring implications of the COVID-19 pandemic for Canada's cities. The COVID pandemic of 2020–2 has not fully resolved, and yet a wide range of its enduring consequences are beginning to become clear. We know that trends underway in Canadian cities even before the pandemic (Ahn et al., 2015) have been reshaped and, in some instances, accelerated and amplified by what has occurred since 13 March 2020, when an unprecedented national lockdown began. Experiences of isolation and hardship have led many urban Canadians into levels of unmitigated distress. In particular, reductions in the proximity of urban Canadians persist even where mobility increases. The impact has been unequal, with some of the most profound problems arising for low-income workers, essential workers, health-care providers, immigrants, children, the elderly, and civil servants. At the same time, our cities proved remarkably adaptable and, in some ways, resilient, even in the face of fundamental challenges to urban ways of life. The conclusion of this chapter emphasizes these strengths to identify emerging opportunities for Canadian cities to become better equipped to confront ongoing, pressing challenges.

The analysis unfolds in a number of sections. First, we briefly review the impact of the pandemic on mobility and proximity in Canadian cities by drawing on several recent reports, including one that we developed late in 2020 in collaboration with colleagues from Cuebiq and the ISI Foundation (Cavalli et al., 2020). Drawing on related research, we describe several dimensions of the impact of changing mobility and proximity among Canadians on individual health, work structure, and economic security. Secondly, the analysis then considers a series of open and critical questions regarding the implications of these changes in

Figure 2.1. Implications of the pandemic: Mobility and proximity, health, structure of work, and socio-economic status

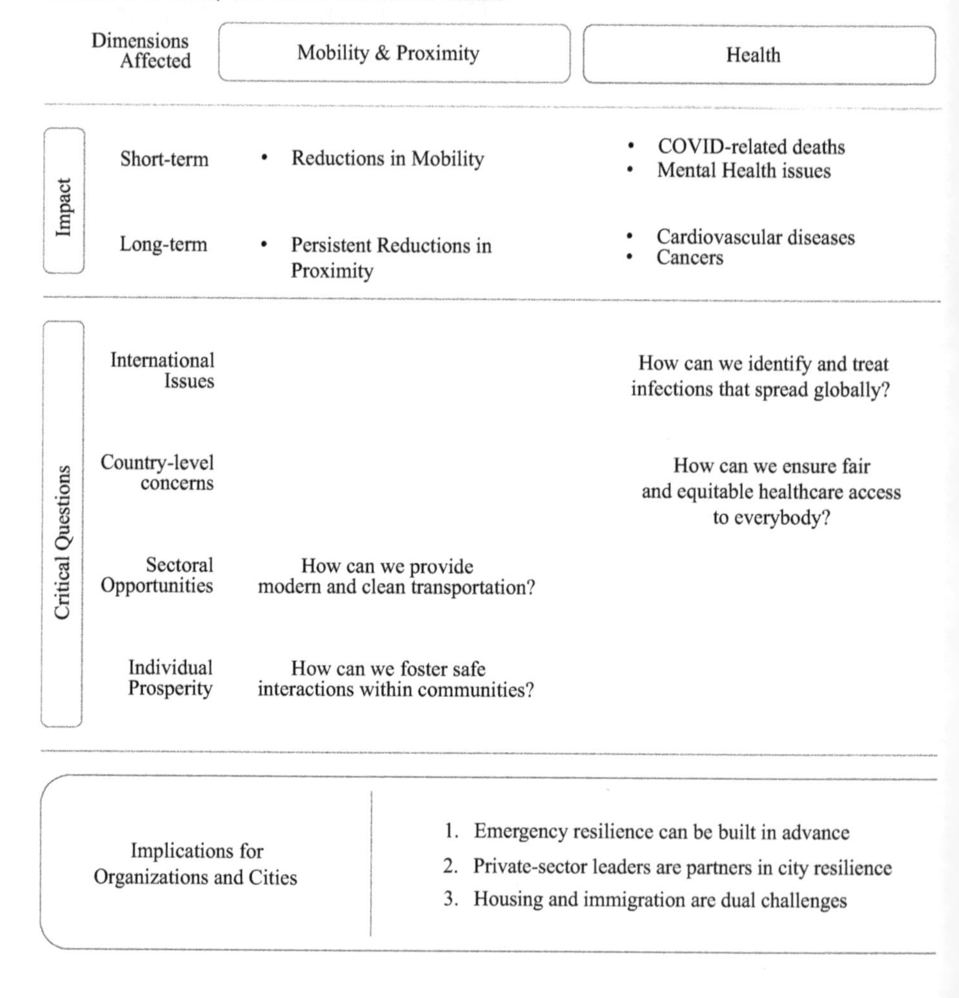

light of trends under way even before the first reported case of COVID was detected in Toronto on 25 January 2020. These trends include climate change, sustainability initiatives, labour issues, and the capacity of the private and public sectors to work together in, for example, vaccine development, and they open up new opportunities for Canadian cities to lead globally in improving quality of urban life. The third section of the paper addresses some immediate and emerging implications for organizations in Canadian cities. The focus is on resolving labour

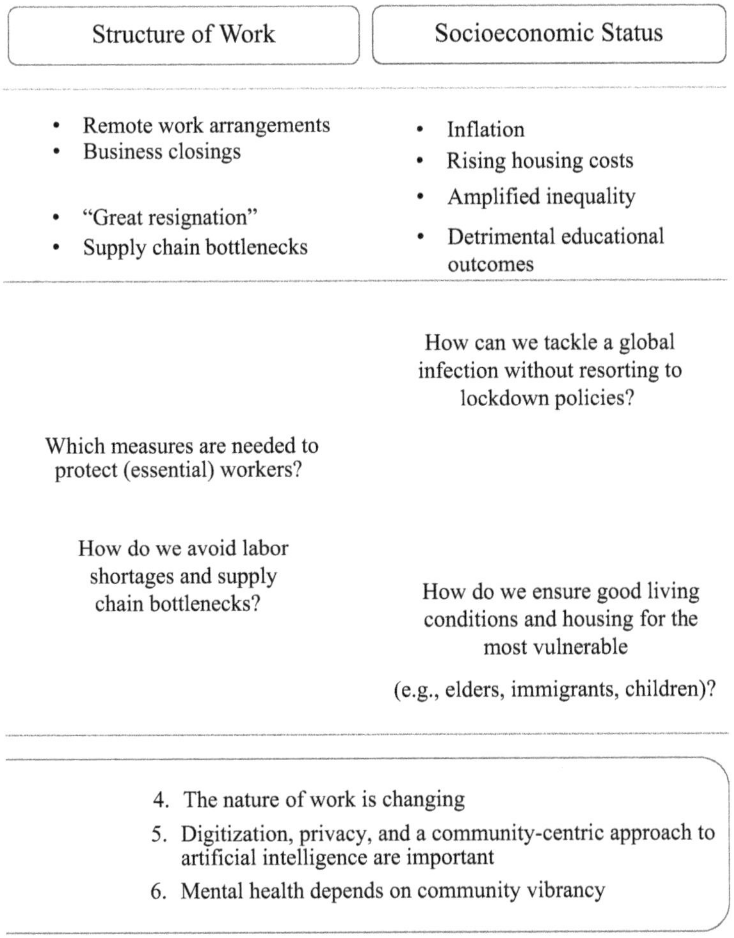

Structure of Work	Socioeconomic Status
• Remote work arrangements • Business closings • "Great resignation" • Supply chain bottlenecks	• Inflation • Rising housing costs • Amplified inequality • Detrimental educational outcomes
Which measures are needed to protect (essential) workers?	How can we tackle a global infection without resorting to lockdown policies?
How do we avoid labor shortages and supply chain bottlenecks?	How do we ensure good living conditions and housing for the most vulnerable (e.g., elders, immigrants, children)?

4. The nature of work is changing
5. Digitization, privacy, and a community-centric approach to artificial intelligence are important
6. Mental health depends on community vibrancy

shortages, dealing with inflation, addressing central problems in the housing stock, and adapting city services to new ways in which urban Canadians interact. The chapter concludes with reflections on the implications of the pandemic for all cities, and points to ways in which Canada's metropolitan areas can lead the world in reconceptualizing city identity, services, and resilience to enable the human interaction required to address the world's most pressing problems, including climate change and sustainable human systems (see Figure 2.1).

COVID's Impacts on Mobility and Proximity in Canadian Cities

Several important studies have documented both a recovery in urban areas after the pandemic, and ongoing reductions in mobility in most cities that may be permanent (Canadian Chamber of Commerce, 2022; Cavalli et al., 2020; Chalise et al., 2022). As Vinodrai (see chapter 4) explains, our understanding of these changes is incomplete at best. Much more research is needed especially to understand in detail the ways in which urban Canadians have changed in their patterns of interaction through the course of the pandemic.

Despite the gaps in data, several important regularities are clear. First, evidence based on cell-phone movements shows conclusively that Canadians responded to stay-at-home orders immediately upon their issuance in March 2020, and that Canadians continued to restrict mobility throughout the pandemic even after stay-at-home orders were modified and lifted (Canadian Chamber of Commerce, 2022; Cavalli et al., 2020). The overall pattern across virtually all Canadian cities as demonstrated by Cavalli et al. (2020) was an initial sharp reduction in mobility after 13 March 2020, followed by some recovery. By June 2020, mobility in the thirty largest Canadian cities had recovered somewhat, and by mid-July 2020, mobility had recovered substantially in many but not all of these cities. Not all types of mobility increased. The results reported by Cavalli et al. (2020) suggested that the rebound was strongly driven by local ground travel within cities and between neighbouring cities (such as Toronto and Peel, or Vancouver and Victoria), rather than by air travel. The analysis showed significant and persistent reductions in the presence of cell phones at airports throughout 2020. By late 2020, evidence suggests that, across Canadian cities, mobility began to recover (Cavalli et al., 2020).

Second, a critical but less fully appreciated fact is that the proximity of Canadians to one another recovered more slowly than mobility (Cavalli et al., 2020). Like mobility, proximity between people can be identified using privacy-preserving cell-phone data. Proximity measures represent the average number of distinct contacts that each member of a group of anonymized persons has in an average day. They can be derived from the locations over time of the devices of the anonymized cell phones of those in the group (Pepe et al., 2020). Thus, "proximity" serves as an approximation of the interactions that a typical member of a population has within an hour (Pepe et al., 2020).

Imagine two members of a family who, after a long period of lockdown, emerge from a shared home to go for a drive to a local park during the summer of 2020. Suppose that both had opted into allowing their locations to be shared anonymously through their smartphone applications. By anonymously tracking the positions of each person's cell phone, analysts can observe de-identified mobility behaviour to establish that mobility increased for the couple, but that their proximities to other persons did not increase. The difference arises because, during the lockdown, the two cell phones were closely located to each other within the home area, constituting an instance of proximity of the two persons only to each other. In the car, the phones may be equally proximate to each other and similarly distant from other phones, despite the increase in mobility arising from the drive. As a result, Canadian proximity did not recover nearly as fast as Canadian mobility (Cavalli et al., 2020). Across the largest Canadian cities, proximity dropped very significantly in March 2020, and it did not begin to recover until the fourth quarter of 2020. By that time, Canadians were more mobile than early in the pandemic, but overall, they continued to avoid interacting with others, with results varying somewhat by cities. Consistent with these findings, other studies also showed that Canadians significantly adhered to social distancing measures (Brankston et al., 2022).

Third, evidence is emerging that variation in mobility and proximity between cities occurred over the course of the pandemic, and that this variation persists at differential rates. Residents of the cities of Toronto and Montreal, for example, were relatively less mobile than residents of Western Canadian cities early in the pandemic, perhaps in part because of weather-related and density exposures to COVID risk (Cavalli et al., 2020). Overall, even partial recoveries in mobility are highly correlated with improvements in weather. During the summer of 2020, for example, Hamiltonians tended to travel longer distances, with a peak day for mobility on 1 July, when the temperature was particularly high (Coleman, 2020). However, other factors than weather appear to be important as well. According to a study published in November 2022 by the Canadian Chamber of Commerce, mobility in the ten largest downtown areas of cities in Canada was down 33 per cent as compared to just prior to the pandemic in January 2020. This figure is significantly different from the national average of 7 per cent less mobility. Across all major metropolitan areas around Toronto and Montreal as well as Vancouver and Alberta, some of the most significant recoveries in mobility occurred in nearby mid-sized cities, suggesting that commuting from areas with lower housing costs into the city core recovered relatively fast. Much more research is needed to understand this variation and its implications.

Fourth, declines in mobility and proximity have significantly affected various health indicators among Canadians. This impact is evident not only among those staying at home but also extends to essential workers, including health-care professionals. Several studies indicate that

Canadians report adverse mental-health impacts of the pandemic arising from a range of factors, including premature death of loved ones, illness, and isolation (Angus Reid Institute, 2022; Buajitti et al., 2022; Chu et al., 2022; Cost et al., 2022; Crowe et al., 2022; Frounfelker et al., 2022). The Angus Reid Institute report indicates that 82 per cent of Canadians surveyed indicate that "the pandemic has pulled people further apart"; 79 per cent say that "the pandemic has brought out the worst in people"; and 61 per cent say that "Canadians' level of compassion for one another has grown weaker" (Angus Reid Institute, 2022). Crowe et al. (2022) report that 100 per cent of Canadian critical care nurses under study suffered "moderate to high burnout" as a result of the pandemic, with 74 per cent dealing with PTSD, 70 per cent with depression, 61 per cent with stress, and 57 per cent with anxiety. These effects were made even more complicated by the way they were treated in the media (Gordon et al., 2022). Authors have also documented concern with the long-term implications of the pandemic for a wide range of health conditions, including cardiovascular disease and cancers, as direct and indirect consequences of COVID-19 (Kite et al., 2022; Martin et al., 2021; McAlister et al., 2022; Yusuf and Tisler, 2020).

Fifth, the decrease in mobility has brought about significant changes in how jobs are designed and geographically distributed. In turn, these changes will have lasting effects not only on urban planning but also on employers' ability to fill vacant positions. The pandemic had a profound and continuing impact on the structure of work (Teodorovicz et al., 2021). Telework, work-from-home, work-from-anywhere, and hybrid work arrangements were widely implemented, tested, improved, and – in many instances – made permanent even after restrictions were lifted (Choudhury, 2020). The consequences of these arrangements for the downtown core of large cities were made immediately evident as commercial vacancy rates increased, servicing stores became shuttered, and furloughs occurred (Cavalli et al., 2020; Rowe, 2020). By the third quarter of 2022, analysts reported that "Canada still boasts three of the five lowest downtown vacancy rates in North America" in Vancouver (7.1%), Ottawa (11.5%) and Toronto (11.8%) (CBRE, 2022; McLean, 2022). At the same time, some Canadian cities had high vacancy rates, including Calgary (32.9%), Waterloo (23.6%), Halifax (18.8%), and Montreal (16.1%) (CBRE, 2022; McLean, 2022). As restrictions on interaction lifted, the "great resignation" and elevated retirements led to labour shortages with consequences for job redesign that amplify the prevalence and extent of attractive remote-work arrangements that are popular with skilled workers (Kniffin et al., 2021). At the same time, the demands for physical presence on essential workers in health care, policing, education, and other sectors have intensified, leading to burnout, distress, and a parallel wave of resignations and retirements (Campbell & Gavett, 2021). The long-term consequences of these changes for cities must be studied intensively.

Sixth, evidence suggests that pandemic-induced reductions in mobility and proximity were tied to changes in several socio-economic indicators. The pandemic has been tied directly to inflation, rising costs of housing, amplified inequality, and detrimental educational outcomes, particularly among children (Chen et al., 2022; UNDP, 2023). Ample evidence exists tying inflation to both supply-chain shortages and labour shortages arising from limitations on mobility, and to monetary infusions that occurred to address job loss during the early part of the pandemic (Helper & Soltas, 2021). Increases in housing prices arose in part from safe-haven investing and shifting needs that arose from remote-work arrangements aiming at reducing worker proximity (Statistics Canada, 2020). Low-paid essential workers, laid-off contract workers, recent immigrants, and those in need of social services all were made more vulnerable by the dual challenges of relatively greater exposure to COVID-19 and lockdown restrictions in cities with inadequate childcare support (Conway et al., 2020; Schembri, 2020). Children also suffered from school lockdowns – both during the pandemic and subsequently as their development slowed – despite evidence that the closing of schools did relatively little to stop COVID transmission (Wibbens et al., 2020).

The bottom line is that the impact of the pandemic is nuanced, variable, and profound across Canadian cities. Evidence suggests that the impact is not limited to the immediate reductions in mobility and proximity that occurred with the implementation of lockdown policies to contain the spread of COVID. The next section describes critical questions that must be answered for our cities to recover innovatively in ways that accelerate human health, well-being, and prosperity rather than recreate the conditions that led to the pandemic in the first place. The pandemic challenges the comparative advantages of Canadian cities and confronts urban leaders with unprecedented opportunities for system-level change in attracting businesses, talent, innovative housing development, transportation innovations, and conveyance of essential service. The stakes have been raised for city leaders who seek to reinvigorate Canada's urban centres to meet the post-pandemic needs of Canadians.

Critical Questions

The rapid spread of COVID-19 in the first few months of 2020 after the disease's emergence in Wuhan, China, made vivid the interdependency of people across the planet. At the same time, it also revealed weaknesses in international institutions and national political systems (Mussachio et al., 2020). The Wikipedia entry entitled "World Health Organization's Response to the COVID-19 Pandemic" contains a sentence saying that "The WHO's handling of the initial outbreak required a 'diplomatic balancing act' between member states, in particular between the United States and China" (Wikipedia, 2022), which points to a shocking set of constraints on the very international institution that leading countries established to endorse scientifically based recommendations regardless of their political content. The politicization of COVID-19 both internationally and nationally in countries such as the United States clearly led to excess deaths at levels that were unimaginable even a few months earlier (Kerr et al., 2021; Pickup et al., 2020). Many of the reasons why international and national responses were impeded relate to other challenges, such as climate change and its implications, that were central concerns prior to the pandemic.

Canada was affected by this backdrop of activity. The federal system of government in Canada led to decentralized responses across Canadian provinces on social-distancing and social-mobility restrictions (McCoy et al., 2020). As national leaders strove to get access to vaccines for Canadians from outside the country, questions arose domestically regarding the locus of authority for action on the health challenge (McCoy et al., 2020). The country's position on access to vaccines in low-income countries became complicated by Canada's frustratingly low capacity to manufacture vaccines domestically (Government of Canada, 2023; Prime Minister of Canada, 2022). Much more research is needed to draw forth lessons for Canadians, but one clear implication of the pandemic for federal and provincial leaders is that domestic strengths are essential to Canada's international standing during times of emergency (see Hutton, chapter 3). Those strengths emerge from across Canadian society, including from activities in cities. There are other lessons as well. As city leaders in Canada worked with provincial, territorial, and national authorities to discern how to address the pandemic, the local relevance of international and national concerns became increasingly clear. The impact of the pandemic was wrapped up in issues and problems that may have seemed remote to many Canadians before 2020. But the future of Canada's cities depends on international and national issues as much as on local and individual concerns.

International Issues

Well before the pandemic, the international focus on climate change carried implications for Canadian cities (Leach, 2019). Discussions of heat waves, intensifying storms, and climate-driven migration occurred in many different venues. What the pandemic demonstrated was the immediate relevance of international emergencies for Canadian cities. Infectious diseases proliferate with climate change (CDC, 2022). Canadian cities are connected through international travel, commerce, economic exchange, and public-policy commitments to places that are especially vulnerable to them. The capacity of Canada's large cities to contain infections and protect local populations from both health threats and the economic devastation of lockdowns is directly tied to the global risk of pandemics. Critical questions arise regarding the capacity in Canadian cities to identify and treat infection without resorting to full-scale lockdowns that carry the kinds of consequences for human health that occurred during the COVID-19 pandemic.

Country-Level Concerns

Canada's cities share challenges that were made evident during the COVID-19 pandemic (McGahan & Sukhram, 2020). These include problems of assuring health-care access and delivery to relatively marginalized and vulnerable populations of immigrants, low-income persons, and members of otherwise vulnerable groups that may be better identified by city governments than by provincial or national authorities.

The COVID-19 pandemic made clear that Canadian cities must adapt to better models of eldercare that keep elders in communities and attend to their needs more humanely and effectively (Government of Canada, 2021). Canadian cities also face common challenges in emergency preparedness construed broadly to include effective access to energy, health care, and priority services for members of vulnerable groups that may not be visible at other levels of government or to private-sector providers of critical services (Benton-Short, 2013).

Canadian cities also share opportunities for collaboration to develop better systems of emergency response enabled by digital technologies, such as in firefighting and policing. Innovations in the arts across cities are central opportunities both to protect workers in the sector, many of whom cannot readily sustain periods of unemployment, and to preserve the vibrancy of local culture (Brail et al., 2020).

Critical questions arise about the ways in which leaders in Canadian cities can learn together from the pandemic about the kinds of common capacities that generate resilience to health and other emergencies.

Sectoral Opportunities

The COVID-19 pandemic revealed that manufacturing efficiencies in the private sector could rapidly become rigidities in periods of emergency

(Ardolino et al., 2022). This problem became particularly evident with supply-chain and labour bottlenecks that originated with mobility restrictions, and that intensified with labour shortages as restrictions were lifted. City resources enabling communication among local leaders during periods of shortage may be warranted by this experience and need further study.

COVID-19 accelerated digitization to the point where some cities, such as Tulsa, Oklahoma, have implemented public projects designed to attract remote workers through incentives (Choudhury et al., 2021). Others have made public internet a city service. Yet others have invested to understand the resilience of communications systems that enable information-sharing during times of public emergency (Abreu et al., 2017; de Jong et al., 2015). Some cities were much more effective than others at delivering childhood education during the pandemic in ways that may be instructive. More research is needed to understand how Canadian cities can learn from and adapt to innovations that could improve quality of life, outcomes, and cost-effectiveness of core systems dependent on IT infrastructure (Brail et al., 2021).

The organizational structure of city government was strained during the pandemic in ways that are important to study and to learn from. Across Canada, some city services were shut down (or nearly shut down) during the pandemic (Bula et al., 2020). In some cases, functions such as licensing and social-services enrolment were suspended (Bowden, 2020). Post hoc reporting of the consequences of stopgap measures as well as closures would inform more effective responses for the future. Such an effort may even point towards opportunities for redesign of organizational arrangements within government for periods of non-emergency.

Innovation in retailing, food retailing, and restaurants during the pandemic was extensive (Opute et al., 2020; see Brail and Donald, chapter 10). The opening of city spaces such as parking spots to restaurants for outside dining exploded in popularity in many Canadian cities (Restaurants Canada, 2020). In the US, removal on restrictions for food distribution to retail customers through commercial challenges facilitated access to food for many families. Removal of reporting requirements facilitated rapid distribution of food and beverages in some communities (PBS News Hour, 2021). Extensive analysis is needed to understand and document the effectiveness of various arrangements, and to promote their widespread adoption if warranted.

The impact of the pandemic on the transportation systems in cities should be studied to cultivate understanding across Canadian cities of transformation opportunities (see Palm & Farber, chapter 6; McNee and Miller, chapter 8; Donald and Brail, chapter 15). As the air cleared during the 2020 lockdowns, Canadians were provided with irrefutable evidence that changes in community and transportation arrangements could significantly and immediately reduce air pollution (University of British Columbia, 2021). Many Canadians reported that the experience of not commuting was central to decisions to resign or retire from employment post-pandemic (Paddon, 2021). Patterns in the usage of roads, subways, trolleys, and train systems during the pandemic reveal important opportunities for city managers to redirect investment towards high-impact reforms with permanent impact.

One of the most important and enduring impacts of the pandemic is in real estate. The transformation of the downtown cores of Canadian cities, discussed above, is already under way, and yet significant opportunity exists for study of how downtown spaces can become revitalized and improved (Tang, 2020). Learning across cities is a major opportunity, as breweries and biotech take hold across the country in urban centres. Much more creative and innovative initiatives are warranted, with experimentation and learning as the objective. The implications of the pandemic for small- and medium-sized businesses across Canadian metropolitan areas also warrants further research as we know little about whether transitions to remote work have led to their growth and prosperity.

Residential real estate and finance are among the largest of Canadian industries. Much more research is needed on Canadian housing values and their distribution. The impact of the pandemic on housing costs is confounded by trends that had been in place prior to 2020. Research is needed on whether and how shifts such as foreign investment, condominium projects, and urban sprawl were affected by the consequences of the pandemic, particularly through inflation and changing wage agreements. The consequences for innovations in housing design, including the construction using climate-friendly materials, should be considered comprehensively.

Individual Prosperity

The famous urban planner and Torontonian, Jane Jacobs, saw cities as places where neighbourhoods and communities thrive (Jacobs, 1961). One of the most poignant and concerning impacts of the pandemic is on the health of Canada's urban dwellers. Too many Canadian urbanites died from COVID. A vast number of Canadian children are behind on educational outcomes (Subramanian, 2021). The epidemic across Canada of mental health challenges, depression, stress, and anxiety arising from the death, illness, and isolation of the pandemic continue to challenge the vibrancy of our neighbourhoods and communities (Findlay et al., 2020; Findlay et al., 2021).

More research is needed to address these problems. We must understand how the neighbourhoods and communities in Canadian cities support the individuals within them. Questions regarding the anger and antisocial behaviour of drivers and merchants and residents in neighbourhoods must be raised and addressed. We must find ways of understanding the stories of individuals who have chosen to resign their jobs and yet stay in urban communities as vital members. Similarly, we must understand the challenges facing those who are not thriving, and the underlying unmet needs that city governments might address.

The COVID-19 pandemic uniquely threatened cities as places that exist to foster interaction and to cultivate community. Understanding both the nature of the threats and the ways that they were overcome is central to our collective recovery.

Implications for Organizations

The rebuilding of organizations after the pandemic cannot fully occur without answers to the critical questions raised earlier, and yet already some successful initiatives have emerged from which leaders in Canadian cities can learn.

Emergency Resilience Can Be Built in Advance

One of the most vivid distinguishing features of cities around the world in pandemic response was their pre-planned capacities to respond effectively once the emergency became clear. Cities that had established emergency response protocols that conferred authority for action on key leaders, and that had developed committee structures that tapped both private-sector and academic authorities, generally were able to marshal an effective response relatively quickly as compared to cities without such capacity (Dzigbede et al., 2020; Gilbert et al., 2020). Post-pandemic retrospectives are beginning to reveal that quick deployment of emergency services at levels appropriate to local needs was important to preserve public trust and to maintain orderly citizen response in the face of the emergency (Gilbert et al., 2020; Goniewicz et al., 2020; Sasangohar, 2020). Specific areas of importance include managing emergency declarations, appropriate enforcement of mandates, services for children, eldercare services, and homelessness.

Private-Sector Leaders Are Partners in City Resilience

By engaging private-sector leaders as partners in pandemic abatement and recovery, city leaders marshalled greater resources and capacity for a coordinated and effective response (Billedeau & Wilson, 2021; Billedeau et al., 2022). Cities such as Montreal sought greater engagement by private-sector organizations (Laframboise, 2021; see Hashemi, Motaghi, and Tremblay, chapter 9). A central opportunity to accomplish this is to engage private-sector leaders in decisions about the resource requirements to amplify, for example, talent retention and community development. This type of engagement may be especially effective when it rewards increasing commitment by private-sector leaders over time.

Housing and Immigration Are Dual Challenges

Evidence reported by Statistics Canada indicates that, for at least the early months of the pandemic, immigrants were likely more vulnerable to COVID-19 due in part to living in crowded and multigenerational housing (Ng, 2021). Even as the pandemic abates, 75–82 per cent of Canada's population growth is from immigration. This percentage is projected to increase over the coming decade (Singer, 2021), and is associated with an escalating housing crisis, particularly in Canadian cities, that is characterized by rapidly rising home prices and the potential for large increases in rents, especially in Western provinces and territories where the increase is likely to be concentrated (Flanagan, 2021). As these developments unfold, the opportunity for pandemic-informed innovations in the housing stock is substantial.

The Nature of Work Is Changing

The remote-work movement that began prior to the pandemic and that was amplified by it is in its infancy (Choudhury, 2020). Much more sophisticated systems of remote work that account for transportation challenges, childcare needs, unstructured interchange between employees, and productivity measurement are in development (Choudhury, 2020; Choudhury et al., 2021). Each of these changes has implications for the structure of cities and city services. Given that these shifts are set against the backdrop of labour shortages, particularly of workers with specific types of skills, opportunities arise for creative city leaders to invest with private-sector leaders to develop specialized programs that differentiate their cities. For example, workers who value flexible hours, affordable childcare, and high-outcome primary education systems that are close to attractive housing may be attracted to both a city and an employer within that city simultaneously.

Digitization, Privacy, and a Community-Centric Approach to Artificial Intelligence Are Important

During the pandemic, many online workers reported subjugation to painful and painstaking productivity-management systems (Jensen et al., 2020). Stories arose of workers choosing to work off the clock rather than to risk that random monitoring through computer systems by an employer would result in an adverse performance appraisal if, for instance, the worker took a few minutes offline to change a child's diaper (Kurkowski, 2021). In some communities, such appraisals were not tolerated through either formal or informal restrictions. The character of a city as a safe and supportive place of residence has now extended into the digital sphere. Opportunities exist for Canadian cities to protect resident privacy while encouraging digital exchange and remote work. Building community in the face of artificially intelligent analytics and measurement systems carries the potential to distinguish a city in the wake of the pandemic. This idea has been tested through the Tulsa Remote program in which prospective residents – in this case with remote jobs – are offered ten thousand dollars (US) in various payments together with high-quality access to Internet connectivity, access to remote workspace, and priority consideration for high-quality urban housing. All that's required is the plan to stay for at least a year. Innovation programs such as Tulsa Remote serve as test beds for comprehensive innovations in attracting and retaining community members.

Mental Health Depends on Community Vibrancy

The pandemic revealed that isolation leads to profound mental health challenges (Pietromonaco & Overall, 2022). Vibrant communities that offer residents opportunities for social exchange, support, entertainment, and meaningful interaction are increasingly valued as a result of the experience of the pandemic (Barbarossa, 2020; Eltarabily and Elghezanwy, 2020). Commitments to local institutions that create great experiences of community such as in the arts, sports, after-school programs, education, and civic engagement are characteristic of vibrant cities (Brail et al., 2020).

Implications for Cities

In this early period of recovery after the COVID-19 pandemic, we do not yet have answers to critical questions about how Canadian cities will develop. What we have learned so far is that proximity of Canadians to one another is not recovering as fast as our mobility (Cavalli et al, 2020). The isolation that we have experienced and continue to experience has profound implications for our health, and especially mental health. Work is restructuring both in essential sectors, where a great resignation and retirement are occurring, and in sectors where remote work is possible. Economic challenges arising from inflation and housing costs have redoubled the challenge for many Canadians. Problems in educational development of many urban children are profoundly consequential. And, importantly, all of the adversity of the pandemic – from excess deaths to economic hardship – have hit the poor and vulnerable among us hardest.

Yet despite the challenges, there is much to learn from the pandemic that can shape investments in our cities. In this chapter, we have argued that Canadian cities were challenged during the pandemic by international, national, sectoral-level, and even individual concerns. Many of the trends and problems that cities faced prior to the pandemic were amplified and extended during COVID-19. We know, for example, that climate change and immigration are both profoundly important for city leaders as well as provincial and federal officials. Cities need strategies for resiliency to emergencies, and for housing immigrants who seek work and who will constitute the major source of population growth in Canadian communities. At the same time, Canadian cities must contend with housing shortages, underdeveloped transportation systems, and the diverse needs of remote workers. The examples described in this chapter suggest that innovation occurred during the pandemic in a range of sectors, including especially retail and essential services. Yet these examples also suggest that much more innovation is needed, particularly on such practices as the monitoring of remote workers and on the provision of safe and humane eldercare.

Already, some opportunities for city development are becoming clear. The pandemic made evident that planning for resilience in the face of emergency is imperative for all Canadian cities. City leaders and managers must be able to respond quickly and effectively – and in partnership with private-sector leaders and those from other sectors – when emergency strikes. These capacities are likely to be tested as climate change progresses and as Canada attracts a large number of immigrants over the coming decade. The implications for innovation in the housing stock are increasingly evident. Building communities in which immigrants and other Canadians can thrive will require a thoughtful commitment to a wide range of activities and social services. Accomplishing this may involve establishing priorities for digital enablement and privacy enhancements through surveillance abatement, for example, that communities can pursue as features of their collective identity.

At the same time as the COVID-19 pandemic challenged Canadian cities at their core, it also revealed the extraordinary strengths of Canadians in our capacity for mutual, coordinated commitment to our health and well-being. As the pandemic ends, we now face the challenge of rebuilding our communities for greater resilience and vibrancy. The challenge is compounded by long-term and perhaps permanent reductions in the types of proximity that Jacobs (1961) emphasized as central to the identities of cities. Even as we have become more mobile, Canadians remain in relatively less proximity. We must find ways to collaborate together even more effectively despite this loss of proximity. Each Canadian city must find a way to extend its identity into a future in which public emergencies emerging from climate change and other threats to sustainability only mount. By investing with solidarity to deepen city engagement, recalibrate core systems, spend wisely, and reconstitute innovation priorities, Canadian cities can emerge into the mid-twenty-first century as places characterized first by resilience and adaptability. We also have the capacity to deepen our communities so as to lead on the world stage in the development of urban systems that encourage immigration, retain members, and support creative exchange that brings new solutions to the world's most pressing problems.

REFERENCES

Abreu, D.P., Valesquez, K., Curado, M., & Monteiro, E. (2017). A resilient Internet of Things architecture for smart cities. *Annals of Telecommunications* 72, 19–30. https://doi.org/10.1007/s12243-016-0530-y.

Ahn, R., Burke, T.F., & McGahan, A.M. (Eds.). (2015). *Innovating for Healthy Urbanization*. Springer.

Angus Reid Institute. (2022, 10 March). *Vast majorities say that the pandemic has pulled Canadians apart, brought out the worst in people*. Available at https://angusreid.org/covid-19-two-year-anniversary/.

Ardolino, M., Bacchetti, A., & Ivanov, D. (2022). Analysis of COVID-19 pandemic's impacts on manufacturing: a systematic literature review and future research agenda. *Operations Management Research 15*, 551–66. https://doi.org/10.1007/s12063-021-00225-9.

Barbarossa, L. (2020). The post pandemic city: Challenges and opportunities for a non-motorized urban environment. An overview of Italian cases. *Sustainability*, *12*(17), 7172. https://doi.org/10.3390/su12177172.

Benton-Short, L. (Ed.). (2013)., *Cities of North America: Contemporary Challenges in US and Canadian Cities*. Rowman & Littlefield.

Billedeau, D.B., & Wilson, J. (2021). COVID-19 and corporate social responsibility: A Canadian perspective. In *COVID-19: Paving the Way for a More Sustainable World* (pp. 23–41). Springer.

Billedeau, D.B., Wilson, J., & Samuel, N. (2022). From responsibility to requirement: COVID, cars, and the future of corporate social responsibility in Canada. *Sustainability*, *14*(11), Article 6658. https://doi.org/10.3390/su14116658.

Bowden, D. (2020). *Federal court of Canada further extends COVID-19 suspension period*. CWilson.com. Available at https://www.mondaq.com/canada/litigation-contracts-and-force-majeure/946702/federal-court-of-canada-further-extends-covid-19-suspension-period.

Brail, S., Des Rosiers, N., & McGahan, A.M. (2020, 19 June). For the arts, the show must go on after COVID-19. *Policy Options*. https://policyoptions.irpp.org/magazines/june-2020/for-the-arts-the-show-must-go-on-after-covid-19/.

Brail, S., Heblich, S., & McGahan, A.M. (2021, 10 March). The innovation imperative: How the pandemic has changed the urban environment where innovation thrives. City workers are likely to continue to work from home at least part-time, which has implications for innovation. *The Financial Post*. https://financialpost.com/technology/the-innovation-imperative-how-the-pandemic-has-changed-the-urban-environment-where-innovation-thrives.

Brankston, G., Merkley, E., Loewen, PJ., Avery, B.P., Carson, C.A., Dougherty, B.P., Fisman, D.N., Tuite, A.R., Poljak, Z., & Greer, A.L. (2022). Pandemic fatigue or enduring precautionary behaviours? Canadians' long-term response to COVID-19 public health measures. *Preventive Medicine Reports*, *30*, Article 101993. https://doi.org/10.1016/j.pmedr.2022.101993.

Buajitti, E., Rosella, L.C., Bryan, K., Giesinger, I., & Goel, V. (2022). Downstream health impacts of employment losses during the COVID-19 pandemic. *Can J Public Health*, *113*, 135–46. https://doi.org/10.17269/s41997-021-00588-3.

Bula, F., Moore, O., & Van Praet, N. (2020, 13 April). Cities face massive financial shortfalls because of COVID. How can they cope when they can't run deficits? *The Globe and Mail*. Available at https://www.theglobeandmail.com/canada/article-cities-face-massive-financial-shortfalls-because-of-covid-how-can/.

Campbell, M., and Gavett, G. (2021, 10 February) *What COVID-19 has done to our well-being, in 12 charts*. HBR.org. Available at https://hbr.org/2021/02/what-covid-19-has-done-to-our-well-being-in-12-charts.

Canadian Chamber of Commerce. (2022, November). *Canada's new workplace mobility trends*. Canadian Chamber of Commerce Business Data Lab. https://bdl-lde.ca/wp-content/uploads/2022/11/Canadas_New_Workplace_Mobility_Report_FINAL_Rankings_EN.pdf.

Cavalli, G., Lake, B., McGahan, A.M., & Pepe, E. (2020, October). Mobility and proximity in Canada during the COVID-19 pandemic. *Innovation Policy Lab, Munk School of Global Affairs & Public Policy*. https://munkschool.utoronto.ca/research/mobility-and-proximity-canada-during-covid-19-pandemic.

CBRE. (2022). *National vacancy contracts by 10 bps, inching Canada's office market closer to equilibrium*. Available at https://cbre.vo.llnwd.net/grgservices/secure/Canada_Office_Figures_Q3_2022.pdf?e=1674598702&h=a839cf7145250028eb798fa852900e53.

CDC (Center for Disease Control). (2022). *Climate change and infectious diseases*. Available at https://www.cdc.gov/ncezid/what-we-do/climate-change-and-infectious-diseases/index.html.

Chalise, N., Gutkowski, V., & Kaplan, H. (2022, 15 November). Perspectives from main street: The impact of COVID-19 on communities and the entities serving them [2022]." *Fed Communities*.

Chen, H., Engert, W., Huynh, K.P., O'Habib, D., Wu, J., & Zhu, H. (2022). *Cash and COVID-19: What happened in 2021*. Bank of Canada Staff Discussion Paper No. 2022-8. https://www.bankofcanada.ca/wp-content/uploads/2022/04/sdp2022-8.pdf.

Choudhury, P. (2020, November-December). Our work-from anywhere future. *Harvard Business Review, 98*(6). https://hbr.org/2020/11/our-work-from-anywhere-future.

Choudhury, P., Foroughi, C., & Larson, B. (2021, April). Work-from-anywhere: The productivity effects of geographic flexibility. *Strategic Management Journal, 42*(4), 655–83. https://doi.org/10.1002/smj.3251.

Chu, C.H., Yee, A.V., & Stamatopoulos, V. (2022). "It's the worst thing I've ever been put through in my life": The trauma experienced by essential family caregivers of loved ones in long-term care during the COVID-19 pandemic in Canada. *International Journal of Qualitative Studies on Health and Well-being, 17*(1). https://doi.org/10.1080/17482631.2022.2075532.

Coleman, J. (2020). *Hamilton is distancing well: UToronto report*. The Public Record. Available at https://www.thepublicrecord.ca/2020/10/hamilton-is-distancing-well-utoronto-report/.

Conway, B., Truong, D., & Wuerth, K. (2020, June). COVID-19 in homeless populations: unique challenges and opportunities. *Future Medicine, 15*(6), 331–4. https://doi.org/10.2217/fvl-2020-0156.

Cost, K.T., Crosbie, J., Anagnostou, E., Birken, C.S., Charach, A., Monga, S., Kelley, E., Nicolson, R., Maguire, J.L., Burton, C.L., Schachar, R.J., Arnold, P.D., & Korczak, D.P. (2022). Mostly worse, occasionally better: Impact of COVID-19 pandemic on the mental health of Canadian children and adolescents *Eur Child Adolesc Psychiatry, 31*, 671–84. https://doi.org/10.1007/s00787-021-01744-3.

Crowe, S., Fuschia Howard, A., & Vanderspank, B. (2022, August). The mental health impact of the COVID-19 pandemic on Canadian critical care nurses." *Intensive and Critical Care Nursing, 71*, Article 103241. https://doi.org/10.1016/j.iccn.2022.103241.

de Jong, M., Joss, S., Schraven, D., Zhan, C., & Weijnen, M. (2015). Sustainable–smart–resilient–low carbon–eco–knowledge cities; making sense of a multitude of concepts promoting sustainable urbanization. *Journal of Cleaner Production, 109*, 25–38. https://doi.org/10.1016/j.jclepro.2015.02.004.

Dzigbede, K.D., Gehl, S.B., & Willoughby, K. (2020, May). Disaster resiliency of US local governments: Insights to strengthen local response and recovery from the COVID-19 pandemic. *Public Administration Review, 80*(4), 634–43. https://doi.org/10.1111/puar.13249.

Eltarabily, S., & Elghezanwy, D. (2020). Post-pandemic cities–The impact of COVID-19 on cities and urban design. *Architecture Research, 10*(3), 75–84. doi:10.5923/j.arch.20201003.02.

Findlay, L.C., & Arim, R. (2020). Canadians report lower self-perceived mental health during the COVID-19 pandemic. *Statcan COVID-19: Data to Insights for a Better Canada*. Available at https://epe.lac-bac.gc.ca/100/201/301/weekly_acquisitions_list-ef/2020/20-17/publications.gc.ca/collections/collection_2020/statcan/45-28/CS45-28-1-2020-3-eng.pdf.

Findlay, L.C., Arim, R., & Kohen, D. (2021). Understanding the perceived mental health of Canadians during the COVID-19 pandemic. *Health Reports, 31*(4), 22–7. https://www.doi.org/10.25318/82-003-x202000400003-eng.

Flanagan, R. (2021, 10 June). *Millions of people will move to Canada in the next 20 years, but where will they live?* CTV News. Available at https://www.ctvnews.ca/canada/millions-of-people-will-move-to-canada-in-the-next-20-years-but-where-will-they-live-1.5464205.

Frounfelker, R.L., Li, Z.Y., Santavicca, T., Miconi, D., & Rousseau, C. (2022). Latent class analysis of COVID-19 experiences, social distancing, and mental health. *American Journal of Orthopsychiatry, 92*(1), 121–32. https://doi.org/10.1037/ort0000593.

Gilbert, M., Pullano, G., Pinotti, F., Valdano, E., Poletto, C., Boëlle, P.-Y., D'Ortenzio, E., Yazdanpanah, Y., Eholie, S.P., Altmann, M., Gutierrez, B., Kraemer, M.U.G., & Colizza, V. (2020). Preparedness and vulnerability of African countries against importations of COVID-19: a modelling study. *The Lancet, 395*(10227), 871–7. https://doi.org/10.1016/S0140-6736(20)30411-6.

Goniewicz, K., Khorram-Manesh, A., Hertelendy, A.J., Goniewicz, M., Naylor, K., & Burkle, Jr, F.M. (2020) Current response and management decisions of the European Union to the COVID-19 outbreak: A review. *Sustainability, 12*(9), Article 3838. https://doi.org/10.3390/su12093838.

Gordon, M., Jayakumar, N., Schaffer, D., Vije, M., Schaffer, A., Niederkrotenthaler, T., Pirkis, J., and Sinyor, M. (2022). An observational study of suicide-related media reports during the early months of the coronavirus d 2019 pandemic in Canada. *The Canadian Journal of Psychiatry, 66*(1). https://doi.org/10.1177/07067437221125316.

Government of Canada. (2021). *COVID-19 and deaths in older Canadians: Excess mortality and the impacts of age and comorbidity."* Available at https://www.canada.ca/en/public-health/services/diseases/coronavirus-disease-covid-19/epidemiological-economic-research-data/excess-mortality-impacts-age-comorbidity.html.

Government of Canada. (2023). *Canada's aid and development assistance in response to the COVID-19 pandemic.* Available at https://www.international.gc.ca/world-monde/issues_development-enjeux_developpement/global_health-sante_mondiale/response_covid-19_reponse.aspx?lang=eng.

Helper, S., & Soultas, S. (2021, 17 June). *Why the pandemic has disrupted supply chains.* White House Council of Economic Advisors. Available at https://www.whitehouse.gov/cea/written-materials/2021/06/17/why-the-pandemic-has-disrupted-supply-chains/.

Jacobs, J. (1961). *The Death and Life of Great American Cities.* Random House.

Jensen, N., Lyons, E. Chebelyon, E., Le Bras, R., & Gomes, C. (2020). Conspicuous monitoring and remote work. *Journal of Economic Behavior & Organization, 176*, 489–511. https://doi.org/10.1016/j.jebo.2020.05.010.

Kerr, J., Panagopoulos, C., & van der Linden, S. (2021). Political polarization on COVID-19 pandemic response in the United States. *Personality and Individual Differences, 179*, Article 110892. https://doi.org/10.1016/j.paid.2021.110892.

Kite, T.A., Pallikadavath, S., Gale, C.P. Curzen, N., & Ladwiniec, A. (2022, August). The Direct and Indirect Effects of COVID-19 on Acute Coronary Syndromes. *Cardiology Clinics, 40*(3), 309–20. https://doi.org/10.1016/j.ccl.2022.03.002.

Kniffin, K.M., Narayanan, J., Anseel, F., Antonakis, J., Ashford, S.P., Bakker, A.B., Bamberger, P., Bapuji, H., Bhave, D.P., Choi, V.K., Creary, S.J., Demerouti, E., Flynn, F.J., Gelfand, M.J., Greer, L.L., Johns, G., Kesebir, S., Klein, P.G., Lee, S.Y., ... Vugt, M.v. (2021). COVID-19 and the workplace: Implications, insights and action for future research and action. *American Psychologist, 76*(1), 63–77. https://doi.org/10.1037/amp0000716.

Kurkowski, H. (2021, 8 December). *Monitoring remote workers: The good, the bad and the ugly. Forbes.* Available at https://www.forbes.com/sites

/forbesagencycouncil/2021/12/08/monitoring-remote-workers-the-good-the-bad-and-the-ugly/?sh=408b61db1da8.

Laframboise, K. (2021, 19 March). *Quebec turns to private companies for COVID-19 vaccination blitz. Global News*. Available at https://globalnews.ca/news/7706615/quebec-covid-19-vaccine-companies-expansion/.

Leach, A. (2019, 19 October). *Opinion election 2019: A national reckoning on climate change*. CBC News. Available at https://www.cbc.ca/news/canada/calgary/election-2019-climate-change-andrew-leach-1.5321425.

Martin, G.P., Curzen, N., Goodwin, A.T., Nolan, J., Balacumaraswami, L., Ludman, P.F., Kontopantelis, E., Wu, J., Gale, C.P., de Belder, M.A., & Mamas, M.A. (2021, May), Indirect impact of the COVID-19 pandemic on activity and outcomes of transcatheter and surgical treatment of aortic stenosis in England. *Circulation: Cardiovascular Interventions, 14*(5). https://doi.org/10.1161/CIRCINTERVENTIONS.120.010413.

McAlister, F.A., Parikh, H., Lee, D.S., & Wijeysundera, H.C. (2022). Health care implications of the COVID-19 pandemic for the cardiovascular practitioner. *Canadian Journal of Cardiology*. https://doi.org/10.1016/j.cjca.2022.11.014.

McCoy, L.G., Smith, J., Anchuri, K., Berry, I., Pineda, J., Harish, V., Lam, A.T., Yi, S.E., Hu, S., Rosella, L., & Fine, B. (2020). Characterizing early Canadian federal, provincial, territorial and municipal nonpharmaceutical interventions in response to COVID-19: A descriptive analysis. *Canadian Medical Association Journal, 8*(3), E545–E553. https://doi.org/10.9778/cmajo.20200100.

McGahan, A.M. (2020, 25 June). How contact tracing could change the economics of innovation. *Policy Options*. https://policyoptions.irpp.org/magazines/june-2020/how-contact-tracing-by-employers-could-change-the-economics-of-innovation/.

McGahan, A.M., & Suhkram, J. (2020, Fall). No going back: Challenges and opportunities after COVID-19. *Rotman Magazine*. https://www-2.rotman.utoronto.ca/insightshub/strategy-value-creation/no-going-back-challenges-oppor.

McLean, S. (2022). *Canada's national office vacancy rate dips in Q3: CBRE*. Available at https://renx.ca/canada-office-vacancy-rate-decline-cbre-q3-2022.

Musacchio, A., McGahan, A.M., Lazzarini, S., & Wong, J. (2020, July). Global health coordination necessary in a pandemic. *Policy Options*.

Ng, E. (2021, 19 June). *COVID-19 deaths among immigrants: Evidence from the early months of the pandemic*. Statistics Canada. Available at https://www150.statcan.gc.ca/n1/pub/45-28-0001/2021001/article/00017-eng.htm.

Opute, A.P., Iwu, C.G., Adeola, O., Mugobo, V.V., Okeke-Uzodike, O.E., Fagbola, O., & Jaiyeoba, O. (2020). The COVID-19 pandemic and implications for business: Innovative retail marketing viewpoint. *The Retail and Marketing Review, 16*(3). Available at https://journals.co.za/doi/abs/10.10520/ejc-irmr1-v16-n3-a8.

Paddon, D. (2021, 12 July). *Retirement levels expected to surge after pandemic-era slump: RBC report*. CTV News. Available at https://www.ctvnews.ca/business/retirement-levels-expected-to-surge-after-pandemic-era-slump-rbc-report-1.5506197.

PBS News Hour. (2021). *How restaurants have innovated to face the pandemic*. PBS. Available at https://www.pbs.org/newshour/economy/how-restaurants-have-innovated-to-face-the-pandemic.

Pepe, E., Bajardi, P., Gauvin, L., Privitera, F., Lake, B., Cattuto, C., & Tizzoni, M. (2020). COVID-19 outbreak response, a dataset to assess mobility changes in Italy following national lockdown. *Scientific data, 7*(1), 230. https://doi.org/10.1038/s41597-020-00575-2.

Pickup, M., Stecula, M., & van der Linden, C. (2020). Novel coronavirus, old partisanship: COVID-19 attitudes and behaviours in the United States and Canada. *Canadian Journal of Political Science, 53*(2), 357–64. https://doi.org/10.1017/S0008423920000463.

Pietromonaco, P.R., & Overall, N.C. (2022). Implications of social isolation, separation, and loss during the COVID-19 pandemic for couples' relationships. *Current Opinion in Psychology, 43*, 189–94. https://doi.org/10.1016/j.copsyc.2021.07.014.

Prime Minister of Canada. (2022). .*Producing made-in-Canada vaccines and creating hundreds of good jobs*. Available at https://pm.gc.ca/en/news/news-releases/2022/04/29/producing-made-canada-vaccines-and-creating-hundreds-good-jobs.

Restaurants Canada. (2020). *Navigating COVID-19: Updates and resources for foodservice operators*. Available at https://www.restaurantscanada.org/industry-news/navigating-coronavirus-covid-19-resources-for-foodservice-operators/.

Rowe, D.J. (2020). *The COVID-19 pandemic has forced a major rethink of downtown cores: Survey*. CTV News. Available at https://montreal.ctvnews.ca/the-covid-19-pandemic-has-forced-a-major-rethink-of-downtown-cores-survey-1.5121592.

Sasangohar, F., Moats, J., Mehta, R., & Peres, S.C. (2020). Disaster ergonomics: Human factors in COVID-19 pandemic emergency management. *Human Factors, 62*(7), 1061–8. https://doi.org/10.1177/0018720820939428.

Schembri, L. (2020, 18 June). *Living with limits: Household behaviour in Canada in the time of COVID-19*. Speech at the Greater Saskatoon Chamber of Commerce. Saskatoon, Saskatchewan. Available at https://www.bis.org/review/r200807g.pdf.

Singer, C.R. (2021, 10 May). *Canada to dramatically increase immigration to more than 400,000 per year. Immigration.ca*. Available at https://www.immigration.

ca/canada-to-dramatically-increase-immigration-to-more-than-400000-per-year/.
Statistics Canada. (2020). *The impact of COVID-19 on key housing markets.* Available at https://www150.statcan.gc.ca/n1/pub/11-627-m/11-627-m2020050-eng.htm.
Subramanian, S. (2021, 4 June). The lost year in education. *Macleans.* Available at https://www.macleans.ca/longforms/covid-19-pandemic-disrupted-schooling-impact/.
Tang, J. (2020, 29 April). How to build more resilient cities post-coronavirus. *The Conversation.* Available at https://theconversation.com/how-to-build-more-resilient-cities-post-coronavirus-136162.
Teodorovicz, T., Sadun, R., Kun, A.L., & Shaer, O. (2021). Working from home during COVID-19: Evidence from time-use studies. *Harvard Business School Working Paper 094.* Available at https://scholar.harvard.edu/files/rsadun/files/working_from_home_during_covid-19-_evidence_from_time-use_studies.pdf.
UNDP (United Nations Development Program). (2023). *COVID-19 socio-economic impact.* Available at https://www.undp.org/coronavirus/socio-economic-impact-covid-19.
University of British Columbia. (2021). COVID-19 lockdowns only reduced air pollution in some parts of the world. Available at https://news.ubc.ca/2021/04/06/covid-19-lockdowns-only-reduced-air-pollution-in-some-parts-of-the-world/.
Wibbens, P., Koo, W., & McGahan, A.M. (2020). Which COVID policies are most effective? A Bayesian analysis of COVID-19 by jurisdiction. *PLOS One, 15*(12), Article e0244177. https://doi.org/10.1371/journal.pone.0244177.
Wikipedia. (2022). The World Health Organization's response to the COVID-19 pandemic. Retrieved 27 November 2022 from https://en.wikipedia.org/wiki/World_Health_Organization%27s_response_to_the_COVID-19_pandemic.
Yusuf, E., & Tisler, A. (2020). The mortality and psychological burden caused by response to COVID-19 outbreak. *Medical Hypotheses, 143,* Article 110069. https://doi.org/10.1016/j.mehy.2020.110069.

3 Implications for Urban Structure and Mobility in the City

TOM HUTTON

Introduction: The Changing Field of Urban Circulation and Mobility

Patterns of circulation and mobility in the city are associated in large part with the evolution of urban economies, and more particularly with the spatial relations (and disjuncture) between the loci of employment and housing. The critical relationships between land use and employment still form defining features of transportation planning and investment in cities situated across diverse urban systems and forms of governance.

As I argue in this chapter, the twenty-first century has seen important and, in some respects, redefining change in how populations and diverse social groups express travel preference, opening new vistas of choice at the localized scale. These include as exemplars short-term car rentals, ride-hailing in jurisdictions which permit this practice, and the introduction of e-bikes and e-scooters as well as including innovative measures to enhance safety within multimodal circulation environments (Pagnucco, 2021). Cities have also experimented with investments in natural amenity including imaginative landscaping of heritage infrastructure to encourage pedestrian travel, exemplified by New York's High Line, emulated in several other cities.

Since the advent of the industrial city in the nineteenth century, and continuing into the twentieth century, transportation, circulation, and mobility have been associated with employment. Major investment in transportation infrastructure and systems facilitated the flow of workers to the zones and places of employment, initially to sites of manufacturing and factory work, then to the service economy of the twentieth-century city, concentrated within the central business district. As an example, the Department of Transport in Britain calculated that over half (57%) of those travelling to London on public transport were "for either commuting or business purposes" (Department for Transport, 2020).

But these largely progressive innovations in urban circulation and mobility have been counterposed against more problematic trends and developments. These include the dire socio-economic consequences of the COVID-19 pandemic and, increasingly, clear empirical evidence of climate change across global space, presenting an existential challenge to the habitation of the earth by humans and myriad other life forms and species.

I follow this introductory commentary with a concise description of what I see as critical aspects of change (and challenge) associated with circulation in the city, followed by a descriptive outline of principal realms of innovation, change, and disjuncture. I then present a discussion of "financialization," which I argue is a profoundly impactful agent of urban change (and more particularly, increasing inequality) across diverse city forms and systems of governance and planning. Relatedly, I present a discussion of urban change calibrated at the localized scale, and which I propose includes complex synergy (and disjuncture) between culture, technology, and the built environment.

I then expand the parameters of change to address implications of the corona virus, now in its fourth year, and emphasize how the pandemic has impacted the social practices of governance, business, and community relations. A final substantive section of this chapter identifies exemplary aspects of experimentation in business location and company operating characteristics as early responses to climate change, including the widespread practice of working from home where this is feasible, but concluding with acknowledgement that these are at best incremental measures in a much more difficult and demanding pathway to sustainable development.

Tracing Lineaments of Innovation in Urban Circulation and Mobility

Over the last century innovation in transportation systems has been associated with growth and change in the city, notably with respect to the structures of the economy and related patterns of employment and population change. An example (and influential precursor) is the Metropolitan Line (1863) as the pioneering project of London's Underground, designed to facilitate the flow of suburban commuters to the City. Another classic case is Chicago, where Robert Lewis (2008) has written about the strategic impacts of urban infrastructure investment in that American city, including goods movement and labour circulation associated with industrial development.

In Canada, pioneering systems of public transit within the city, notably streetcars in Montreal and Toronto (1861 for each), were followed by rail service connecting growing suburban populations with the development of the central city economy of intermediate services, including finance, head offices, and business services. The exceptional growth of the Greater Toronto Area (GTA), Canada's global city, has been shaped in part by an extensive system of fixed-rail and bus service to serve an expanding regional metropolis, but also by high volumes of auto commuters.

In Vancouver, the pioneering Livable Region Plan (1986) established the need for regional transit (fixed-rail and expanded bus services) as key to a sustainable future. In East and Southeast Asia the spectacular growth of business centres such as Tokyo, Seoul, Shanghai, and Singapore has been accompanied (and to a significant extent facilitated) by massive public investment in both fixed-rail transit systems and bus service.

In each case the argument for investment in public transportation systems, whether anticipatory or simply in response to established demand patterns, is embedded within a matrix of economic rationales and environmental values. There is every prospect that these rationales will continue to inform decision-making on urban-regional transportation options and choices.

But in this chapter, I argue that the mix of factors shaping transportation innovation has expanded to include both larger issues, and more nuanced behavioural factors. First, the weight of scientific evidence demonstrating the dire environmental consequences of climate change constitutes a powerful argument for investing in more sustainable transport infrastructure and travel choices at the strategic scale of decision-making, as well as accommodating more localized options including bicycles and e-scooters. Relatedly, an increasing number of local authorities are introducing forms of road pricing both to disclose, and effectively address, the full costs of private auto use in cities and urban regions.

Second, recurrent cycles of innovation in communications technology in this century, including personal computers, the Internet, and more recently Smartphones, have reconfigured networks of commercial, retail, and interpersonal connectivity. "Virtual shopping," exemplified by the growth of Amazon as well as UPS and other warehouse-to-home delivery networks, demonstrates another facet of the increasingly complex transportation-communications nexus with its service provider price signals and, for consumers, trade-offs between cost, convenience, and experience.

Third, a scalar recalibration of issues and opportunities associated with the costs of travel (environmental as well as financial) has stimulated innovation at the localized (district and community) scale. These include the provision of bicycles for short-term hire at the local scale sponsored by local governments, by advocates for cycling, and in some cases local business associations, and form part of the branding and marketing of cities. There are also options for short-term travel in the form of ride-hailing, such as Uber, among other options.

A fourth aspect of transport innovation incorporates a generational dimension, exemplified in the use of e-bikes and e-scooters as choices on the part of (mostly) younger travellers in the city, and reflecting a recalibration of the scalar features of space, place, and mobility at the local level.

Finally, any contemporary assessment of change in travel modes must acknowledge implications of the COVID-19 pandemic. The first-order impacts of COVID include of course massive human costs in terms of death, disease, welfare, and income. There are also serious implications of the pandemic for many worldwide who have experienced damaging – and in many cases debilitating – social isolation and alienation from family and friends. Debates over responses to COVID and public health treatments, including vaccination as well as isolating protocols, have proven toxic in many states and societies, notably the US and UK among others. These conflicts over values and choices have also extended to debates over transportation preferences, which are parts of the larger landscapes of bitter contestation that characterize the age.

As I discuss in this chapter, the pandemic has influenced the broader patterns of enterprise location and internal operating dynamics of the urban economy, hollowing out the office economy and much of the retail and consumption sectors.

Processes and Consequences of Urban Change

Over the past half-century, the form and structure of the city have been reconfigured by internal processes but more forcibly by exogenous factors and influences. These have included the globalization of production and consumer markets, and localized imprints of an evolving international division of labour; political factors incorporating the pervasive influence of neo-liberalism on the regulation of property development and more particularly housing; and, increasingly, technological factors exemplified by innovation in communications and production technologies that yield productivity enhancement but also labour-shedding.

In this century there has been if anything a deeper and more pervasive array of forces acting on the city and its constituent spaces and territories. Compelling scientific evidence of climate change and its devastating impacts on the natural environment and its constituent ecological systems and life forms has stimulated a more urgent enquiry into remedial programmes. But it seems clear that even in the most progressive societies there remains a disconnect between current policy practice, on the one hand, and, on the other, what is required to bring human values and behaviour into alignment with the conservation of critical ecological systems (Stephens, 2022). At another scale of causality there is the heightened competition among cities for capital, skilled labour, and market advantage across key sectors, producing inequality as well as opportunity for urban populations.

What characterizes urban change in the present era is not only a forceful array of exogenous factors but also local aspects that include the quality of space in the city. In many cases these represent not simply an inert factor of development but rather a rich (and often conflictual or contested) source of prompts for both private and public actors and interests. At one level there is the basic geometry of space, including location, scale, and dimensionality – each of which presents both possibilities and limitation for various classes of development. But in many cases, there are physical, historical, and semiotic qualities of place and territory that evoke memory and the urban imaginary as cues for recurrent cycles of redevelopment and capital relayering in the city.

Important framing devices for study include functional dynamics of growth and change and, relatedly, typologies of space in the city. In each case there are complex implications for modal travel preference and patterns of mobility and circulation. I start first with financialization, which narrowly pertains to the massive increase in capital deployed from domestic and (increasingly) exogenous sources for high-margin development, but more strategically can be situated as a principal paradigmatic force for change: in the words of some scholars, nothing less than a successor to the post-industrial city of the late twentieth century. In this reading, the power of capital shapes political values, policy preferences, and local planning, overriding the needs and values of local communities and neighbourhoods.

A second and related aspect of redevelopment in the city comprises very large capital sites characterized as "urban mega-projects," which bring together multiple agents, factors, and narratives of change and displacement. There are of course differences in scope, scale, and ambition from place to place, but in many cases they embody expressions of civic aspiration and involve large pools of capital sourced from local and more frequently external actors.

These development sites typically project massive scale, employ star architects, shape in ways large and small contours of urban morphology, and are (re)deployed by the local state as evidence of commitment to competitive advantage in a globalizing world. These capital projects typically entail major investment in transportation and communications systems, and in some cases experimentation in forms of mobility and circulation.

The third key factor of urban change comprises the multifaceted domain of "culture," comprising both semiotic and tangible expressions of identity, history, and memory, values, tastes, and preferences. We can readily identify aspects of revealed cultural values in local design idioms, architecture, and the myriad performative arts and allied institutions that increasingly comprise not fringe aspects of the city but defining qualities deployed in place-making and external marketing. Relatedly, innovation in progressive forms of mobility represent contemporary expressions of cultural values in the city.

Fourth, I identify technology as a critical factor in urban development across increasingly broad and complex domains. At the urban systems scale, technology facilitates the transmission of capital, design, and other factor inputs, while within cities advanced communications technology can expedite the connectivity of elements of the regional space-economy. Cities have actively promoted technology districts as lead developmental features, observed in the MARS project in Toronto, Seattle's Fremont and South Lake Union districts, and Shoreditch in London. Within each, the built environment, the semiotic aspects of place, and talent and creativity are recombinant features of innovation across a spectrum of industries and product sectors. Further, many of these urban innovation districts have been designed to benefit from (and promote) "walkability" in terms of access and movement: a key tenet of the green city movement.

Relatedly, I discuss circulation as a key aspect of urban development, including both the everyday patterns of movement in the metropolis and its contribution to the social life of cities as well as critical field of innovation in transportation systems, and also the field of experimentation in mobility enhancement. These include for the purposes of example ride-sharing and ride-hailing, and the use of e-scooters and e-bikes. Innovation in metropolitan transportation systems is typified by the recent opening of the Elizabeth Line serving London and its larger region, some 160 years after the pioneering Metropolitan Line in Britain's capital.

Financialization as Programme, Policy Value, and Practice: A Focus on Vancouver

The power of capital and allied business cohorts was identified as a critical force for urban development and associated upgrading and displacement by Harvey Molotch almost a half-century ago in his depiction of the urban growth machine (Molotch, 1976). David Harvey's penetrating account of the power of capital in the reproduction of space and affiliated social dislocations in Paris offers a counter-narrative to the standard imagery of preservationist practices in the French capital relative to (notably) London and New York, where redevelopment is the dominant modality of city planning (Harvey, 2006).

In this century scholars have advanced our understanding of the power of capital to transform the city and its constituent spaces and sites. Greta Krippner describes financialization as "a pattern of accumulation in which profits accrue primarily through financial channels rather than through trade and commodity production" (Krippner, 2005, 174), tendencies observed in many cities and societies within states notionally characterized by differing political systems.

In a paper titled "Making Sense of Financialization" Natascha van der Zwan describes the parameters of an "increasingly autonomous realm of global finance" (van der Zwan, 2014, 99), within which finance entails "not the neutral allocation of capital, but rather an expression of class, a control mechanism . . . a rationality associated with late twentieth century capitalism" (102). Coupled with the growth machine values and practices of the state, financialization represents a potent force for serial upgrading and dislocation at the local level in the city.

The potential of capital to transform urban territory in the late modern era was actualised first by Margaret Thatcher's neo-liberal revolution following her election as British prime minister in 1979, mimicked by Ronald Reagan in the US and subsequently in a diluted form by Canadian prime minister Brian Mulroney, among other international exemplars. There were to be sure salient differences from place to place, but the defining values included a political willingness to displace long-established communities in the interests of redevelopment in the city, with the Canary Wharf megaproject situated in London's Isle of Dogs a paradigmatic example.

Vancouver

Capital, both domestic and increasingly foreign-sourced, has been instrumental to the reproduction of space and territory in Canadian cities, and involving in each case investment and innovation in transportation. Expo '86 (originally "Transpo 86") in Vancouver was an instructive example, with exhibitions arrayed on the north shore of False Creek depicting transportation innovation sourced from domestic and international places. The construction of Greater Vancouver's first fixed-rail rapid transit system linking the downtown with south-eastern suburbs was a key affiliated feature of the project.

The City of Vancouver's planning for the former Expo site took the form of new high-rise residential development situated on the north shore of False Creek, adjacent to Vancouver's central area, and directly across from the earlier (and paradigmatic) False Creek South community, a hallmark of multi-level government planning and investment in the 1970s (Ley, 1996).

The city planning department's project plans, endorsed by Vancouver City Council and involving a path-breaking program of public consultation (as well as a decision to forego possibilities of industrial renewal in False Creek), attracted local investment but more decisively offshore capital, notably from Hong Kong and other Asia-Pacific markets, marking the introduction of a new development model for Vancouver, widely emulated in megaprojects situated elsewhere.

While planning for megaprojects includes in most cases at least a nod to public consultation with local communities, the scale of both public funding and market investment tends to drive the agenda. As an example, the planning and development of the Broadway Line linking Vancouver's central area with the University of British Columbia involves an extensive high-rise housing program situated within the larger transportation corridor, with revenues accruing to private developers but also supporting public infrastructural investment.

Such is the power of capital operating within Vancouver's housing market, increasingly oriented to affluent first-time buyers, that it effectively limits other developmental possibilities. The city's space-economy includes tertiary education, the visitor economy, and pockets of innovation in places like Mount Pleasant and the areas fringing the central waterfront, and niches within the largely hollowed-out central business district, following the secular decline of the financial and business industry sectors.

But overall, lucrative revenue streams accruing from high-rise residential development in Vancouver, including both private-sector profits and taxes and fees for government, suggest the operational features of a latter-day urban growth machine far exceeding the parameters and force of those described by Harvey Molotch almost a half-century ago.

Condominium development, and the adaptive reuse of older office buildings for residential use, are features of many cities worldwide, although there are of course significant aspects of contingency from place to place.

Culture and Technology in the Spaces of the City: Synergy, Succession, and Dislocation

While capital, sourced both from domestic and international actors and agencies, represents a first-order factor of urban growth and change, the motive forces and palette of urban activity in many cities present a more complex morphology, accommodating a diversity of use types, although upgrading is pervasive and disruptive for established neighbourhoods and social groups.

As is well known, gentrification within the city was shaped primarily by the rise of a new middle class of professionals and managers who eschewed the culturally anodyne suburbs in favour of the inner city – a principal narrative of late-twentieth-century cities within advanced societies. This social movement incorporated a distinctive cultural aspect, expressed in affinity for the post-industrial built environment, in the semiotic value of neighbourhood environments embodying cultural capital, and the elevated consumption preferences of professionals and managers.

The gentrification narrative in this century has evolved, incorporating new actors, agencies, and motive forces, and including in globalizing cities large inflows of those with access to very large pools of capital. Allied to this inflection of social change is a separate but related narrative of industrial succession at the localized scale in the city, reflecting the city's role as a crucible of innovation. The original cohorts of gentrifiers have been accompanied by new, more affluent social actors, and a vanguard of activities, including cultural industries and institutions, and innovative technology firms, representing a critical new field of urban change.

The first wave of change involved artists who in many cases comprised the initial cohort of gentrifiers, and more recently have incorporated professional design firms and cultural institutions, clusters of firms engaged in processes of technological innovation, and, increasingly, high-value consumption sites and precincts. As for previous rounds of upgrading in the city, sequences of change take the form of recombinant industries, key actors and entrepreneurs and state agencies committed to high-margin development in the spaces of the city as keys to regeneration.

While there is of course novelty within processes of economic change, a careful review of trends and factors within the spaces of the city disclose significant commonality, underscoring the value of urban districts for industrial innovation over time.

While the city performs as a site of technological innovation, cultural experimentation, and high-amenity habitus for many, for others it represents a place of deprivation and struggle. Within Vancouver's Downtown Eastside, the inflow of middle-class residents within parts of Strathcona and a number of upscale restaurants and high-value product producers (notably Herschel) situated in Railtown have exerted more upgrading pressure on low-income residents and those afflicted with addiction to fentanyl and other drugs (Newman, 2014), publicised in the media as a tent city comprised of very low income residents and characterized by rampant crime.

Just a little way to the west, in Victory Square, a lively economy of arts and cultural experimentation coexists uneasily with a population experiencing poverty (Mckenzie & Hutton, 2015). And beyond the central city, there are less-publicised low-income neighbourhoods situated within outer suburban municipalities, such as Surrey and Coquitlam.

Even in cities considered liberal in broad political terms, and successful in capturing large shares of high-growth, high-value industries and employment, there are populations of deprived communities struggling with low incomes and access to services and basic life needs.

The Coronavirus Pandemic and Its Impacts on Business, City Planning, and Sustainability

The coronavirus pandemic has generated enormous human costs in terms of infection, illness, disability, and death. Important spillover effects include massive dislocations of labour markets, company closures and rising unemployment, and sharply reduced income streams for firms, families, and governments at all levels.

In this more hopeful post-2023 stage of the pandemic there is of course speculation on how economies, labour markets, and business practices may evolve. At a broader level, research has addressed how businesses and institutions – private and public – respond to the "constrained opportunity" environment of infection, in an economic landscape already reconfigured by new global actors, rivalries, and conflicts as well as the impacts of telecommunications on the space-economy described earlier in this chapter. What follows is a necessarily selective review of trendlines and consequences.

The expansive range of possibilities of the pandemic on the organizational structures and operating systems of business is the subject of a paper published by Nicolai J. Foss in the *Journal of Management Studies*. Underscoring the degree of uncertainty associated with the pandemic, Foss suggests that processes of scenario-building "may mean everything from a complete reversal of the pre-pandemic situation to a more or less permanent situation of sporadic outbreaks and lockdowns that require more social distancing" (Foss, 2021, 270).

In a similar vein, the World Economic Forum has acknowledged uncertainties in the employment and labour management environment under COVID but also identified five ways in which the pandemic has changed workforce management and operations. These include (1) a "rapid reskilling" of the workforce to enable the rapid adoption of new technologies and methodologies, requiring in return acceptance of a "learning mindset"; (2) changing leadership and management competencies, including effective (and sensitive) ways of addressing fear and uncertainty; (3) increased emphasis on trust and transparency, rather than "control," in order to facilitate employee performance with less direct oversight; (4) a renewed focus on employee well-being in a time of pandemic, acknowledging the "four pillars" of mental, social, physical, and financial well-being; and (5) a focus on learning to work in a more "agile" way, as the pandemic forces change in the pace of workplace innovation (World Economic Forum, 2020).

The exigencies of operating in competitive local and global markets has stimulated private-sector responses to the pandemic, incorporating in many cases those enumerated above. But government agencies and public institutions have advanced an extensive range of remedial programs and recommendations. As example, the Government of Ontario implemented changes to a range of operations and programs. These include flexibility around meetings and business operations during the pandemic, including "virtual" meetings and deferred annual meetings, and new procedures for filing key reporting and financial documents to minimize the need for proximate contact (Government of Ontario, 2021).

Ashley Stahl has developed an outline of likely changes in the conduct of business in space shaped by the pandemic. She offers a nostalgic description of life in the pre-pandemic office, including "brainstorming on communal couches, buffet lines of corporate cafeterias, shared cubicles, and, for the lucky few, a keg of beer or nitro coffee at the ready" (Stahl, 2021). Typical of corporate investment strategies for office space pre-pandemic included Pinterest leasing five hundred thousand square feet of office space for its workforce, including a "reported termination fee of $89 million," and Twitter leasing five hundred thousand square feet of office space in San Francisco. But by May 2020 both Twitter and Facebook approved working remotely for most of their respective workforces, while Shopify CEO Toby Lutke approved permanent working from home (Stahl, 2021).

A legacy of the pandemic includes nuances of change in employment location and travel within the city and urban region. In the case of the Seattle region, in 2023 a survey conducted by a local news program found that 52 per cent of employees sampled worked remotely Mondays, and 57 per cent on Fridays: part of a hybrid business and employment model that included over half the workforce travelling to work Tuesday-Thursday (King County Newscast, 2023). What this case demonstrates is that while the pandemic has stimulated widespread working from home, enabled by telecommunications innovation, there is still value for many companies in having employees travel to places of work for contact-intensive forms of communication, deliberation, and exchange.

While uncertainties abound, we can conjecture that the effects of the pandemic on business, the economy, and labour markets have in important respects potentiated trendlines already evident pre-COVID. These include the substitution of capital for labour both in services and goods production among advanced and, increasingly, developing economies, and the convenience of online retail over personal shopping. Relatedly, the downsizing of office employment, the principal form of labour market growth from the 1970s onward in major business centres – accelerated by the effects of the pandemic, which also impacts retail and consumption industries and labour – has effectively "hollowed out" the central city workforce in many cities.

Climate Change and Urban Sustainability

As in other cases, crisis tends to generate major socio-economic costs and hardship, but also stimulates innovation. The experience of the last three years has generated new conversations and discourses concerning the future of cities, encompassing modes and patterns of mobility, circulation, and consumption, as evidenced in the massive growth of Amazon's warehouse-to-household delivery service.

The growth of Uber and myriad other forms of ride-hailing was impacted by the contagion vectors of COVID, and most restaurants were forced to close or offer only takeout orders. But mutations stimulated by the pandemic include growth in the companies engaged in point of food production to home consumers, including Uber Eats,

DoorDash, and many other variants (see chapter 10). For some portion of the consumer market, these practices are likely to survive as a means of convenience.

COVID has also stimulated conversations about how the quality of urban living can be enhanced by the learning process generated from the last three years. At this early stage, a fully developed and coherent narrative is still in its nascent stage, but some observations and suggestions have been advanced. To illustrate, city planning professor Jordi Honey-Roses has offered observations about how the COVID experience can stimulate city and neighbourhood planning innovation for a more sustainable city: "As people were in enforced lockdown with restricted mobility, urban planning concepts like the 15-minute city came to the fore, with more people becoming intrigued by the promise of a more local and slow pace of life where you can go to school, work and shop within a 15-minute walk or cycle from your home" (Honey Roses, 2021).

As examples drawn from cities situated in different countries and cultures, Honey-Roses cites Paris, which has restricted vehicles in the city centre, while adding bike lanes and creating more green space; Bogota, which has added 130 kilometres of bike lanes to the urban transportation system; Barcelona, which has established a network of "green streets" with reduced speed limits; and Vancouver, which has implemented a program of street calming, bike lanes, and sidewalk extensions. Entrepreneurial companies in Vancouver offer a range of sustainable travel options, including short-term rental of Tesla cars (see Figure 3.1).

Relatedly, cities are also focal points for research and policy experimentation associated with the consequences of climate change. There is a vigorous and sometimes conflictual discourse around the timelines of climate change, and an unhelpful narrative of denial on the part of some political figures. But there is evidence that scientific panels, regulatory bodies, and media channels are gaining traction in establishing what will be very demanding protocols for managing growth and change.

The urgency of the climate change problematic was recently affirmed in a statement issued by United Nations secretary-general António Guterres at the UN COP27 Climate Change Summit. As Guterres declared during the summit, "our planet is fast approaching tipping points that will make climate chaos irreversible ... [w]e are on a highway to hell with our foot on the accelerator" (*Guardian*, 2022). In framing the temporal dimension, Secretary-General Guterres declared that the battle for managing climate change will be won (or lost) in the current

Figure 3.1. "Zerocar" electric vehicle for rent, Kitsilano, Vancouver

decade (*Guardian*, 2022). If this prognosis is correct, the future for viable life on this planet appears dire as such a reversal of long-established human practices seems a remote prospect at best.

Conclusion: Parameters of Urban Change, Crisis, and Opportunity

For this chapter, I have endeavoured to trace the lineaments of urban change, which include key internal economic, social, climate, and spatial factors. In this concluding section I offer a synopsis of change and its implications for cities derived from this reading, including the effects of change in mobility and circulation in the city.

A key reference point for understanding both the origins of contemporary change and clear departure in the present points is the globalization experience of the last two decades of the twentieth century: notably, a new international division of labour favouring rapid industrialization in East and Southeast Asia, and a corresponding collapse of Fordist industry among the economies of the North Atlantic realm.

The corollary rise of a new middle class of professionals and managers reshaped the social order and housing markets of cities in the West, and increasingly in East and Southeast Asia (Ley, 2010).

The remarkable growth and socio-economic transformation of major cities was a defining spatial effect of capitalism and its national/regional variants, and appeared to presage an era of managed competition and cooperation. State commitments to education, housing, and public transit at the regional and local levels offered both efficiency gains and a more sustainable urban future.

The discussion presented here offers a necessarily selective review of innovation in approaches to problems (and opportunities) of transportation in the city, incorporating concepts of circulation and mobility as framing features. Broadly, the last two decades or so of the last century was characterized in many societies by debates over private automobile use and (alternatively) public transportation, the latter including a range of rapid transit modes, both fixed rail and the generally more economical bus lines.

In part the discourse included not only narrow cost-benefit parameters, but also arguments around which form of transit provided the most effective approach to larger urban planning and regional development, with fixed rail seen as more strategic in terms of shaping sustainable urban form, and with buses as more flexible and generally cheaper. The field of circulation and mobility has become more complex, incorporating new actors, technologies, and travel modalities. Innovation in communications technology has enabled in many business, government, and education settings practical alternatives to meeting face to face. This practice will likely continue even as the pandemic becomes more manageable.

As I have discussed in this chapter, a proliferation of mobility options at the localized scale, including e-bikes and e-scooters, offers convenient means of travelling shorter distances within more compact spaces of the city, enriching the experience of neighbourhoods, cultural quarters, and consumption districts that comprise important features of the convivial city. As cities pursue densification strategies, these mobility innovations are likely be useful complements.

To these we can add ride-hailing, such as Uber, and short-term auto rental, including (in British Columbia) Evo, operated by the BC Automobile Association. These innovations appeal especially to younger riders, and therefore are imbued with salient demographic and cultural signifiers, forming in turn the potential for an enrichment of the human experience of the city and its diverse spaces.

REFERENCES

Department for Transport (UK). (2020). *Transport solutions for a greener Britain*. Government of the United Kingdom.

Foss, N.J. (2021, January). The impact of the Covid-19 pandemic on firms' organizational design. *Journal of Management Studies, 58*(1), 270–4. https:doi .org/10.1111/joms.12643.

Government of Ontario. (2021, 22 October). Responses to COVID. News release. Ontario releases plan to safely reopen Ontario and manage COVID-19 for the long term. News release. Office of the Premier.

Guardian, The. (2022). The global climate fight will be won or lost in this decade'.https://www.theguardian.com/environment/2022/nov/07/cop27 -climate-summit-un-secretary-general-antonio-guterres/accessed/2022 /nov/07.

Harvey, D. (2006). *Paris, capital of modernity*. Routledge.

Honey-Rosés, J. (2021). *Our cities may never look the same again after the pandemic*. School of Community and Regional Planning Research Paper.

King County Newscast. (2023, 16 March). Travel trends in King County.

Krippner, G.R. (2005). The financialization of the American economy. *Socio-Economic Review, 3*(2), 173–208. https://doi.org/10.1093/SER/mwi008.

Lewis, R. (2008). *Chicago made: Factory networks in the industrial metropolis*. University of Chicago Press.

Ley, D. (1996). *The new middle class and the remaking of the central city*. Oxford University Press.

Ley, D. (2010). *Millionaire migrants: Transpacific lifelines*. Blackwell.

Mckenzie, M., and Hutton, T. (2015). Culture-led regeneration in the post-industrial built environment: Complements and contradictions in Victory Square. *Journal of Urban Design, 20*(1), 8–27. https://doi.org/10.1080/13574809.2014.974149.

Molotch, H. (1976). The city as growth machine: Toward a political economy of place. *American Journal of Sociology, 82*, 309–32.

Newman, K. (2014). Commodifying poverty: Gentrification and consumption in Vancouver's Downtown Eastside. *Urban Geography, 35*(2), 157–76. https://doi.org/10.1080/02723638.2013.867669.

Pagnucco, B. (2021). *Scoot over: Strategies for facilitating safe interaction between e-scooter riders and pedestrians*. Capstone Professional Report: Plan 528, Masters in Community & Regional Planning. University of British Columbia.

Stahl, A. (2021). *The future of offices and workspace, post-pandemic*.Forbes. https://www.forbes.com/sites/ashleystahl/2021/04/16/the-future-of-offices-and -workspaces-post-pandemic/?sh=258ffd5e6442. Accessed 5 November.

Stephens, B. (2022, 28 October). Yes, Greenland's ice is melting. *New York Times*. https//www.nytimes.com/interactive/2022/28/opinion/climate-change-bret-stephens.html.
van der Zwan (2014). Making sense of financialization. *Socio-economic Review, 12*(1), 99–129. https://doi.org/10.1093/ser/mwt020.
World Economic Forum (2020, 2 June). Five ways Covid-19 has changed workforce management. Retrieved 7 November 2022 from https:/www.org/agenda/2020/06/covid-homeworking-systems-ofchange-face-of-workforce-management/.

4 Urban Mobility and the Digital Economy: Capturing the Experiences of Canadian Cities

TARA VINODRAI

Introduction

For those of a certain age growing up in Canadian cities, a familiar sound of summer was the ringing of bells as the Dickie Dee ice cream cart passed by, its driver pedalling hard to propel a modified tricycle to parks and playgrounds to serve on-demand cool treats. The Dickie Dee Ice Cream Company was founded in the 1950s in Winnipeg, Manitoba, and was purchased by two brothers, who subsequently grew a sizeable ice cream delivery business with a fleet of modified tricycles, scooters, and trucks, with operations in over three hundred cities across North America. In 1992, the Dickie Dee Ice Cream Company was acquired by Unilever, and in the early 2000s the urban ice cream delivery business was closed (Dunne, 2015). Fast forward to the present day, and in some Canadian cities it is not unusual to see a cargo bike or e-bike passing by on its way to deliver packages to homes and local businesses. And ice cream trucks still circle around city neighbourhoods on hot summer days.

While this chapter is not about ice cream, the story of the Dickie Dee Ice Cream Company reminds the reader that mobility-enabled services have been part of the urban fabric of Canadian cities for a long time. Other scholars similarly note that much of what has been celebrated as new or disruptive has been decidedly ahistorical in perspective (Biber et al., 2017), missing the broader context in which these so-called disruptive technologies and business models have emerged and taken hold. The circulation of people and goods within cities is an enduring feature of the city. And the infrastructure, institutions, and systems that facilitate these movements help to form and define the built environments in which we live (Moos et al., 2020). Capturing the characteristics of – and changes in – urban mobility in Canadian cities is difficult.

The next section discusses the challenges in systematically measuring and capturing urban mobility with an eye to comparing the experiences of Canadian cities from coast to coast to coast. It highlights the limitations of both traditional, government data sources and the growing numbers of other sources of data that help us to understand various facets of urban mobility. Following this discussion, I explore three specific areas of contemporary urban mobility research and policy interest: sustainable commuting and micromobility; urban logistics and delivery systems; and the rapid rise of remote work and its impact on daily mobility patterns in Canadian cities. In each case, I deploy evidence from both traditional and non-traditional data sources. In doing so, I demonstrate the data challenges and issues inherent in understanding mobility in Canadian cities, while also highlighting the differential experiences of Canadian cities. I conclude by revisiting the benefits and drawbacks of existing data to understanding urban mobility in Canada, and advocate for high-quality, publicly accessible data.

Measuring Urban Mobility in Canadian Cities

Urban mobility encompasses a range of activities related to the movement of goods and people in the city (Brail & Donald, 2020). Yet, measuring mobility-related activity, whether it be the movement of goods or people in cities, can be a substantive challenge. This is especially true when considering the emerging digital technologies and platforms that are emboldening the growth of urban mobility (see Rojas and Miller, chapter 5). In 2017, Statistics Canada released its first (and – to date – only) estimates related to the use of digitally enabled platforms. Statistics Canada asked a subset of Labour Force Survey (LFS) panel respondents seven questions related to the use of ride-hailing services (e.g., Uber, Lyft) and short-term rental services (e.g., AirBnB, FlipKey). The results showed that 7 per cent of Canadian adults used ride-hailing services, with variation by age and location. Urban data were only available for the largest eight census metropolitan areas (CMAs) in Canada. Amongst these CMAs, the largest proportions of adults using ride services were in Ottawa-Gatineau (17.6%), Toronto (14.8%), and Edmonton (9.8%) (Statistics Canada, 2017). The survey also provided limited information on spending on ride-hailing services, suggesting that Canadians had spent an average of $122 per person and a total of $241 million nationwide (Statistics Canada, 2017). These data provide very limited (and dated) information on the state of ride-hailing, which is only one aspect of urban mobility.

More recently, Statistics Canada (2019) released a framework for assessing the digital economy, which distinguishes between digitally enabled infrastructure, digitally ordered transactions (i.e., e-commerce), and digitally delivered products. In their report, the authors concede, however, that it is difficult to separate platform-enabled transactions from other forms of e-commerce. In practice, this means that the framework does not adequately capture – or even include – many activities such as ride-hailing or food-hailing, which are central to understanding mobility in the contemporary Canadian city.

The public data described above provide very limited insights into urban mobility in Canada, only scratching the surface of our understanding (see Palm and Farber, chapter 6; Kong and Leszczynski, chapter 7; McNee and Miller, chapter 8). Faced with such limited information and data, it is important to consider other ways to capture and measure urban mobility in Canadian cities. Elsewhere, efforts to assess the dynamics of local and regional economies have increasingly relied upon digital third-party data sources at the individual, firm, and institutional level (Fang et al., 2022; Feldman & Lowe, 2015; Holicka & Vinodrai, 2022; Spigel & Vinodrai, 2021). As Feldman and Lowe (2015) note, such novel digital data are often drawn from "accessible and frequently updated records, scraped from the web, pulled from voluminous documents through text analysis, found on open data platforms, [or] purchased from third parties" (1791). Examples include data scraped from social media platforms, such as Twitter, or collected via cell phones or loyalty programs (see also Brail, 2022a; Brail & Donald, 2020; Fang et al., 2022).

However, analysts face a series of interrelated trade-offs in using or relying upon novel digital datasets developed by private (and normally, for-profit) interests rather than those developed for the broader public good by governmental or not-for-profit agencies. This section identifies a series of interrelated issues that one encounters in selecting traditional public data sources or other digital data sources. Table 4.1 summarizes the key issues in using different types of data, including issues related to classification, temporality, privacy, access and availability, granularity, and geography. I discuss each of these issues in greater detail below.

First, there are issues related to classification. Traditional data sources often rely upon strict, standard classification systems that provide a consistent framework for analysis and endure over time. In contrast, private data providers often use their own classification systems to suit their own purposes. Although, as Fang et al. (2022) note, many emerging digital sources involve big data, with millions of unclassified,

Table 4.1. Differences between traditional and unconventional digital data sources

	Traditional sources	Unconventional sources
Source	Government Public agencies Not-for-profit organizations	Private
Classification	Standard classification systems that are updated or modified infrequently	Categories designed by provider User-driven classifications
Temporality	Less frequent collection and reporting of data (e.g., monthly, quarterly, or annually)	Real-time data collection and reporting often possible
Privacy	Strict protocols governing privacy of individuals and firms	Providers devise their own end-user agreements and standards
Data quality	Transparency related to data collection and processing	Little information provided regarding data collection and processing
Access and availability	Costs are variable with some data available for free Data often available via repositories over long periods of time	Costs are variable, with most data being expensive. User agreements and licensing arrangements limit access
Granularity	Ability to parse data using an array of socio-economic and demographic information	Limited information available for socio-economic and demographic groups
Geography/ Spatial boundaries	Standard geographical units and/or administrative boundaries Data may be geocoded to align with administrative/ analytical spatial units Often available for small-scale geographies, including socio-economic and demographic information	Data often reported for non-standard geographical units Data may be geocoded to align with administrative/analytical spatial units May be available for small-scale geographies, without socio-economic and demographic information

individual observations where it is possible to classify or code data in a flexible manner. This, in some cases, makes it possible to categorize data to align with standard classification systems. However, rigid classification systems can make it difficult to identify new or emerging activities (Feldman & Lowe, 2015). One of the biggest challenges associated with any form of new economic activity, including platform-based activity, is the ability of researchers or policymakers to estimate even basic indicators, such as the number of firms or employment, in these emerging areas. Ride-hailing and other digital platform activities that permeate our cities are a prime example. In the Canadian context, platform firms that offer ride-hailing services could be classified as either technology companies or taxi companies (see Brail and Donald, chapter 10; Spicer, Zwick, and Young, chapter 13). The classification of these firms was a point used by the firms themselves in early disputes with jurisdictions over regulation and governance (Spicer et al., 2019; Woodside et al., 2021). In other words, the consequences of classification extend beyond simply measurement.

A second issue that challenges researchers in using digital data to understand urban mobility is temporality. Unconventional data sources can provide more fine-grained temporal and time-sensitive data, either through the frequency of data collection (monthly, weekly, daily, even to the minute or second) or through its immediacy in availability. In other words, there are substantial analytical gains to be made by drawing upon a broader range of unconventional data sources, including being able to access data in real time, rather than waiting long periods of time for datasets to be released (see also Vinodrai and Brail, 2023). For example, census data – often the gold standard for much scholarly research – is collected every five years in Canada and often is not released for several years after its collection. This issue is acutely challenging in the current context given that the Canadian census was collected during May 2021, when Canada and its many jurisdictions still had restrictions and emergency measures in place. Absent other data sources, we will not have detailed information to understand fully the impact of the pandemic on mobility in Canadian cities until quite possibly 2026 or 2027, after the next census.

A third issue relates to privacy. Public datasets are normally subject to strict rules governing confidentiality and anonymity. This can lead to data suppression to protect the easy identification of individuals or businesses. By contrast, there is a patchwork of rules governing private data that are determined by the private data providers themselves via end-user agreements or licensing arrangements. This connects to a fourth issue related to data quality and transparency. Unconventional digital data are often produced in a black box and made available "as is" with little guarantee or transparency about how that data might have been collected or processed. Indeed, quantitative evidence is often presented with little to no information on how it was derived making replication nearly (if not entirely) impossible (Vinodrai and Brail, 2023). This extends to other issues related to data quality. For example, there are issues related to both response and non-response bias, meaning there are many unknowns related to quality, especially with private digital data sources.

Another challenge relates to the cost and ease of accessing data and its broad and long-term availability. While some private companies share (limited amounts) of their data, most unconventional digital data are costly to access even for non-commercial purposes. In addition, data may not be available to allow for future studies, as there are no promises that data will continue to be collected, released, or shared. Digital data are often collected within the confines of a specific context or application, limiting their broader and long-term use. For example, an organization may change what data and how much data they share via their website or digital dashboard (Vinodrai and Brail, 2023). For researchers interested in questions of replicability, whether to confirm results or to apply a specific approach to a different industrial, organizational, geographical, or other context, the lack of guarantees about future data availability is a substantial drawback.

There are often trade-offs related to the granularity of data. A major shortcoming of many newer digital data sources, such as cell phone or loyalty program data, is the lack of information about key variables that explain differences, such as race, gender, class, sexuality, ethnicity, and other markers of identity. For this reason, scholars, decision-makers, and business leaders often rely on publicly available datasets, such as those derived from large-scale government datasets, like the census. The benefit of sources of data such as the census is that they allow for nuanced analysis that can take into account socio-economic and demographic dimensions such as gender, ethnicity, language, income, industrial and occupational structure, labour market status, housing characteristics, and immigration status.

Related to the above point is the need for detailed geographic information. To understand urban mobility and change, we are often interested in small-scale geographies to facilitate comparisons both between and within cities. Understanding neighbourhood-level change, not just in aggregate, but also for specific groups (e.g., women, low-income populations, racialized populations) and over time, necessarily requires highly detailed data of the kind most readily available from public sources such as the census. This is particularly important in answering critical questions about equity and inclusion related to urban mobility problems. Moreover, similar to the challenges of classification and granularity already discussed, digital data are not always reported using consistent geographic boundaries. In some cases, regardless of its source, individual records with postal code or GPS-derived coordinates allows for data to be aggregated to consistent spatial units.

Finally, a culmination of the challenges outlined above is that it is often problematic to integrate traditional and less conventional data sources (see also Brail, 2022b; Feldman and Lowe, 2015). Moreover, it is hard to make consistent comparisons over time and space, and these are exacerbated even more by data costs and availability, the time and scale of data collection, changes in classification schemes, and other related factors. This is not to suggest that traditional, public data sources are without flaws. Indeed, some of these challenges persist, although normally to a lesser degree. However, there remains a need for high-quality, publicly available data (see also Brail, 2022b; Vinodrai & Brail, 2023; Vinodrai & Moos, 2015), provided with a high degree of transparency regarding how it is collected, processed, and reported.

Using Data to Explore Urban Mobility in Canadian Cities

Clearly, this chapter cannot resolve the issues related to the absence of high-quality, consistent, comparable data to understand the shifts and changes to urban mobility in Canadian cities. Nonetheless, this section explores several areas of urban mobility research and policy using existing data to understand the experiences of Canadian cities. In doing so, it highlights the empirical data challenges associated with understanding contemporary urban mobility. First, I examine the relationship between more sustainable forms of commuting (e.g., active or public transit-oriented commutes) and the growing interest in micromobility solutions, such as e-scooters and e-bikes. Second, I examine the impact of the on-demand economy, which requires vast warehousing capacity and last-mile delivery solutions. Moving goods through the city has implications for land use and employment. Third, I consider the implications of the rise of remote work, accelerated due to the COVID-19 global pandemic, and its impact on Canadian cities. Each example draws upon a combination of traditional, public sources of data alongside evidence from other data sources with the goal of illustrating the promise and pitfalls of integrating traditional data sources alongside new and novel data.

Sustainable Commuting and Micro-Mobility Solutions

People travel in cities to visit family and friends, participate in recreational and leisure activities, engage in civic life, access services, and go to school or work. While many people rely on cars for their commute, growing concerns related to climate change mean that policymakers have increasingly paid attention to what policy levers are available to reduce emissions in cities. Beyond car-oriented solutions, such as high-occupancy vehicle lanes on highways and major arterial roads, intended to promote carpooling, policymakers have looked to reduce automobile dependency and encourage more sustainable forms of commuting. More sustainable commutes include active forms of transportation (e.g., walking and cycling) and the use of public transit systems.

Through the Census of Population, Statistics Canada collects a wide array of information related to commuting, including time, distance, and modal choice. Table 4.2 shows the proportion of urban commuters that use public transit or active transportation across thirty-three CMAs in Canada. Overall, there is a great deal of unevenness in the use of more sustainable forms of commuting across Canadian cities. Generally, the use of more sustainable forms of commuting is much higher in Canada's largest cities. Almost one-third of commuters in the Toronto, Montreal, Vancouver, Ottawa, and Victoria CMAs use a more sustainable commute mode compared to smaller Canadian CMAs, such as St. John's, St. Catharines-Niagara, Windsor, Thunder Bay, and Abbotsford, where workers using sustainable commute modes account for less than 10 per cent of commuters. Notably, the Victoria CMA has the highest proportion of active transport commuters (16.9%) compared to any other Canadian CMA. Certainly, several factors contribute to this, including the presence of supportive public infrastructure, relatively warm weather year-round, and active local policymaking oriented towards improving walkability and investing in cycling infrastructure as well as climate and sustainability issues more broadly. Cities with higher levels of public transit use, such as Toronto (24.3%), Montreal (22.3%), and Vancouver (20.4%), have comparatively well-developed multimodal local and regional transit systems. These large Canadian city-regions register much higher proportions of people commuting via public transit compared to Canada's smaller urban centres. Other factors shaping these patterns include issues related to built form, industrial structure, the location of workplaces, weather conditions, public infrastructure investments, and supportive local policies and programs (Moos et al., 2020).

At the onset of the COVID-19 global pandemic, amidst stay-at-home orders, lockdowns and other (temporary) limits to individual mobility, many urban residents suddenly found themselves no longer travelling to work or school. The short-term impacts were immediately noticeable with eerily quiet and empty downtowns, highways, streetcars, and subways in Canada's largest cities. Public transit ridership in Toronto provides an example of this impact. Figure 4.1 shows the average weekly ridership from January 2016 to June 2022. While there is some seasonal variation in the level of ridership over time, the pandemic led to dramatic decreases in ridership beginning in March 2020. Certainly, these data are instructive in showing both the immediate impact of the

Table 4.2. Sustainable commuting modes in Canadian CMAs, 2016

	Active transport		Public transport		All commuters
	#	%	#	%	#
Toronto	183,450	6.7	667,255	24.3	2,747,055
Montreal	136,135	7.2	419,765	22.3	1,883,815
Vancouver	105,065	9.1	235,985	20.4	1,159,210
Calgary	42,310	6.2	98,510	14.4	684,215
Ottawa-Gatineau	54,355	8.7	115,005	18.3	627,570
Edmonton	30,610	4.7	73,660	11.3	653,740
Quebec City	29,920	7.6	43,540	11.1	392,930
Winnipeg	23,610	6.2	51,390	13.6	377,845
Hamilton	17,600	5.1	33,710	9.8	342,520
Kitchener-Waterloo-Cambridge	13,925	5.5	15,280	6.0	253,445
London	14,870	6.7	16,145	7.2	222,815
St. Catharines-Niagara	9,395	5.4	4,735	2.7	174,605
Halifax	17,900	9.2	22,975	11.8	194,800
Oshawa	5,955	3.4	16,635	9.5	174,200
Victoria	28,885	16.9	18,610	10.9	170,830
Windsor	5,885	4.3	4,595	3.4	136,940
Saskatoon	8,940	6.1	6,325	4.3	145,810
Regina	6,095	5.1	6,040	5.1	119,575
Sherbrooke	5,885	6.3	3,925	4.2	93,465
St. John's	4,705	4.8	3,045	3.1	97,920
Barrie	3,875	4.1	4,135	4.3	95,540
Kelowna	6,300	7.2	3,365	3.9	87,160
Abbotsford	3,010	3.7	1,995	2.5	80,900
Sudbury	3,670	4.9	3,635	4.9	74,740
Kingston	6,835	9.5	4,890	6.8	71,980
Saguenay	2,960	4.3	1,500	2.2	69,105
Trois-Rivières	3,520	5.3	1,505	2.3	66,490
Guelph	5,255	6.9	4,860	6.4	76,090
Moncton	4,105	6.0	2,345	3.4	68,285
Brantford	2,845	4.6	1,915	3.1	61,550
Saint John	2,995	5.4	2,315	4.1	55,965
Peterborough	4,250	8.3	2,005	3.9	51,375
Thunder Bay	2,950	5.4	2,120	3.9	54,635
Lethbridge	2,945	5.4	1,585	2.9	54,825
Belleville	2,785	6.2	1,045	2.3	44,885

Source: Authors' calculations based on custom tabulations of the Canadian Census of Population, 2016.

Figure 4.1. Toronto Transit Commission average weekly ridership, January 2016 to June 2022

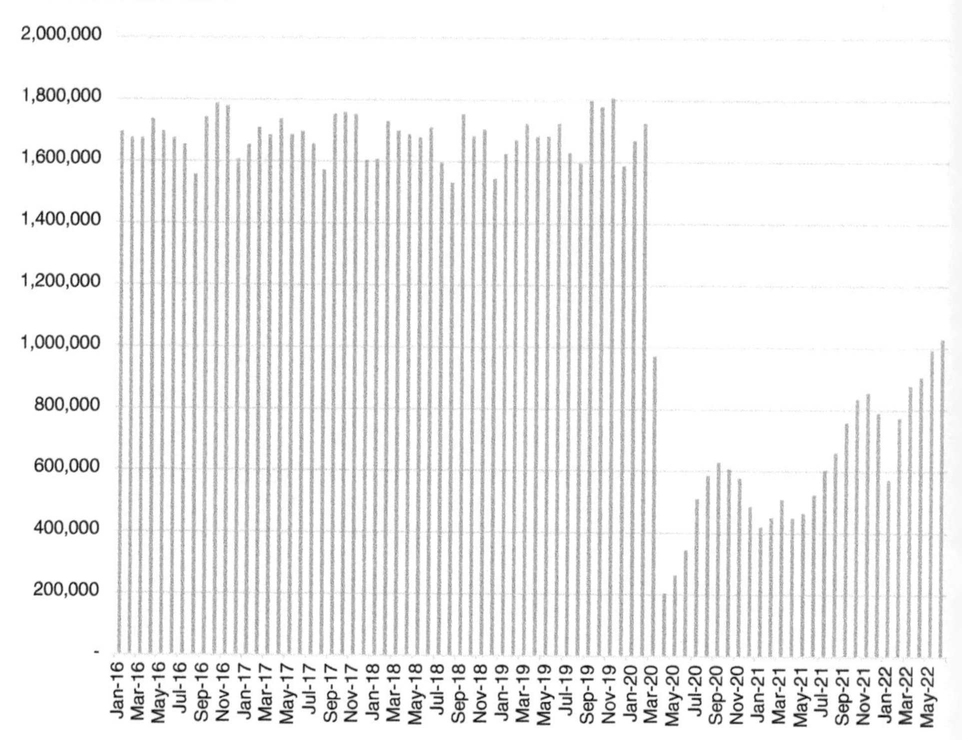

Source: City of Toronto (2022) Toronto Dashboard. https://www.toronto.ca/city-government/data-research-maps/toronto-progress-portal/

pandemic on commuting and mobility, but also the slow and halting recovery of ridership levels with downstream impacts on the urban landscape, such as the vibrancy of downtowns and the employment in businesses that support office work.

Coincident with reduced transit ridership in Toronto and other Canadian cities due to the COVID-19 pandemic, observers suggested that other forms of mobility were growing in popularity. While car ownership and use increased due to people avoiding poorly ventilated, crowded indoor spaces, there was also growing interest in other forms of urban mobility including e-scooters and e-bikes. E-scooters and e-bikes are not without controversy. On the one hand, they offer another potentially more sustainable commute mode, along with allowing greater flexibility to users and longer travel distances. Indeed, e-bikes may provide some populations, including older adults, additional modal choices (Dean & Donato, 2021; Leger et al., 2019). On the other hand, there are potential issues related to safety (e.g., helmets, speed limits) and regulatory challenges related to where e-bikes and e-scooters operate (e.g., bike lanes, trails, sidewalks, or roads).

E-scooters provide an interesting example of how micromobility solutions introduce challenges in Canadian cities. E-scooters are not new urban technologies and already had a foothold in the Canadian mobility landscape prior to the pandemic. Localities face issues related to managing how dockless e-scooter programs operate, including designated zones in the city, parking, numeric restrictions, and storage. The City of Waterloo was one of the first municipalities in Canada to work with Lime, a global shared electric vehicle company, on a pilot project to introduce e-scooters to the University of Waterloo campus (University Relations, 2019). Canadian cities have used pilots to test different policy solutions, regulatory changes, and new technologies to address urban mobility challenges. As Brail & Vinodrai (2020) note, "Pilots (or trials) are an increasingly popular way for governments to test out whether and how an idea might work in practice, in a low-threat environment. They can be limited in terms of geography, and also can have a limited period of time in which testing is conducted" (46). In the case of Lime's e-scooter pilot program in Waterloo, the pilot was not renewed. The company cited legislative barriers as an impediment, since – at the time – the Province of Ontario did not allow the use of these electrified vehicles on public streets. However, Lime does have a presence in several cities in Western Canada, including Edmonton, Calgary, Kelowna, North Vancouver, and Richmond.

In 2020, legislative changes allowed Ontario's municipalities to participate in a five-year pilot e-scooter program that permits municipalities to decide where and how e-scooters can be used. In fact, the Region of Waterloo has recently allowed the use of e-scooters, whereas in Toronto, the use of e-scooters remain illegal. However, it will be very difficult to know how much of an impact the availability and use of e-scooters and e-bikes will have on mobility patterns. Understanding the impact of these shifts to newer forms of urban mobility on aggregate sustainable commute levels across Canadian cities remains an open question but will be hard to capture without changes to the questions asked in the census long form or identifying alternative data sources.

Logistics, Local Delivery, and the Last Mile

The distribution of goods within and between urban regions relies on a complex web of supply chains, logistics, and distribution networks. Firms involved in these activities face the dual challenge of having to store goods so they are at the ready when purchased with an expectation of on-demand, just-in-time delivery, but also having to deliver those goods to specific home and business addresses across a complex internal geography of the city. This latter challenge, last-mile delivery, requires firms to identify new urban mobility solutions to ensure timely local delivery and – increasingly – such solutions rely on digitally enabled, app-based delivery services, the popularity of which has been amplified over the course of the pandemic (see Brail and Donald, chapter 10). In combination, these forces result in two distinct geographic problems: 1) sprawl of distribution centres and employment along major highways and on the edges of cities; and 2) last-mile delivery, especially in denser urban neighbourhoods where parking, traffic congestion, and other infrastructural challenges abound.

To capture the expansion of warehousing, I examine the changes in employment in the warehousing and storage industries over the period between 2006 and 2021 for Canada's three largest CMAs using LFS data (see Figure 4.2). In each of the Toronto, Vancouver, and Montreal CMAs, there is a consistent upward trend in the growth of employment in the warehousing and storage industries over the past fifteen years. The growth trajectory of the Toronto CMA is especially notable, as growth appears to be on a steeper trajectory compared to Montreal and Vancouver. This may be because the Toronto region serves as a hub for a much larger population of households and firms across Southern Ontario compared to other urban regions in Canada. Two issues might confound the employment patterns in the warehousing and storage industries observed here. The first relates to measurement. The level of industrial and geography detail involved means there is fluctuation in the annual LFS estimates related to sample weighting; however, it is still possible to examine overall trends. Indeed, it would be challenging to examine other Canadian cities due to data suppression rules to maintain confidentiality and privacy. Second, it is hard to ascertain if the use of automation and workplace robotics systems, which have replaced labour in many warehousing operations, influences some of the observed variation.

One company that both employs large numbers of workers and relies heavily on robotics is Amazon. Unquestionably, Amazon, with its large market share and presence, has fundamentally shaped the logistics industry (MacGillis, 2021). Amazon has an expansive footprint associated with its warehousing and fulfilment centres. It has a large – and growing – number of warehouses located within Canada's urban regions, including in the suburban parts of the Toronto and Vancouver CMAs. For example, Brampton has multiple Amazon facilities, including the first of Amazon's robotics-intensive fulfilment centres, which are a source of local

Figure 4.2. Employment in the warehousing and storage industries in Toronto, Montreal, and Vancouver CMAs, 2006–21

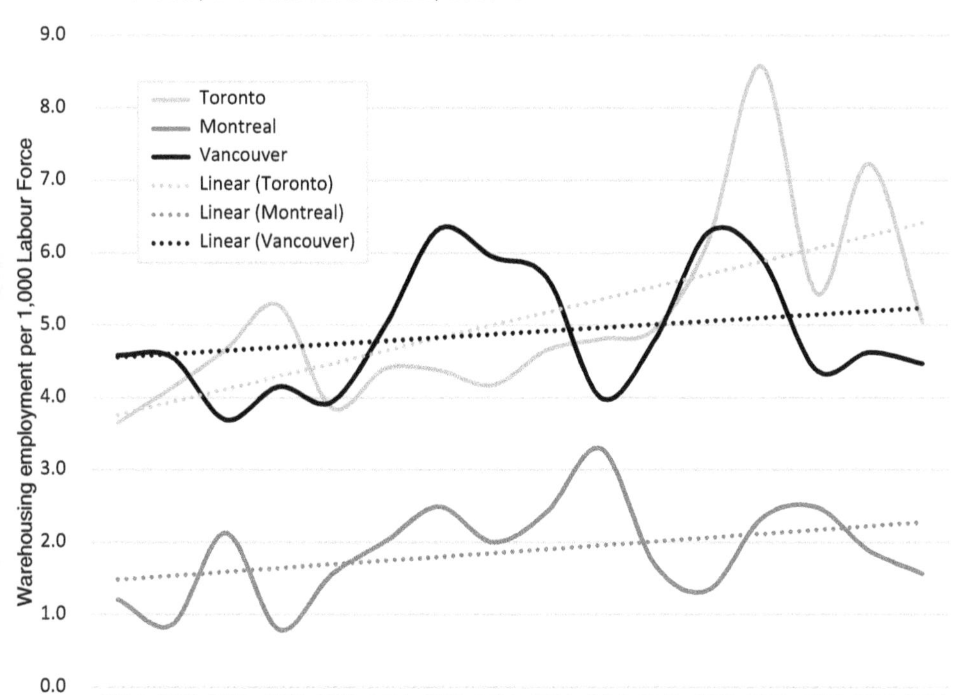

Source: Authors' calculations based on custom tabulations of Statistics Canada's Labour Force Survey

jobs, albeit generally low skill, low wage positions. Amazon also recently opened a warehouse in Hamilton, with promises of approximately 1,500 jobs; although there are concerns about the quality and wage levels associated with these jobs (Hristova, 2022).

Amazon warehouses were the sites of controversy during the early days of the global COVID-19 pandemic. While many workers stayed home and avoided entering grocery stores, restaurants, and other retail spaces, local delivery services experienced unusually high demand. Amazon's warehouses were extremely busy and their warehouses in the Greater Toronto Area (GTA) experienced some of the highest rates of COVID-19 infections. Moreover, the employees at these specific facilities in Brampton and Mississauga are often racialized, new immigrants, primarily of South Asian descent, along with other younger workers (including international students) with precarious work status

(Grant, 2021). In other words, there are social consequences to urban mobility services, which contribute to urban inequalities. In another controversy related to Amazon's ongoing expansion, the City of Cambridge has allowed for the development of a new 1 million square foot fulfilment facility that will open in 2025 (Pickel & Sandstrom, 2022). Local residents have objected to the development, despite the promised jobs and economic benefits, due to concerns over noise, traffic, and the environmental impacts given the large numbers of delivery trucks and personal vehicles that will enter and exit the facility on a daily basis. Issues related to warehouse expansion will continue to reverberate across Canadian cities, as cities contend with the lack of suitable spaces for logistics providers (CBRE, 2021; Spivacek, 2022).

While the warehousing elements of the on-demand economy appear to promote sprawl, encourage automobile dependence, and contribute to emissions, delivery companies also face challenges associated with the last mile. In these cases, delivery companies have turned to experiments and pilot projects using urban mobility solutions intended to reduce emissions, avoid congestion, and improve the speed of delivery with the aim of better efficiency in urban delivery. For example, during the second summer of the pandemic, FedEx launched its first North American pilot project involving cargo e-bikes. The company was able to take advantage of Toronto's extended cycling networks introduced by local authorities during the pandemic, while also addressing challenges it faced related to congestion and other last-mile issues. According to FedEx, the cargo e-bike pilot program was successful; the result has been that the company is expanding its program to a number of other Canadian cities including Ottawa, Gatineau, Montreal, Calgary, Richmond, Vancouver, and Windsor. The company reports that these locations were selected because local governments were amenable to developing programs and there was support for cycling-based solutions (Hansen-Gillis, 2020; Yakub, 2022). The company plans to expand their fleet to include other electric vehicles.

Another global courier company, Purolator, has similarly introduced new urban mobility solutions in Canadian cities. Purolator was the first courier company in Canada to operate a fully electric curbside delivery fleet in Vancouver, noting that home delivery grew by approximately 50 per cent during the pandemic (Purolator, 2021). This shift to electric vehicles builds on their usage of e-cargo bikes in Toronto and Montreal. More recently, Purolator announced a new pilot program on the University of Toronto's downtown campus, which aims to create a small distribution centre made from a shipping container and replacing local delivery trucks with cargo e-bikes (Purolator, 2022). While the initiative is intended to address last-mile issues, the pilot project also involves

Figure 4.3. Employment in the local messenger and local delivery industries in the Toronto CMA, June 2017 to May 2022

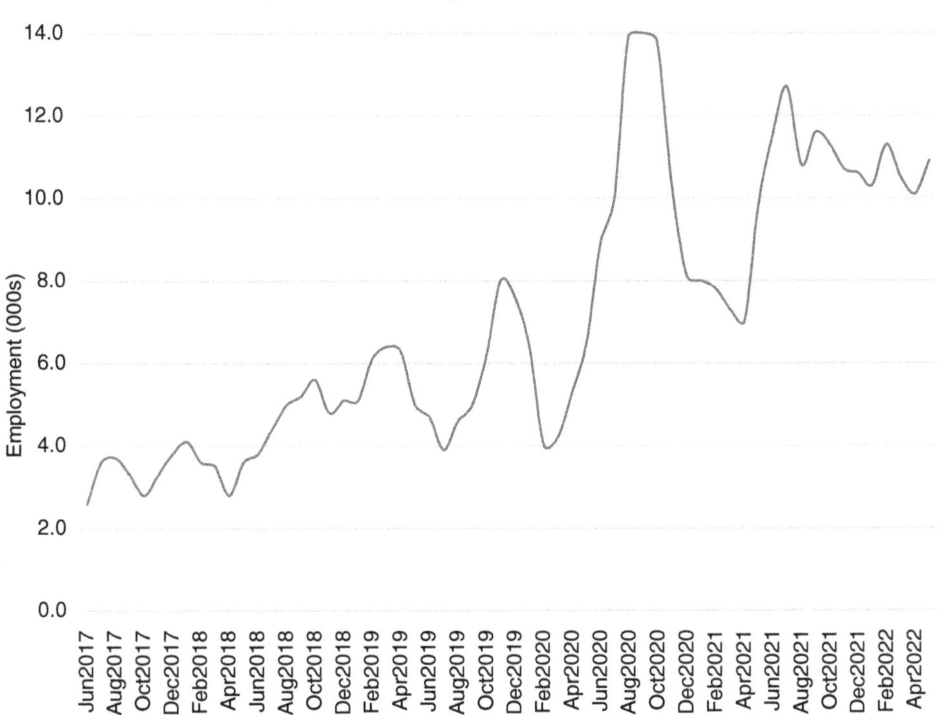

Note: Monthly data represent three-month moving averages.
Source: Authors' calculations based on custom tabulations of Statistics Canada's Labour Force Survey

researchers investigating various facets of urban mobility using sensors and other smart technologies.

While these cases are illustrative of how the COVID-19 pandemic has accelerated change in particular industries, it is also helpful to understand the impact of the pandemic in a quantitative fashion. Much like in the case of warehousing (above), there are limited data available to systematically explore this issue. Nonetheless, Figure 4.3 uses monthly LFS data to show employment in the local messengers and delivery industries in the Toronto CMA between June 2017 and May 2022. Over this five-year period, there is a steady increase in employment in these local delivery industries. Most notably, however, there are steep increases in employment levels coincident with the initial shutdowns associated with the pandemic followed by continued growth, with some fluctuations.

Some of the fluctuation observed in these data are due to the challenges of the LFS itself, which relies on a nationwide monthly panel survey. For this reason, detailed data, such as looking at one specific industry in one particular city-region, should be treated with some caution and such analysis is only possible for Canada's largest cities due to data quality issues. However, for understanding trends in Canada's largest cities, these data are incredibly useful and provide empirical confirmation of growth in local delivery services employment at the onset of the COVID-19 pandemic.

Remote Work and Its Impact on Daily Mobility

As noted above, with the advent of the pandemic, many urban workers found themselves commuting less (if at all) and working from home for the first time. Businesses, especially those that have relied on offices, shifted to remote operations involving various types of telework and work-from-home (WFH) arrangements. Firms used this opportunity to assess the extent to which remote work may become a more permanent organizational feature, with many firms experimenting with fully remote and hybrid work arrangements (see Cavalli and McGahan, chapter 2). For cities, an open question remains as to the impact of this shift given that urban downtowns are the prime locations of high proportions of office workers (Brail, 2022a; Chapple et al., 2022; Florida et al., 2021; Nathan & Overman, 2020).

WFH is not new and was already prevalent in some industries and occupations prior to the COVID-19 pandemic; indeed, there has been long-standing interest in the consequences of telework on office location and organizational dynamics (c.f., Di Martino & Wirth, 1990; Huws et al., 1990; Johnson et al., 2007). Shearmur (2018) makes the case for the extension of the workplace stretching beyond offices and reaching into many corners of the everyday life of urban professional workers, including airport lounges, coffee shops, co-working spaces, and public transit. In other words, work has always taken place outside of offices and other traditional workplaces with a complex relationship to urban mobility. People may work while being mobile or in transit, but there are also shifts in destinations as people change how and where they work, shaping the daily rhythm of the city.

Much of the statistical work estimating the potential for the more extensive practice of remote work in the economy relies on a timely study by Dingel & Neiman (2020), which classified occupations in the United States based on the likelihood that the work could be performed remotely. In Canada, Deng et al. (2020) adapted this methodology and applied it to the Canadian context. Their analysis indicated that almost four in ten jobs could be conducted remotely. However, the same analysis indicated that there was considerable variation across different sectors and demographics. Specifically, the study showed that sectors such as professional services and finance had much higher remote work potential. This suggests that places that skew towards these knowledge-intensive activities (cities) may have to contend more directly with the influence of organizational decisions related to remote work.

I draw on previous work to estimate the remote work capacity of Canadian cities to understand historic WFH levels in Canadian cities (Vinodrai et al., 2020; see also Toronto Region Board of Trade, 2020). Using industry-level estimates of remote worker capacity (Deng et al., 2020), I calculate the proportion of potential remote workers for each Canadian CMA, thereby accounting for local industrial structure in our analysis. I also present data that help to provide a context for understanding how the potential for remote work has an uneven geography by looking at where people worked prior to the COVID-19 pandemic.

Table 4.3 shows place of work data using the 2016 Canadian census, as well as our estimates of remote work capacity, for the thirty-five largest CMAs in Canada. Overall, the data are revealing of an already uneven pattern related to where people worked. The proportion of the workforce that had a regular place of work ranged from 73.9 per cent (Abbotsford) to 87.9 per cent (Saguenay) across Canadian CMAs. Correspondingly, the proportion of the workforce that has no fixed workplace ranged from a low of 7.5 per cent in Saguenay to a high of 18.2 per cent in Abbotsford. Overall, 1 per cent or fewer urban residents regularly worked outside of Canada across Canadian CMAs, except for Windsor (4.3%). Windsor's position as a border town helps to explain the relatively high proportion in comparison to other Canadian cities. Windsor's role as a border town came into sharper focus during the earlier periods of the COVID-19 pandemic when workers who could normally go to work, whether in health care or other settings, had their commutes suddenly confounded by layers of rules and regulations dictating border crossings. Finally, and of relevance to the discussion at hand, the proportion of the employed labour force that regularly worked from home ranges from 4.1 per cent in Windsor and 4.2 per cent in Sudbury to a high of 9.3 per cent in Kelowna and 8.4 per cent in Victoria. These higher proportions in places like Kelowna and Victoria may reflect self-employed workers who have chosen to live in these

Table 4.3. Place of work and remote work capacity in Canadian CMAs, 2016

	Total employed	Usual place of work		Worked at home		Worked outside Canada		No fixed workplace		Remote work Capacity
	#	#	%	#	%	#	%	#	%	%
Toronto	2,984,995	2,400,850	80.4	218,365	7.3	19,575	0.7	346,205	11.6	43.7
Montreal	2,026,050	1,704,040	84.1	135,480	6.7	6,755	0.3	179,780	8.9	41.2
Vancouver	1,276,900	989,890	77.5	104,865	8.2	12,825	1.0	169,325	13.3	41.2
Calgary	740,235	582,020	78.6	52,835	7.1	3,185	0.4	102,190	13.8	38.8
Ottawa-Gatineau	694,180	554,660	79.9	38,745	5.6	1,695	0.2	99,085	14.3	45.0
Edmonton	676,480	567,645	83.9	46,420	6.9	2,490	0.4	59,925	8.9	37.6
Quebec City	417,200	359,810	86.2	23,490	5.6	775	0.2	33,120	7.9	41.7
Winnipeg	397,545	337,320	84.9	18,325	4.6	1,375	0.3	40,525	10.2	39.2
Hamilton	369,070	303,105	82.1	25,040	6.8	1,515	0.4	39,415	10.7	39.5
Kitchener-Waterloo-Cambridge	271,875	227,585	83.7	17,320	6.4	1,105	0.4	25,865	9.5	40.0
London	239,940	197,045	82.1	16,065	6.7	1,050	0.4	25,770	10.7	38.9
Halifax	209,420	172,585	82.4	13,580	6.5	1,045	0.5	22,215	10.6	34.7
St. Catharines-Niagara	188,220	154,300	82.0	12,030	6.4	1,580	0.8	20,305	10.8	42.0
Victoria	187,335	148,165	79.1	15,755	8.4	755	0.4	22,665	12.1	39.8
Oshawa	186,205	152,050	81.7	11,635	6.2	370	0.2	22,145	11.9	41.0
Saskatoon	155,345	125,960	81.1	9,220	5.9	315	0.2	19,845	12.8	35.2
Windsor	149,385	123,710	82.8	6,100	4.1	6,350	4.3	13,230	8.9	38.0
Regina	126,100	106,185	84.2	6,275	5.0	260	0.2	13,390	10.6	40.1
St. John's	103,000	87,705	85.2	4,640	4.5	440	0.4	10,215	9.9	37.5
Barrie	102,610	81,880	79.8	6,740	6.6	330	0.3	13,660	13.3	39.4
Sherbrooke	100,425	85,575	85.2	6,660	6.6	300	0.3	7,885	7.9	36.9
Kelowna	96,495	72,150	74.8	9,010	9.3	325	0.3	15,010	15.6	35.4
Abbotsford	87,815	64,910	73.9	6,550	7.5	360	0.4	15,990	18.2	32.9

Guelph	82,325	68,570	83.3	5,930	7.2	310	0.4	7,530	9.1	37.3
Sudbury	78,105	66,230	84.8	3,255	4.2	115	0.1	8,510	10.9	41.0
Kingston	77,330	64,395	83.3	5,010	6.5	330	0.4	7,590	9.8	36.7
Moncton	72,870	61,535	84.4	4,335	5.9	250	0.3	6,750	9.3	36.7
Saguenay	72,470	63,665	87.9	3,285	4.5	80	0.1	5,440	7.5	39.9
Trois-Rivières	70,515	60,760	86.2	3,870	5.5	155	0.2	5,730	8.1	39.1
Brantford	65,765	54,460	82.8	4,010	6.1	205	0.3	7,095	10.8	34.1
Lethbridge	59,330	46,890	79.0	4,375	7.4	125	0.2	7,940	13.4	38.1
Saint John	59,065	49,995	84.6	2,895	4.9	205	0.3	5,975	10.1	37.1
Thunder Bay	56,945	48,390	85.0	2,190	3.8	120	0.2	6,250	11.0	37.4
Peterborough	55,800	44,660	80.0	4,310	7.7	115	0.2	6,715	12.0	35.2
Belleville	47,715	39,860	83.5	2,695	5.6	135	0.3	5,020	10.5	35.7

Source: Authors' calculations based on custom tabulations from the Canadian Census of Population, 2016

smaller urban centres for lifestyle related reasons and therefore operate their businesses from home.

While these data show unevenness in where remote work was already prevalent, as discussed above, estimates of remote work capacity can account for local industrial structure. The final column in Table 4.3 shows remote work capacity, or the proportion of jobs that have the *potential* to be conducted remotely, for each Canadian CMA. Again, there is variation across Canada. Larger centres like Toronto (43.7%), Montreal (41.2%), Vancouver (41.2%), and Ottawa (45%), which all have high proportions of employment in the professional and financial services industries, rank much higher in terms of remote work capacity.

Certainly, the actual amount of remote work that is taking place in Canadian cities due to the pandemic remains difficult to estimate. A report from Statistics Canada estimates that 22.2 per cent of the Canadian labour force was working from home in December 2021, almost double the United States (Clarke & Hardy, 2022). Data from the United States suggests that many workplaces have adopted a hybrid work strategy. For example, Barrero et al. (2021) finds that – for now – a relatively steady state pattern has emerged; 50 per cent of US workers are fully in person, 30 per cent are working in hybrid arrangements, and the remainder are working fully remotely. Yet, if WFH rates are markedly divergent between the US and Canada, it is reasonable to expect different patterns in Canada and its cities.

Private companies have also derived various indices using their own proprietary systems to estimate the return to the office, and by extension, the potential recovery and vitality of urban downtowns (see Brail, 2022b for discussion). In Canada, two recent studies leverage private data that shed light on Canadian cities during various stages of the pandemic. Cavalli et al. (2020) use Cuebiq data (derived from cell-phone usage) to estimate daily and weekly mobility as well as proximity to understand changing mobility patterns. Their study, limited to the initial months of the pandemic, found that after dramatic decreases in mobility, Canadians returned to pre-pandemic levels of mobility, all while maintaining distance and interacting with fewer individuals. People in Toronto and Montreal appeared to interact less than in Vancouver, Edmonton, and Calgary, which the authors attribute to different local conditions. They also found that while travel resumed, there was greater travel between neighbouring regions (e.g., between Mississauga and Toronto or Vancouver and Victoria). In another study, Chapple et al. (2022) use Safegraph cell-phone data to understand downtown recovery trajectories across sixty-two metropolitan regions in North America, including ten Canadian city-regions. They find that North American downtown recovery has been slow, especially in comparison to other parts of the city-region. Moreover, downtowns in larger metropolitan regions that are denser, transit-oriented, older, and rely on professional and tech-oriented employers have faced the largest struggles.

The pandemic-accelerated shift to remote work and its mobility-related consequences related to transit use, commuting, and travel patterns opens a research agenda that can be both theoretically informed and empirically rich. However, we lack conclusive data and evidence on the full impact of pandemic-induced or pandemic-accelerated work and mobility patterns. At the time of writing, the COVID-19 pandemic continues to shape individual behaviours and the locational decisions of people and firms with profound influence on cities and their rhythms. It remains unknown as to the extent to which changes under way are permanent or evolving, finding a new equilibrium or steady state after the initial exogenous shock of, and the subsequent responses to, the global pandemic. Many of the earliest predictions calling for the death of cities or the end of the office ring false; there is a much more nuanced and complex entanglement between history, geography, institutions, and place. In some cases, the pandemic accelerated or exacerbated underlying trends and conditions, rather than halting them in their tracks. There remain open questions about the futures of cities and their downtowns, which is at once shaping and shaped by urban mobility. The complex interplay between workplace organization, mobility, and cities means that we still know relatively little, especially in systematic and comprehensive ways, about how Canadian cities have fared during the pandemic.

Coda: Pandemics, Platforms, and Public Data

This chapter illustrated the data issues apparent in understanding the contours of urban mobility, using examples from three areas of contemporary urban mobility research and policy interest. In doing so, it provided a window into the uneven geographies of urban mobility across Canadian cities using publicly collected data as well as data from other sources. It also demonstrated that there is a lack of consistent and comparable quantitative data to understand urban mobility across Canadian cities. In the case of sustainable commuting, the most reliable and consistent measures come from the Canadian Census of Population, conducted every five years. The chapter notes that it is incredibly difficult to conduct comparative analysis of transit ridership across Canadian cities. Yet, doing so is useful for many reasons, including in understanding the impact of the COVID-19 pandemic on daily travel patterns via public transit (see Palm and Farber, chapter 6). There is no single data repository; different agencies and jurisdictions disseminate their data in different formats, over different time scales, and using different geographies that do not align neatly with existing classifications systems. And our existing public datasets do not adequately capture new forms of micromobility that may be used by commuters and in last-mile delivery solutions, often enabled by various digital platforms. While it is possible to trace some dimensions of employment related to the growth of local delivery services and warehousing implicated in the growth of digitally enabled retail and e-commerce platforms, it is difficult to do so beyond the largest urban centres due to data suppression. Finally, in understanding the impact of the pandemic on remote work (with its knock-on effects on mobility), the chapter reports on the findings of several studies that rely on private data sources to tell compelling stories of mobility change in Canadian cities. These efforts are commendable in marshalling data to explore urban mobility. However, private commercial interests govern the collection and access to these data. This may well limit researchers' ability to conduct comparable studies or follow-up studies due to access issues as well as changes in methodologies in how (and what) data is collected and aggregated and for how long. Moreover, it is difficult to parse out variations amongst different socio-economic and demographic groups.

As noted at the outset of this chapter, there is a series of interrelated issues that leads to trade-offs between traditional data sources and the vast array of emerging non-traditional digital data sources. In most cases this boils down to accessing slow public data, with guarantees around quality, privacy, and transparency, or fast private data, which captures new and emerging trends, but suffers from being costly and/or inaccessible, lacking in transparency, and being produced in a black box. Due to these issues, it remains entirely too difficult to access timely, comparable, and detailed quality data for Canadian cities to understand differences between and within them.

In present-day Toronto and other Canadian cities, on a hot summer day you can sometimes hear the music emanating from a truck as it rolls slowly to a stop at the edge of a splash pad or playground. It is an ice cream truck in search of exuberant young children desperate for sugary treats. The city has always been the site for the flows of people and goods, and the technologies that enable mobility are always changing. As scholars, we continue to puzzle through how cities in Canada and elsewhere change and transform over time. High-quality, publicly available data are necessary for that task.

Acknowledgments

The author is grateful for funding from the Social Sciences and Humanities Research Council of Canada and the University of Toronto Mississauga Mobility Network. The author would also like to thank David Attema (School of Planning, University of Waterloo) for research assistance, as well as Shauna Brail, Betsy Donald, and the reviewers for comments and suggestions that helped to inform and refine this chapter.

REFERENCES

Barrero, J.M., Bloom, N., & Davis, S.J. (2021). *Why working from home will stick* (National Bureau of Economic Research Working Paper 28731). https://doi.org/10.3386/w28731.

Biber, E., Light, S.E., Ruhl, J., & Salzman, J. (2017). Regulating business innovation as policy disruption: From the Model T to Airbnb. *Vanderbilt Law Review*, *70*(5), 1561–626.

Brail, S. (2022a). COVID-19 and the future of urban policy and planning. *Current History*, *121*(838), 298–303. https://doi.org/10.1525/curh.2022.121.838.298.

Brail, S. (2022b, December 13). *Visualizing the impact of COVID-19 on Toronto.* Spacing. http://spacing.ca/toronto/2022/12/13/visualizing-the-impact-of-covid-19-on-toronto/.

Brail, S., & Donald, B. (2020). Digital cities: Contemporary issues in urban policy and planning. In M. Moos, T. Vinodrai, & R. Walker (Eds.), *Canadian cities in transition: Understanding contemporary urbanism* (6th ed., pp. 70–86). Oxford University Press.

Brail, S., & Vinodrai, T. (2020). The elusive, inclusive city: Toronto at a crossroads. In S. Bunce, N. Livingstone, L. March, S. Moore, and A. Walks (Eds.), *Critical dialogues of urban governance, development and activism: London and Toronto* (pp. 38–53). UCL Press.

Cavalli, G, Lake, B., McGahan, A.M., & Pepe, E. (2020). *Mobility and proximity in Canada during the COVID-19 pandemic* (Innovation Policy Lab Working Paper, Munk School of Global Affairs, University of Toronto).

CBRE. (2021). So much demand, so little space. Retrieved from https://www.cbre.ca/insights/articles/so-much-demand-so-little-space.

Chapple, K., Leong, M., Huang, D., Moore, H., Schmahmann, L., & Wang, J. (2022). *The death of downtown? Pandemic recovery trajectories across 62 North American cities* (Research Brief, Institute of Governmental Studies, University of California Berkeley). https://downtownrecovery.com/death_of_downtown_policy_brief.pdf.

Clarke, S., & Hardy, V. (2022, 24 August). Working from home during the COVID-19 pandemic: How rates in Canada and the United States compare. https://www150.statcan.gc.ca/n1/pub/36-28-0001/2022008/article/00001-eng.htm.

Dean, J., & Donato, E. (2021). New micro-mobilities and aging in the suburbs. In M. Hartt, S. Biglieri, M.W. Rosenberg, & S.E. Nelson. *Aging people, aging places: Experiences, opportunities, and challenges of growing older in Canada* (pp. 115–32). Bristol University Press.

Deng, Z., Morissette, R., & Messacar, D. (2020, 28 May). Running the economy remotely: Potential for working from home during and after COVID-19. *StatCan COVID-19: Data to Insights for a Better Canada*. https://www150.statcan.gc.ca/n1/pub/45-28-0001/2020001/article/00026-eng.htm.

Di Martino, V., & Wirth, L. (1990). Telework: A new way of working and living. *International Labour Review, 129*(5), 529–54.

Dingel, J.I., & Neiman, B. (2020). *How many jobs can be done at home?* (National Bureau of Economic Research Working Paper 26948). https://doi.org/10.3386/w26948.

Dunne, M. (2015, 3 August). Dickie Dee men still pedalling bikes, ringing bells, selling treats. Retrieved from https://www.cbc.ca/news/canada/windsor/dickie-dee-men-still-pedalling-bikes-ringing-bells-selling-treats-1.3175873.

Fang, L., Green, J., Ye, X., & Shi, W. (2022). Time to upgrade our tools: Integrating urban data science into economic development research and curriculum. *Journal of Planning Education and Research*. https://doi.org/10.1177/0739456X221128501.

Feldman, M., & Lowe, N. (2015). Triangulating regional economies: Realizing the promise of digital data. *Research Policy, 44*(9), 1785–93. https://doi.org/10.1016/j.respol.2015.01.015.

Florida, R., Rodríguez-Pose, A., & Storper, M. (2021). Cities in a post-COVID world. *Urban Studies, 60*(8), 1509–31. https://doi.org/10.1177/00420980211018072.

Grant, T. (2021, 10 January). Business is booming at Amazon Canada, but workers say the pandemic is adding to safety concerns. *Globe and Mail*. https://www.theglobeandmail.com/canada/article-business-is-booming-at-amazon-canada-but-workers-say-the-pandemic-is/.

Hansen-Gillis, L. (2020, 10 November). Why FedEx chose a Canadian city to launch its unique ebike pilot project. *Cycling Magazine*. https://cyclingmagazine.ca/sections/feature/fedex/.

Holicka, M., & Vinodrai, T. (2022). The global geography of investment in emerging technologies: The case of blockchain firms. *Regional Studies, Regional Science, 9*(1), 177–9. https://doi.org/10.1080/21681376.2022.2047769.

Hristova, B. (2022, 19 April). *Amazon opens Hamilton warehouse, announces 3 more in Ontario*. CBC News. Retrieved from https://www.cbc.ca/news/canada/hamilton/amazon-warehouse-hamilton-1.6423335.

Huws, U., Korte, W.B., & Robinson, S. (1990). *Telework: Towards the elusive office*. Wiley.

Johnson, L.C., Andrey, J., & Shaw, S.M. (2007). Mr. Dithers comes to dinner: Telework and the merging of women's work and home domains in Canada. *Gender, Place and Culture: a Journal of Feminist Geography, 14*(2), 141–61. https://doi.org/10.1080/09663690701213701.

Leger, S.J., Dean, J.L., Edge, S., & Casello, J.M. (2019, May). "If I had a regular bicycle, I wouldn't be out riding anymore": Perspectives on the potential of e-bikes to support active living and independent mobility among older adults in Waterloo, Canada. *Transportation Research Part A: Policy and Practice, 123*, 240–54. https://doi.org/10.1016/j.tra.2018.10.009.

MacGillis, A. (2021). *Fulfillment: Winning and losing in one-click America*. Farrar, Straus and Giroux.

Moos, M., Vinodrai, T., & Walker, R. (Eds.). (2020). *Canadian cities in transition: Understanding contemporary urbanism* (6th ed.). Oxford University Press.

Nathan, M., & Overman, H. (2020). Will coronavirus cause a big city exodus? *Environment and Planning B: Urban Analytics and City Science, 47*(9), 1537–42. https://doi.org/10.1177/2399808320971910.

Pickel, J., & Sandstrom, A. (2022). *Controversial Blair warehouse will be Amazon fulfillment centre*. CTV News. Retrieved from https://kitchener.ctvnews.ca/controversial-blair-warehouse-will-be-amazon-fulfillment-centre-1.6086537.

Purolator. (2021, 29 March). *Purolator hits the road as first national courier to deploy fully electric delivery vehicles*. Newswire. Retrieved from https://www.newswire.ca/news-releases/purolator-hits-the-road-as-first-national-courier-to-deploy-fully-electric-delivery-vehicles-889913704.html.

Purolator. (2022, 12 October). *Purolator, University of Toronto and City of Toronto launch Urban Quick Stop on campus*. Retrieved from https://www.purolator.com/en/articles/purolator-university-toronto-and-city-toronto-launch-urban-quick-stop-campus.

Shearmur, R. (2018). The Millennial urban space economy: Dissolving workplaces and the de-localization of economic value-creation. In M. Moos, D. Pfeiffer, and T. Vinodrai (Eds.), *The millennial city: Trends, implications, and prospects for urban planning and policy* (pp. 65–79). Routledge. https://doi.org/10.4324/9781315295657-6.

Spicer, Z., Eidelman, G., & Zwick, A. (2019). Patterns of local policy disruption: Regulatory responses to Uber in ten North American cities. *The Review of Policy Research, 36*(2), 146–67. https://doi.org/10.1111/ropr.12325.

Spigel, B., & Vinodrai, T. (2021). Meeting its Waterloo? Recycling in entrepreneurial ecosystems after anchor firm collapse. *Entrepreneurship and Regional Development, 33*(7–8), 599–620. https://doi.org/10.1080/08985626.2020.1734262.

Spivacek, M.S. (2022, 1 February). Warehouse space is the latest thing being hoarded. *New York Times*. https://www.nytimes.com/2022/02/01/business/warehouses-supply-chain.html.

Statistics Canada. (2017, 28 February). *The sharing economy*. The Daily. https://www150.statcan.gc.ca/n1/daily-quotidien/170228/dq170228b-eng.htm.
Statistics Canada. (2019). *Measuring digital economic activities in Canada: Initial estimates*. (Catalogue no.13-605-X). Statistics Canada. https://www150.statcan.gc.ca/n1/pub/13-605-x/2019001/article/00002-eng.htm.
Toronto Region Board of Trade. (2020). *Shaping our future: Playbook for rebooting and reimagining the regional economy in Ontario's Innovation Corridor*. Toronto Region Board of Trade.
University Relations. (2019, 24 May). Waterloo students are the first in Canada to pilot Lime e-scooters. Retrieved from https://uwaterloo.ca/news/waterloo-students-are-first-canada-pilot-lime-e-scooters.
Vinodrai, T., Attema, D., & Moos, M. (2020). *COVID-19 and the economy: Exploring potential vulnerabilities in the Region of Waterloo*. Report for the Region of Waterloo. https://t.co/SJCEyhLO0O?amp=1.
Vinodrai, T., & Brail, S. (2023). Cities, COVID-19, and counting. *Big Data & Society*, *10*(2). https://doi.org/10.1177/20539517231188724.
Vinodrai, T., & Moos, M. (2015). Do we still have quality data to study Canadian cities? In P. Filion, M. Moos, T. Vinodrai, and R. Walker. (Eds.), *Canadian cities in transition: Perspectives for an urban age* (5th ed., pp. 436–9). Oxford University Press.
Woodside, J., Moos, M., & Vinodrai, T. (2021). Private car, public oversight: Municipal regulation of ride-hailing platforms in Toronto and the Greater Golden Horseshoe. *Canadian Planning and Policy/Aménagement et politique au Canada*, *2021*(1), 146–65. https://doi.org/10.24908/cpp-apc.v2021i01.14362.
Yakub, M. (2022, 2 August). E-bikes and BrightDrop electric vans helping FedEx Express Canada meet its corporate zero-emission goals. Retrieved from https://electricautonomy.ca/2022/08/02/e-bikes-brightdrop-fedex-express-canada/.

5 Emerging Mobility Technologies and Transportation Systems in Canadian Cities

LISA L. LOSADA-ROJAS AND ERIC J. MILLER

Introduction

Canadian cities, like cities worldwide, have faced changes in the mobility services that first disrupted the transportation landscape ten years ago. When ride-sharing services began in Toronto in 2012, they opened an array of mobility options in Canada that have endured despite regulations in different cities. Today, various mobility services are in operation in many Canadian cities, including car-sharing, bike-sharing, ride-hailing, and demand-responsive transit. Some services include different modalities (i.e., car-sharing as one-way or two-way, ride-hailing as pool service or individual service), which define their operating characteristics and capabilities. Some mobility services have also transitioned or expanded to integrate other services, such as food delivery, e-commerce, and, during the COVID-19 pandemic, access to vaccination sites and other health services. However, this chapter focuses only on services that cater to passengers, either concurrently (where passengers share the vehicle's space, such as public transit and ride-hailing) or non-concurrently (where passengers use the service individually).

Experts classify emerging mobility options in different ways. For instance, shared mobility typology outlines four types: public transit, micromobility, automobile-based modes, and commute-based modes (Shared Use Mobility Center, 2018). On a similar note, Calderón & Miller (2020, 2022) establish a taxonomy of mobility services within the mobility as a service (MaaS) paradigm that is based on various combinations of operational features of the mobility services. The authors argue that these services generally share functional features such as matching, rebalancing, pricing, and pooling, but differ in terms of vehicle technology, service attributes, etc. Ride-matching is the problem of finding the best match between rides and available vehicles in real time. Rebalancing is a problem for free-floating or docked bike-sharing and car-sharing systems. Calderón and Miller identify individual services, such as ride-hailing, carpooling, bike-sharing, e-scooters, and demand-response transit, as existing operational mobility service implementations within this generalized framework. In our analysis, we use Calderón and Miller's taxonomy and focus on these particular services.

The State of New Mobility Services in Canadian Cities

Various academic and non-academic studies predict that changing mobility patterns will reduce the vehicle ownership share of the transportation market (Nicoll & Armstrong, 2016). However, the most recent Canadian vehicle ownership statistics do not support this expectation. New vehicle registrations increased by 6.5 per cent between 2020 and 2021, with an average of 4 per cent annually since 2017 (Government of Canada, 2021a). Pre-pandemic concerns existed regarding possible stagnation or even decline in public transit usage in Canadian cities (Miller et al., 2018). Pandemic lockdowns, the "aftershocks" of increased working from home, and concerns about transit safety have resulted in major decreases in transit usage that are only slowly being recovered. Cities are mainly affected by this trend, given their robust and connected public transportation systems. In recent years, new technologies have complemented (and sometimes replaced) the use of public transit. Below, we briefly describe each emerging service and the state of mobility in Canada's main urban regions. Taking population size, availability of services, and geographical representation into consideration, we selected the following cities: Toronto; Greater Toronto-Hamilton Area (GTHA); Montreal; Vancouver; Calgary; Winnipeg; Quebec City; Halifax; and Saskatoon.

Individual Mobility (Ride-Hailing)

Ride-hailing services use passenger vehicles to provide trips reserved and paid (in most cases) via an app (Shaheen & Chan, 2015). The rider requests the service, and within minutes, the app uses GPS-based platforms to match the rider with an available driver. Most apps today allow you to rate the service, share your ride with a trusted contact, and reserve trips in advance. In Canada, the most popular ride-hailing companies are Uber, Lyft, Steer, TappCar, and Uride. Ride-hailing competes with a long-existing and regulated industry: taxis. Some scholars have also argued that ride-hailing competes with public transportation in specific markets while complementing it in others (Habib, 2019; Hall et al., 2018).

Uber arrived in Canada in 2012, with its first ride booked in Toronto. Since then, the company has expanded to serve 140 municipalities across nine provinces as of 2022 (Uber, 2022). Uber is available in all cities considered in this study; however, in some, the service has faced possible discontinuation, mainly because provinces and municipalities have expressed regulatory concerns (see Zwick, Young, and Spicer, chapter 13). For example, the service in Edmonton was discontinued and then reimplemented, while in Vancouver, it was not until January 2020 that the service was implemented due to delays in establishing an approved regulatory environment (Nair, 2020). The second biggest ride-hailing service in Canada is Lyft, which first arrived in 2017 (Milenkovic, 2022). Toronto was the first international city served by this San Francisco–based mobility provider. Since launching in Canada, Lyft has offered services in ten cities, some of which Uber does not cover.

SHARED RIDE-HAILING AND OTHER SERVICES

Before the COVID-19 pandemic, ride hailing services provided shared-ride options. These shared/pooled trips happened when a driver picked up two or more unconnected passengers with similar routes over the course of a trip. However, these services, provided mainly by Uber and Lyft, were suspended due to public health concerns, a suspension that continues as of August 2022 (for other changes in ride hailing in response to COVID-19 see Brail and Donald, chapter 10). Other ride hailing services available in the cities examined target long hauls (Poparide) or serve a specific market segment, such as women and girls (DriveHer). These companies have also evolved to provide food delivery, e-commerce, and healthcare access.

Car-Share

Car-sharing services provide a network of cars to pre-screened members for short-term rentals based on hours rather than days. This service is ideal for trips ranging from eight kilometres to more than thirty kilometres, especially when members are shopping or carrying other belongings (Shaheen et al., 2018). There are three different car-sharing services: round-trip, one-way, and free-floating. Fixed stations provide round-trip services within predefined service areas, and drivers must return the vehicle to the same station where they picked it up (Calderón & Miller, 2020). Unlike round-trip car-sharing, one-way allows users to return the vehicle to a different station. Lastly, the free-floating service lets the user pick up and drop off cars at any parking spot within a specific service area.

Car-sharing services in Canada are also classified according to their business models. For instance, the co-op car-sharing model allows people to buy shares in the organization, and the organization uses that funding to subsidize maintenance costs and other fees of the car fleet. Peer-to-peer car-sharing lets car owners make their vehicles available for others to rent for short periods. Finally, private car-sharing companies offer car rental services, which can be round-trip, one-way, and free-floating in nature (Government of Canada, 2021b). Car-sharing companies in Canada are Communauto, Zipcar, Enterprise CarShare, Evo, and Modo. The first three companies are available in the GTHA. Communauto is available in Montreal, Quebec City, Calgary, and Halifax, while Evo and Modo operate in Vancouver, a city recognized by many as the car-sharing capital of North America (for a discussion on car-sharing in Montreal, see Hashemi, Motaghi, and Tremblay, chapter 9) (Britten, 2018; movmi, 2020).

Bike-Share

Bike-sharing services provide a fleet of bicycles for self-rental for a few minutes to hours in a predefined area (see McNee and Miller, chapter 8). In docked systems, riders leave the bicycles at a fixed station; in dockless systems, riders can leave the bicycle anywhere within a designated area. Demand for hybrid dock-and-float configurations is increasing (Shared Use Mobility Center, 2018). Toronto was the first Canadian city to implement bike-sharing services in 2001. In this grassroots bike-sharing program, cyclists rented bicycles as needed. Although this initial service stopped in spring 2007, Toronto relaunched the service with Bike Share Toronto in 2011, which is still serving the city with 6,850 bicycles and 625 stations across 200 km^2 (Bike Share Toronto, 2022).

Of the other cities examined, Montreal, Vancouver, Calgary, and Quebec City have bike-sharing services. Bixi, in Montreal, is Canada's first widely deployed bike-sharing system. Its operation model offers an exemplary way for members to escape high fuel prices and reduce car use (Serebrin, 2022). Hamilton, part of the GTHA area, also has a bike-sharing program, one designed mainly for the student population. Bike-sharing has been supported by infrastructure investments and enabling regulations. However, these initiatives can face financial and operational challenges, including vandalism and theft. As of 2021, Ottawa, Kitchener, Victoria, and Edmonton had shut down their bike-share services (Hansen-Gillis, 2021). Additional detail concerning the

Toronto bike-share experience as a specific case study is provided by McNee and Miller in chapter 8.

E-BIKE-SHARES

With the advent of electric bicycles, alternative bike-sharing options have surged. Similar to conventional bike-sharing services, Zygg rents bicycles in Toronto and Vancouver. However, the bicycles are electric, and the company rents them for longer periods (one month or more). Advocates claim that electric bicycles are easier to ride and better for climbing steep streets (Edge & Goodfield, 2017).

E-SCOOTERS

Like electric bicycles, electric scooters (e-scooters) are available for short-term rental and rely on similar technologies to enable services. Because these small vehicles are usually dockless, riders can park them anywhere, and some cities have expressed concern about cluttered sidewalks and blocked wheelchair ramps. However, many consider e-scooters essential for providing first- and last-mile (FMLM) services in North American cities (movmi, 2022). FMLM refers to the start or end of an individual trip before using a primary mode, usually public transportation. Several e-scooter businesses serve Canadian municipalities: Lime, Uber, Lyft, Link, Neuron, Roll, and Bird. Of the cities considered in this study, Calgary, Ottawa, and Edmonton have e-scooter services as of May 2022 (for a case study of e-scooters in Calgary, see Kong and Leszczyznski, chapter 7). Dockless scooters were available in Toronto from 2019 until 2021, when the city opted out of Ontario e-scooter pilot program (Bird Canada, 2021). As of August 2022, e-scooter use is illegal in the city of Toronto, although some people use these vehicles anyway.

Demand-Response Transit

Demand-response transit (DRT) typically includes smaller vehicles than traditional transit vehicles, an approach that suits occasional short trips. Transit experts see DRT as a potential solution to conventional transit's low occupancy in low-density service areas (Calderón & Miller, 2020). Unlike "dial-a-ride" services, which have existed for a long time, DRT services use recent technological advances as part of their operations. DRT can provide both many-to-many or many-to-one services. The first option is more common in small or medium-sized communities that need to serve an entire area. Vehicles take several riders to several locations. The second option refers to trip patterns with origins or destinations at a single stop or "hub." Vehicles either pick up passengers at a shared spot or deliver them to the same location. Cities typically describe these two options as FMLM services.

A recent study in Canada shows that DRT may no longer only be used as a cost-effective alternative to fixed-route buses in areas with very low demand (Klumpenhouwer, 2020). The study surveyed twenty-six organizations scattered through seventeen municipalities in Canada, demonstrating a growing variety of implementation contexts and strategies. As of 2020, Ontario encompassed a significant portion of recent DRT projects, while the survey found no projects east of Montreal. Of the cities examined in this chapter, Calgary, Winnipeg, and Saskatoon either had or had planned a DRT.

Finally, several recent transit-related experiments in smaller communities deserve mention. Although we do not formally include them in this review of mobility services in Canada's largest cities, they represent creative responses to transit issues, and each one uses a new mobility service either to complement conventional transit or to stand in for it. In one notable example, Belleville, ON, replaced fixed-route/schedule transit services with a fully demand-response service during evenings and nights (Zhang et al., 2020). In another example, Innisfil, ON, completely replaced conventional transit with a subsidized Uber ride-hailing service (Uber, 2018). More generally, in response to dramatic COVID-driven ridership declines, a number of communities have converted suburban bus routes to some form of DRT service (Klumpenhouwer, 2020; Yakub, 2022).

How New Technologies Have Changed the Planning Landscape

Mobility services have been evolving, growing, and also sometimes disappearing. According to many experts, changes often take place because of regulatory barriers (or opportunities) within the municipalities or provinces where the services are operating (Équiterre, 2017; movmi et al., 2022). For example, ride-hailing often makes the news, given its competition with the taxi industry. This mobility service has disrupted local and regional economies and faced a variety of regulatory responses during its implementation in different cities (Ditta & Urban, 2016). Although ride-hailing appeared in Canada in 2012, Edmonton was the first city to legalize the service in 2016, rapidly followed by other cities such as Toronto. This legalization represents a significant industry evolution, creating a model that allows the taxi business and private transportation companies (PTCs) to coexist. However, other municipalities, such as Vancouver, did not legalize ride-hailing until 2020. Thus, the experience with ride-hailing (and mobility services more generally) has varied considerably across Canadian cities in terms of timing of adoption, regulatory environment, and impacts.

For planners, this uneven acceptance of ride-hailing services poses challenges. For example, as part of its approval of PTC, Toronto required companies to submit travel records to the city (Équiterre, 2017). Toronto officials can use these data for audit and compliance as well as accessibility, transportation planning, and environmental initiatives relevant to the city's goals. For this reason, cities should always request such crucial data because it helps them to understand the impact of mobility services on existing services such as public transit. This knowledge can help governments not only plan future transportation investments at the municipal, provincial, and federal levels but also integrate all transportation modes in a MaaS scheme.

The pandemic affected ride-hailing immediately. Still, companies quickly evolved to provide other services that allowed them to stay in business, such as providing food delivery or discounted prices for specific events, such as elections (see Brail and Donald, chapter 10). However, one of the most notable drawbacks of ride-hailing is its potential competition with transit (Ditta & Urban, 2016). Especially after COVID-19, experts believe that the public will need several years to regain confidence and trust in public transit, with some users looking for other transportation options (Phillips & Rickmers, 2021). This is an opportunity for ride-hailing to position itself as an active part of a multimodal transportation system, especially in the provision of FMLM solutions that are transit-supportive. However, this position will greatly depend on each city's regulations and incentives.

On a different note, car-sharing in the cities examined in this study has faced fewer barriers than ride-hailing (Ditta & Urban, 2016), with parking regulations as the main obstacle. For instance, Car2Go, a well-known car-sharing company, left Toronto abruptly in 2018, arguing that city rules inhibited the technology's development (Migdal, 2019). The company later withdrew its operations from the entire North American market in 2020, motivated by increased infrastructure costs and volatility in the global mobility industry. This decision left around eighty thousand users in Toronto without Car2Go services. Although other car-sharing businesses exist in the cities we examined, the pandemic affected some of them as well, such as ZipCar, which shut down operations in Vancouver. Car-share revenues have declined significantly due to the pandemic, while operating costs have increased, mainly due to sanitization requirements (Phillips & Rickmers, 2021).

Even before COVID-19, cities began to incorporate carsharing as part of a multimodal transportation system. In Toronto, mobility hubs such as Union Station and GO stations now have designated parking spaces for car-sharing companies, and in other cities, car-sharing parking permits in residential areas are becoming more common. These policies imply a change in land-use rules and might influence travel behaviour (movmi, 2020).

Unlike the other services, bike-sharing usually begins with a city-requested proposal and operates under various public-private partnerships. Because it satisfies the need for more active travel and for FMLM for public transit (Santander et al., 2022), operators need to prioritize the municipality's goals. When they do, the city can integrate bike-sharing into the transportation system more easily. Some cities have recently invested in projects that support bike-sharing. For example, "the minimum grid" is a project in Toronto to create about one hundred kilometres of connected and protected bicycle lanes on main streets and an additional one hundred kilometres of bicycle boulevards on low-traffic streets (Ditta & Urban, 2016). In the past decade, bike-sharing services have influenced transportation planning by demanding new infrastructure and demonstrating changes in users' behaviour. During the pandemic, bike-sharing has been a clear winner as people have shifted away from motorized modes towards micromobility.

Several Canadian communities are exploring the idea of shared micromobility, including the use of e-scooters. Still, the idea faces obstacles such as limited budgets and laws prohibiting the use of e-scooters on public roads as well as concerns about winter operations (movmi et al., 2022). For example, people in Toronto cannot operate e-scooters in public spaces, take them on board a TTC vehicle, or ride them on GO Transit property. On the other hand, several municipalities, such as Calgary, Edmonton, Vancouver, and Montreal, have approved them for personal use. Some of these cities also have issued requests for proposals for shared micromobility. When e-scooter services first entered Canada, their impacts on parking and safety were typically not consistently considered by the service providers and posed a significant challenge to the government with respect to how to deal with these issues. However, this approach has started to change, with more operators assigning specific parking areas for the scooters and providing helmets to riders (movmi et al., 2022).

E-scooters have altered transportation behaviour, mainly in urbanized areas, where they serve as a FMLM option to connect with transit. Many transit experts claim that shared micromobility services could reduce car use, car ownership, and greenhouse gas emissions, benefiting both transportation and the environment. Nevertheless, for these services to succeed, cities need to integrate them with existing transportation and infrastructure (movmi, 2022).

Pathways and Future Challenges for the Evolution of Mobility Services

Emerging mobility services will keep evolving and growing. The pandemic demonstrated that new technologies could provide convenient alternative modes of transportation. With the increasing popularity of these modes and the advent of autonomous vehicles, cities should stop opposing new services and instead look for ways to integrate them into the current transit system. However, before complete integration can take place, municipalities will need to consider several challenges.

Regulation

Regulation of mobility services should consider federal, provincial, and municipal goals. This unified approach will address many cross-sectional issues across regions, including licences, driver/passenger behaviour, data security and safety, and low earnings for ride-hailing drivers. Regulation in a multi-scale context will also facilitate discussions about parking permits in the case of car-sharing, and speed limits and safety in the case of bike-sharing and e-scooters. If these requirements also include clear expectations from the operation side, then regulation will allow big tech companies and local operators to play a role in the transportation transformation. Additionally, cities should encourage public-private partnerships, which can assist communities to reach their goals and depend less on private transportation.

Integration with Other Modes

For the future of mobility, cities must also ensure that riders can plan trips blending different modes (see Gorachinova, Huh, and Wolfe, chapter 11, and Lorinc, chapter 14). This integration will likely depend on policy objectives in each jurisdiction. However, many experts argue that if cities make emerging mobility services part of a larger mobility ecosystem, transportation will become more convenient, the use of single-occupancy vehicles will decrease, and people's travel behaviour will change (Ditta & Urban, 2016). This integration should include planning, infrastructure, and fare payment systems so that riders have timely information about different routes. Planners will also need to consider not only specific issues, such as accessibility for people with disabilities, but also preferences, such as more environmentally friendly options.

Broader Impacts

Emerging mobility technologies have broader social, environmental, and economic impacts beyond the provision of transportation services per se. For example, some analysts claim that ride-hailing services increase congestion because vehicles drive extra kilometres while waiting to pick up another passenger. In the case of car-sharing, critics frequently point to parking and infrastructure misuse. Regarding e-scooters, experts highlight speed limits and safety issues. In response to these concerns, different alternatives have emerged. Specifically, for the ride-hailing and car-sharing problems, road- and land-use pricing may offer a way to discourage single-occupancy vehicle use and to motivate shared mobility and more sustainable modes such as public transit. In the case of e-scooters and bike-sharing, new software and hardware technology could help enhance safety. For instance, when operators use geofencing to limit areas where e-scooters can travel, this technology can help control speed limits, by disabling the capacity to accelerate and reduce conflicts with other road users (movmi et al., 2022). In addition, e-scooter companies could encourage riders to wear helmets and use warning noises when approaching other road users. These new procedures could help change the safety perception of these micromobility services.

Availability of Standard Data

To understand future travel patterns and compare modes, collecting, sharing, and analysing data gathered from emerging mobility technologies is essential. Yet sources of data vary (see Vinodrai, chapter 4). For example, while writing this chapter, we could not find out how many trips took place in each emerging mobility service. While car-sharing services usually report the number of members, public transit reports ridership, and ride-hailing services report app users and drivers. These varied databases also pose challenges for sharing the information in one platform, making MaaS schemes difficult.

Appeal and Access for Riders

The pandemic affected transportation modes and users unequally. Studies have found that services based on emerging mobility technologies attract specific population groups, leaving behind others who might need the service more (see Palm and Farber, chapter 6). For example, Ipsos found that Canadian consumers disliked several features of ride-hailing, including its surge pricing and imprecise insurance policies. They also felt uncertain about data security and safety, while resenting the inability to hail a vehicle (Nicoll & Armstrong, 2016). As well, research has associated ordering these services with tech savviness and economic status, two qualities that some age cohorts and socio-economic groups lack (Shaheen et al., 2017). Given these inequalities, governments must fund research and development pilot projects on accessible shared mobility and provide options for people who do not know how to use the technology.

What Is Next for Emerging Mobility Technologies in Canada?

There is growing recognition that urban transportation needs new mobility technology services. Demand clearly exists for many of these services when they are well-designed to meet a given market need. However, governments at all levels – but particularly municipal – must work proactively to integrate mobility services within the overall transportation system in order to maximize their social, environmental, and economic benefits. Even if a market exists for a service, governments must not assume that private companies will automatically provide a socially attractive solution. Instead, public-private collaborations are essential so that private-sector innovation can support the public good. Further, one size does not fit all. Different services and combinations of new and conventional services will suit different municipalities, depending on each city's size, existing transit, transportation capabilities, local geography, and weather. As the pandemic has revealed, new technologies, well designed and operated, can help people move safely and easily, and cities must plan to incorporate these emerging mobility technologies as part of their future transportation options.

REFERENCES

Bike Share Toronto. (2022). *Convenient, affordable & fun transportation*. Bike Share Toronto. https://bikesharetoronto.com/.

Bird Canada. (2021, 6 May). *Bird Canada disappointed in City of Toronto e-scooter pilot decision*. https://www.businesswire.com/news/home/20210505006243/en/Bird-Canada-Disappointed-in-City-of-Toronto-e-Scooter-Pilot-Decision.

Britten, L. (2018, 26 January). *Vancouver is "car-sharing capital of North America," report says*. CBC News. https://www.cbc.ca/news/canada/british-columbia/vancouver-car-share-car2go-evo-1.4504926.

Calderón, F., & Miller, E.J. (2020). A literature review of mobility services: Definitions, modelling state-of-the-art, and key considerations for a conceptual modelling framework. *Transport Reviews*, *40*(3), 312–32. https://doi.org/10.1080/01441647.2019.1704916.

Calderón, F., & Miller, E.J. (2022). A conceptual framework for modeling the supply side of mobility services within large-scale agent-based travel demand models. *Transportation Letters*, *14*(6), 600–9. https://doi.org/10.1080/19427867.2021.1913303.

Ditta, S., & Urban, M.C. (2016). *Sharing the road: The promise and perils of shared mobility in the GTHA*. Mowat Centre.

Edge, S., & Goodfield, J. (2017). Responses to electric bicycles (e-bikes) amongst stakeholders and decision-makers with influence on transportation reform in Toronto, Canada. *Proceedings of the 52nd Annual Conference*. Canadian Transportation Research Forum. https://ctrf.ca/wp-content/uploads/2017/05/CTRF2017EdgeGoodfieldActiveandGreenTransportation.pdf.

Équiterre. (2017). *SHARED MOBILITY Removing regulatory barriers in Canadian cities*. https://www.equiterre.org/en/resources/news-shared-mobility-removing-regulatory-barriers-in-canadian-cities.

Government of Canada. (2021a). *Automotive statistics*. https://www.statcan.gc.ca/en/topics-start/automotive.

Government of Canada. (2021b). *Car sharing*. Industry Canada; Innovation, Science and Economic Development Canada. https://www.ic.gc.ca/eic/site/Oca-bc.nsf/eng/ca03013.html.

Habib, K.N. (2019). Mode choice modelling for hailable rides: An investigation of the competition of Uber with other modes by using an integrated non-compensatory choice model with probabilistic choice set formation. *Transportation Research Part A: Policy and Practice*, *129*, 205–16. https://doi.org/10.1016/j.tra.2019.08.014.

Hall, J.D., Palsson, C., & Price, J. (2018). Is Uber a substitute or complement for public transit? *Journal of Urban Economics*, *108*, 36–50. https://doi.org/10.1016/j.jue.2018.09.003.

Hansen-Gillis, L. (2021, 12 April). Can you guess which Canadian city has the most affordable bike-share system? *Canadian Cycling Magazine*. https://cyclingmagazine.ca/sections/gear-reviews/bikesframes/bike-share-canadian-cities/.

Klumpenhouwer, W. (2020). *The state of demand-responsive transit in Canada*. (UTTRI Report, p. 48). University of Toronto. https://uttri.utoronto.ca/files/2020/08/The-State-of-Demand-Responsive-Transit-in-Canada-v1.2-FINAL.pdf.

Migdal, A. (2019, 18 December). *Share Now, formerly Car2Go, to end service in North America in February*. CBC News. https://www.cbc.ca/news/canada/british-columbia/car2go-share-now-shutting-down-1.5401113.

Milenkovic, D. (2022, 9 August). *Lyft vs Uber – Key differences between in Canada*. Carsurance Canada. https://carsurance.net/canada/insights/lyft-vs-uber/.

Miller, E.J., Shalaby, A., Diab, E., and Kasraian, D. (2018, October). *Canadian Transit Ridership Trends, Final Report*. Report to the Canadian Urban Transit Association, Canadian Ridership Trends Research Project. University of Toronto Transportation Research Institute.

movmi. (2020). *Carshare City Awards Report* (p. 10). Carsharing association. www.movmi.net.

movmi. (2022, 13 May). *Micromobility update 2022: Shared E-scooters in Canada*. https://movmi.net/blog/shared-e-scooters-in-canada-2022/.

movmi, Phillip, S., & Gopal, V. (2022). *Evaluating share micromobility*.

Nair, R. (2020, 25 January). *Vancouver's long and winding road to ride-hailing*. CBC News. https://www.cbc.ca/news/canada/british-columbia/timeline-uber-vancouver-1.5439522.

Nicoll, E., & Armstrong, S. (2016). *Ride-sharing: The rise of innovative transportation services*. MaRS Discovery District. https://www.marsdd.com/news/ride-sharing-the-rise-of-innovative-transportation-services/.

Phillips, S., & Rickmers, J. (2021). *Rebuild tomorrow's mobility: Reporting back on the impact of COVID-19 in Vancouver, 2020.*. movmi.net. https://movmi.net/wp-content/uploads/2020/08/RebuildTomorrowsMobility_MetroVancouver.pdf.

Santander, C., Phillips, S., Van Audenhove, F.-J., Jouron, A., & Gopal, V. (2022). *Rethinking bike-sharing performance* (p. 8). Arthur D. Little.

Serebrin, J. (2022, 6 June). *Montreal's bike culture a model for beating high gas prices*. Canada's National Observer. https://www.nationalobserver.com/2022/06/06/news/montreals-bike-culture-model-beating-high-gas-prices.

Shaheen, S., & Chan, N.D. (2015). *Mobility and the sharing economy: Impacts Synopsis*. Innovative Mobility. http://innovativemobility.org/wp-content/uploads/Innovative-Mobility-Industry-Outlook_SM-Spring-2015.pdf.

Shaheen, S., Cohen, A., & Jaffee, M. (2018). *Innovative mobility carsharing outlook – Winter 2018*. Innovative Mobility. http://innovativemobility.org/?p=3082.

Shaheen, S., Cohen, A., & Yelchuru, B. (2017). *Travel behavior: Shared mobility and transportation equity* (PL-18-007; p. 66). Federal Highway Administration.

Shared Use Mobility Center. (2018). *What is shared mobility?* Shared-Use Mobility Center. https://sharedusemobilitycenter.org/what-is-shared-mobility/.

Statistics Canada. (2022, 19 May). *Urban public transit, March 2022*. The Daily. https://www150.statcan.gc.ca/n1/daily-quotidien/220519/dq220519d-eng.htm.

Uber. (2018). *Innisfil transit partnership*. https://www.uber.com/ca/en/u/innisfil/.

Uber. (2022). *The impact of Uber in Canada – How Uber has transformed the on-demand economy*. https://ubercanada.publicfirst.co/.

Yakub, M. (2022, 22 June). *On-demand transit could transform the commuter experience in Canada*. Electric Autonomy Canada. https://electricautonomy.ca/2022/06/22/on-demand-transit-canada-autocrypt-webinar/.

Zhang, Y., Farber, S., & Young, M. (2020). *The benefits of on-demand transit in Belleville: Findings from a user survey* (p. 32). University of Toronto. https://pantonium.com/pantonium-on-demand-transit-project-begins-in-belleville-ontario/.

SECTION III: PLACES, PATTERNS, AND CHALLENGES

6 Prospects for Public Transit Equity in the Twenty-First Century: Lessons from COVID-19's Impacts on Canada's Largest Metropolis

MATTHEW PALM AND STEVEN FARBER

Introduction

COVID-19 and societal responses to it pushed Canadian public transit into an existential crisis. Suddenly, the mode of travel that researchers identified as an enabler of agglomeration economies (Chatman & Noland, 2014), robust real estate markets (Welch et al., 2018), and a green future (CUTA-ACTU, 2021b) became a potential disease vector, placing its users and operators at risk of infection. The actual risk of COVID-19 spreading through public transit was not well understood throughout this crisis, leading governments and transit agencies themselves to discourage ridership except for essential purposes (TransLink, 2020; TTC, 2020). Ridership in North America would bottom out at an average of 85 per cent below pre-pandemic levels (Liu et al., 2020). For Canadian agencies that depend heavily on fare revenue, this drop precipitated an apocalyptic loss of funding, a catastrophe that could trigger further ridership decline in a mutually reinforcing feedback loop known as a "ridership spiral" (CUTA-ACTU, 2021c).

As fare-dependent systems struggled, industry voices across North America advocated for lifeline funding and began drawing more heavily on an equity-infused narrative, one that asserted transit was essential for millions of people's well-being. One commentator went so far as to argue that public transit "makes urban civilization possible" because essential riders, "who are overwhelmingly low-income, have always been there, moving around quietly in our transit systems, keeping our cities functioning" (Walker, 2020). Industry bodies calling for sustained funding to carry transit through the crisis frequently drew on this narrative as well, with the Canadian Urban Transit Association reminding the federal government that "public transit enables equity by allowing access to employment, education, and daily life irrespective of income" (CUTA-ACTU, 2021a).

Political theorist John Kingdon once referred to public transit as a solution seeking a problem, "portrayed by its advocates as a solution to the problems of traffic congestion, air pollution, and energy shortages as each of these problems has heated up" (Kingdon, 1993, p. 41). Although Kingdon was referring to public transit in the post-war United States, his insights are relevant to Canada during COVID-19 and its aftermath. COVID-19 will likely change policymakers understand public transit's purposes and benefits. In short, the pandemic reminded policymakers that public transit plays a critical role in ensuring equitable access to destinations in an automobile-dominated country like Canada. Three dynamics of continued transit ridership during the crisis, viewed through this equity lens, underscored the importance of public transit. Each dynamic, expressed as a question, inspires a lesson for what policymakers need to consider when delivering equitable public transit systems in the future:

- Who continued to use transit during the pandemic and why?
- Which transit modes did riders continue to use?
- When did riders continue to ride?

Using a case study analysis of Canada's largest region, the Greater Toronto Area (GTA), this chapter investigates these spatial, modal, and temporal dynamics and the lessons they offer on how public transit can equalize access in Canadian cities. We begin with a summary of the state of transit equity in Canada and the GTA at the onset of the crisis.

The Equity of Canada's Transit Systems on the Eve of the Pandemic

Transportation equity refers to the fair distribution of the benefits and burdens of transportation systems across the populations that those systems serve (Litman, 2002). Academics studying transportation systems have agreed on accessibility as the key metric of transportation system benefits that policymakers should use to evaluate transportation equity (see Kong and Leszczynski, chapter 7) (Martens & Golub, 2021; Pereira & Karner, 2021; Pereira et al., 2017). Most research on the topic defines accessibility as a person's ability to reach a range of potential opportunities from a given location within a given time frame (Martens & Golub, 2021), often the number of jobs that residents can reach within a standard commuting time. The debate now focuses on which destinations are most important to measure access to, and where the distribution of accessibility should stand before a transit system can be declared "equitable." One camp argues forcefully that policymakers should strive to provide every transit rider with a sufficient level of transit service to achieve equity with the level of access provided by the automobile and to stop concentrating on relative disparities between groups (Martens et al., 2022).

Before the pandemic, accessibility research portrayed Canadian transit systems as equitable in allocating service spatially, but still insufficient in providing enough service for all urban populations in need. Notably, this literature focuses almost exclusively on income groups. One study of eleven metros found that vulnerable residents had higher-than-average accessibility to all jobs and low-wage jobs specifically (Deboosere & El-Geneidy, 2018). A similar study comparing residents' accessibility by income in the same eleven metros found that residents on low incomes had better access to jobs than those on high incomes (Cui et al., 2020). However, small average differences between income groups do not mean that all workers on low incomes had adequate transit access. A separate analysis of the country's eight largest metros revealed that nearly a million Canadians with incomes below the low-income cut-off, a threshold for poverty, lived in neighbourhoods in the bottom decile of transit access to jobs (Allen & Farber, 2019). The authors concluded that these households were at risk of transport poverty, a circumstance in which a lack of adequate transportation worsens existing social disadvantages and may prevent people from exiting poverty (Lucas, 2012).

Two interrelated trends in the social geography of Canadian cities contributed to the rise of transport poverty in the country's urban areas: the gentrification of downtown cores and the growth in suburban poverty. Decades of inner-city gentrification threatened access to transit in Canada's largest cities by making neighbourhoods next to heavy rail and commuter rail prohibitively expensive (Grube-Cavers & Patterson, 2015; Kramer, 2018). Concurrently, the number of households on low incomes in post-war inner suburbs increased in most major metros (Ades et al., 2012; Allen, 2022; Grant et al., 2020). In Toronto, these trends led to income polarization spatially arranged around the subways: neighbourhood average incomes along subway lines rose after the 1970s, while average incomes in suburban neighbourhoods far from the lines decreased relative to overall incomes (Hulchanski, 2010). This mismatch between transit service provision and growth in suburban poverty is associated with declining participation in daily life activities and increased commute times in high-poverty suburbs across the region (Allen & Farber, 2021). Decades of planned investment in rapid rail transit failed to keep pace with geographic shifts in need, as the province wielded rail expansion as a tool of regional economic development, often in direct conflict with the City of Toronto's own transit vision (Addie, 2013; Addie & Keil, 2015). These political challenges (see Hutton, chapter 3), compounded by a lack of integration between municipal transit systems, disadvantaged some residents. In particular, the situation left many riders in the region's suburban but densifying "in-between" cities without adequate access to other suburban centres that they needed to reach (Hertel et al., 2016).

The spatial polarization with respect to income also mirrored an increasingly racialized "post-Fordist" city. During this time, a racially diverse population of newcomers entering Canada by way of the GTA settled in outer and inner suburban communities with relatively lower levels of transit (Walks, 2001). By 2016, immigrant settlement in the GTA was more suburbanized and less clustered around rail transit than in the comparable "gateway city" of Sydney, Australia (Allen et al., 2021). These dynamics presented a further challenge to transportation equity, as newcomers rely more on transit for commuting and other travel compared to the general population (Allen et al., 2021; Lo et al., 2010). Perhaps as a result, most immigrant cohorts in Canada are now more likely to face extreme commutes of more than sixty, seventy-five, and ninety minutes compared to commuters born in the country (Allen et al., 2022).

Canadian cities pre-pandemic were at least "vertically equitable," meaning that people with a greater need for transit, such as newcomers and those at risk of poverty, had higher average levels of transit access than their respective reference groups (Canadian-born residents, people with high incomes). However, spatial polarization driven by gentrification and the suburbanization of both poverty and immigration threatened this access. By 2016, nearly a million people on low incomes across the country likely lacked the level of transit access they needed to avoid transport poverty. Furthermore, many people with transit access did not have good access to where they needed to go, particularly for "suburb-to-suburb" travel (Hertel et al., 2016). These historical factors would help explain why transit ridership unravelled during the pandemic and why the return to normal has been slow. In particular, they help to account for the spatial, temporal, and modal dynamics that shaped pandemic transit use.

Data

Synthesizing the equity impacts of a global pandemic on even one region's transit system is a daunting task, and it required us to draw on multiple sources of data, some public and some private. First, to adequately contextualize pandemic-era changes, several of the following sections draw on the 2016 Transportation Tomorrow Survey, or TTS (DMG, 2014). The TTS is the largest one-day household travel survey in Canada, representing 5 per cent of the population of the Greater Golden Horseshoe Area, which includes the GTA, Greater Hamilton, and adjacent cities in southern Ontario such as Niagara Falls. An online data retrieval system, which produced weighted summary statistics, allows users to calculate cross-tabulations without requiring access to raw data files.

Second, we pulled transit ridership statistics from various sources. The Canadian Urban Transit Association provided us with annual ridership data by transit mode for the GTA's major transit agencies. The Toronto Transit Commission (TTC), the region's largest agency serving the City of Toronto, also provided further custom ridership tabulations by time of day.

Third, we drew upon our own Public Transit and COVID-19 Survey (Palm, Allen, et al., 2020), which targeted Torontonians who rode transit at least weekly before the pandemic. That survey asked riders about their continued use of transit during the pandemic, their perceived difficulties in avoiding transit, and their attitudes towards future transit use. We began data collection in May 2020, limiting ourselves to Facebook advertisements and e-mails to community group LISTSERVs because in-person recruitment was not possible (ethics board restrictions). The ads and survey were translated into Simplified and Traditional Chinese. Ads were seen by over seven hundred thousand unique Facebook profiles based in Toronto. We gathered over eighteen hundred usable responses in Toronto that we deploy in this chapter (Palm, Allen, et al., 2021). Despite our convenience sampling, the data successfully replicate known relationships between socio-demographic, built form, and transportation outcomes (Zhang et al., 2020). Where possible, we have referenced our survey findings against studies that asked similar questions but collected survey data through traditional means. Given the unique restrictions we faced on survey fielding, this cross-referencing provided an additional check on our results. The combination of our three distinct data sources allows us to explore the equity implications of COVID-19 transit ridership at the scale of the individual, city, and region.

How COVID-19 Reshaped Equity Considerations

Who Kept Riding, and Why?

For some people, public transit played such an important role in pre-pandemic life that they continued to ride against the advice of medical experts and even transit agencies themselves (Kamga & Eickemeyer, 2021). Early media narratives portrayed these riders as the "working poor" (Spurr, 2020): people still riding because they didn't have a car and they couldn't do their jobs remotely. In the Canadian context, this group included workers in manufacturing, trades, transportation, health, sales, and services (Statistics Canada, 2021). These workers were more likely to be male and newcomers to Canada (Statistics Canada, 2021). The ability to work from home broke heavily along lines of income, with just 7.9 per cent of workers in the bottom income decile working from home compared to 62.5 per cent of workers in the top decile (Statistics Canada, 2021). Although the capacity to work from home was a defining factor in who kept riding transit, particularly along lines of income, industry, and immigration history, it was not the only determinant. Public narratives emphasizing essential workers obscured other factors driving continued ridership.

Many pre-pandemic riders continued to depend on public transit to reach essential activities other than work. In our 2020 survey of Torontonians who used transit regularly before the pandemic, we asked respondents which travel mode they used most often for each of six trip purposes (Palm, Allen, et al., 2020). From that survey, Figure 6.1 shows the May 2020 mode split by trip purpose, but the data in each column are limited to those who used public transit for that respective purpose before the pandemic. The figure thus shows the degree of mode shift by purpose as well as the type of mode that riders shifted to. Mode shift was smallest for grocery trips (40%), followed by irregular or other trips (22%), medical trips (22%), work trips (21%), care or support trips (14%), and finally recreational trips (7%).

This focus on mode shift highlighted a significant fact. Transit was considerably more important for conducting grocery, medical, and other irregular travel compared to work travel, as fewer people shifted modes for those activities. These findings mirror another survey in the region collected using representative panels. That study suggests that during the first six months of the pandemic, just under half of people

Figure 6.1. May 2020 mode split by trip purpose, among Torontonians who usually used transit for this purpose pre-pandemic (Authors' data)

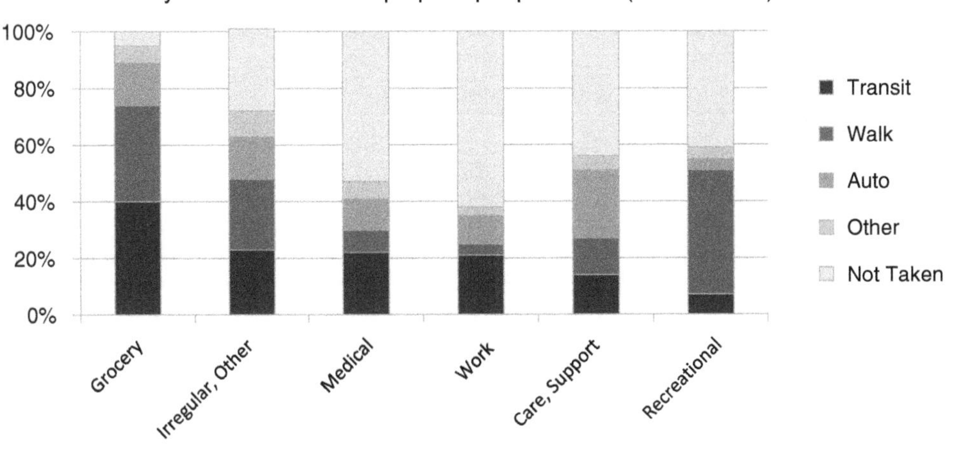

still using transit continued to ride for groceries at least weekly, compared to about a quarter continuing to ride for work as often (Wang et al., 2021a, p. 20). Wang et al. (2021a) also found relatively higher levels of continued transit use for visiting others (42% at least weekly). The difference between our surveys regarding this activity may relate to the wording of questions. We asked about typical mode for "care or support" trips, while Wang et al. (2021a) focused on frequency of transit trips for "visits" to other people over a weekly time frame. By fall 2021, the share of transit users in their survey riding at least weekly for different purposes rose to 57 per cent for shopping, 62 per cent for restaurants, and 44 per cent for visiting others. In contrast, the share of riders commuting to work by transit at least weekly remained at a much lower 26 per cent (Wang et al., 2021b, p. 23).

Not all segments of the pre-pandemic ridership shifted modes to the same degree across different trip purposes. In a concurrent study, we used a two-step selection approach to model who continued to use transit to reach groceries during the pandemic. Step one predicted use of transit to get groceries before the pandemic, and step two modelled continued use of transit for groceries during the pandemic. We found that people over age 64 were more likely than those aged eighteen to twenty-nine to continue using transit for groceries, as were racialized

Table 6.1. Share of people continuing to ride transit for a given purpose in May 2020, by demographic group

	Irregular Medical				Work		Care and support		Recreational	
Sample	923		916		1059		1031		521	
Gender										
• Female	20%		20%		7%		11%		7%	
• Male	31%	**	27%	^	13%	**	19%	***	11%	insig.
• Non-binary	22%		24%		6%		12%		5%	
Age										
• 18–29	15%	***	20%	**	5%	*	11%	**	7%	insig.
• 30–50	22%		19%		9%		14%		6%	
• 50–64	38%		28%		12%		21%		14%	
• 64+	38%		35%		12%		6%		9%	
Immigration										
• No	15%	insig.	20%	0	5%	i	11%	ii	7%	insig.
• Yes	22%		19%		9%		14%		6%	
Race										
• Racialized	25%	insig.	27%	ii	12%	iii	19%	iii	10%	0
• White	21%		19%		5%		10%		5%	
Has a disability										
• No	22%	i	20%	iii	8%	i	13%	i	7%	insig.
• Yes	37%		41%		18%		24%		9%	
Vehicle access										
• Neither	30%	***	31%	***	15%	***	22%	***	11%	***
• Access	16%		15%		3%		8%		4%	
• Own	4%		7%		1%		1%		2%	
Income										
• Under $40k	31%	***	33%	***	15%	***	21%	***	9%	
• $40k - $59k	27%		23%		9%		17%		12%	
• $60k - $99k	22%		17%		5%		8%		2%	insig.
• $100k - $124k	15%		15%		6%		9%		7%	
• $125k and up	9%		10%		2%		2%		2%	

Chi-squared: ^ $p < .10$, * $p < .05$, ** $p < .01$, *** $p < .001$; Difference in means: ^{0} $p < .01$, ^{i} $p < .05$, ^{ii} $p < .01$, ^{iii} $p < .001$. Insig. for insignificant results for either test.

riders relative to white riders (Palm, Allen, et al., 2023). Respondents with longer walking distances to the nearest grocery store were also more likely to continue using transit. All told, our work suggests that older riders living in more suburban environments with fewer walkable destinations remained more dependent on transit for groceries during the pandemic. Notably, in a large UK study, walking distance to food retailers is also negatively associated with continued transit use among older adults (Carney et al., 2022).

The demographic patterns of continued transit use vary substantially among the other trip purposes measured in our survey. Table 6.1 presents the share of people who continued to use transit for each of the other five reasons presented in Figure 6.1. We used chi-squared tests to find differences between groups according to gender, age, income, and vehicle access, while we used difference-in-means tests to find differences based on disability, race, and immigration. Significance levels from tests on each demographic variable appear to the right of the statistics and represent the results for the entire variable. For example, the two asterisks below gender and irregular trips indicate that the three statistics for continued ridership for irregular trips among females, males, and non-binary respondents are significantly different according to a chi-squared test at $p < .01$.

Socio-demographic differences shaped riders' continued use of transit for all purposes, except with respect to gender. Racialized riders were more likely to keep riding for medical, work, and care or support trips, as were recent immigrants for the latter two of those activities. Riders over the age of sixty-four were most likely to continue riding for irregular and medical trips, while those between fifty and sixty-four were most likely to ride for work and care or support trips. Riders without access to a vehicle were more likely to continue using transit for all these purposes, reflecting the overall importance of automobile access in determining continued transit use (Md. Mashrur et al., 2022; Palm, Allen, et al., 2021). Similarly, those with higher incomes were less likely to continue using transit for all trip purposes compared to those with lower incomes, except for recreational trips.

The number of people who used transit for work before the pandemic was much larger than for other purposes, and those trips remained more frequent. Not surprisingly, therefore, policymakers have focused on work trips. However, the socio-demographic make-up of ridership by trip purpose before and during the crisis exposes the limitation of this approach. When developing equitable transit systems, policymakers must also consider non-work travel – a consideration that the pandemic has only elevated.

Figure 6.2. Household income distribution of ridership by transit mode, 2016 Transportation Tomorrow Survey

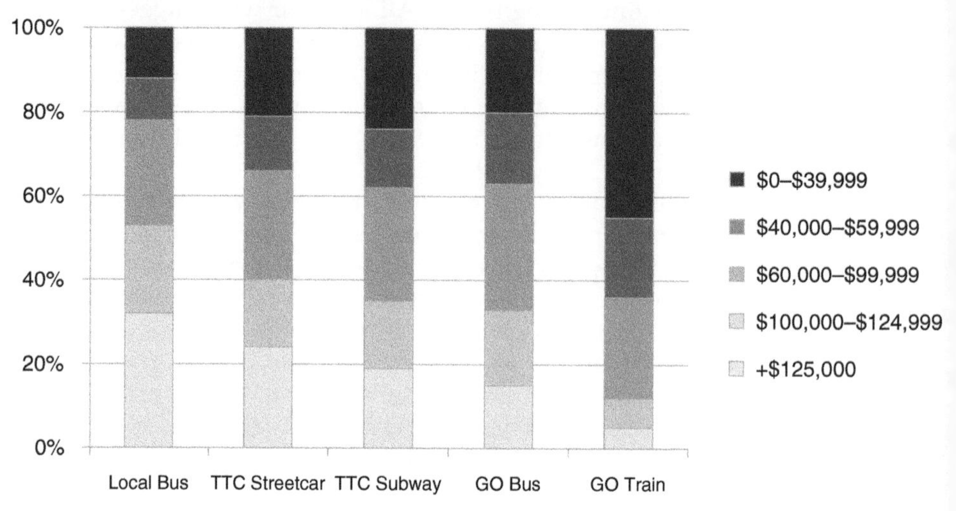

How Did People Travel?

In the period leading up to the pandemic, two factors created a polarization of incomes by transit mode: the socio-spatial polarization of Toronto with respect to the subway lines and the growth of automobile-dependent communities in surrounding localities. The 2016 Transportation Tomorrow Survey, depicted in Figure 6.2 below, reveals these differences in household income distribution by transit mode. We consider each segment of a transit journey separately. For instance, a person taking a bus to a subway and then another bus counts thrice: once for the subway and twice for the bus. This approach helps capture the relative importance of each mode to each income group.

Just over half of riders relying on local bus service across the region lived in households earning less than sixty thousand dollars per year, compared to 40 per cent of streetcar riders and just over a third of subway riders. GO ridership skewed even more heavily towards residents with high household incomes, as over 70 per cent of GO Train riders lived in households earning over one hundred thousand dollars per year. These percentages reflect the government's use of rapid transit investment primarily as a tool of economic competitiveness and development (Addie, 2013; Addie & Keil, 2015). However, this

Figure 6.3. Annual ridership by mode and city relative to 2019, 2019–21 (adapted from CUTA 2022)

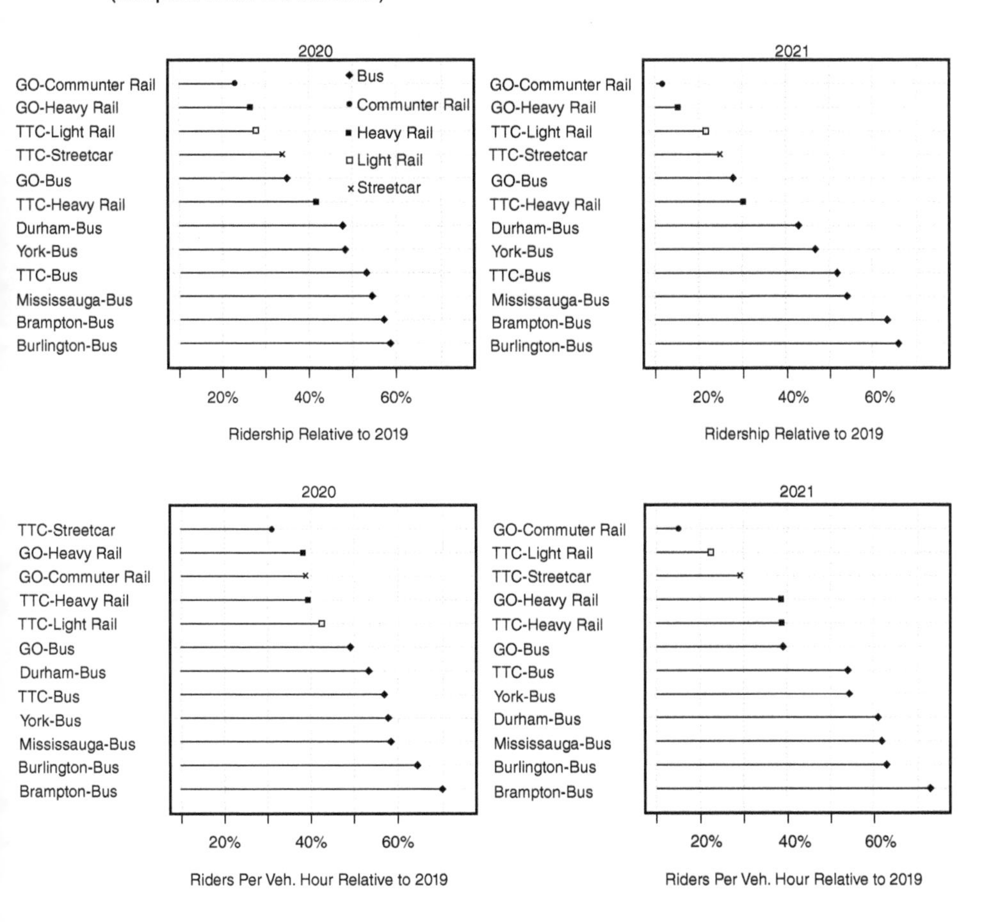

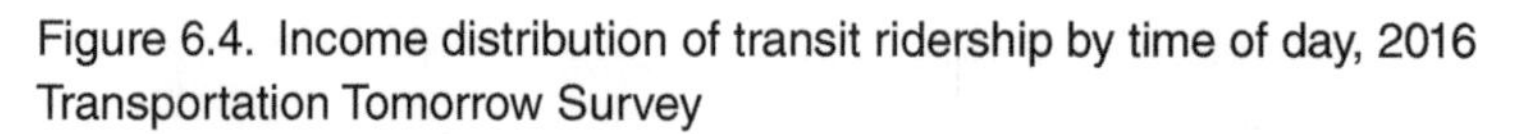

Figure 6.4. Income distribution of transit ridership by time of day, 2016 Transportation Tomorrow Survey

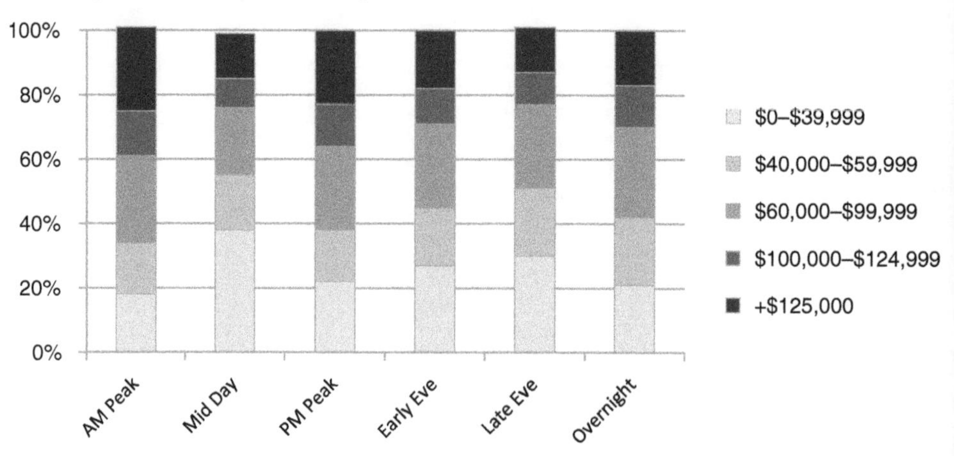

approach has failed to provide fast or affordable options to a growing number of suburb-to-suburb commuters on low incomes who currently rely on local bus networks (Hertel et al., 2016). This failure, combined with unequal capabilities for remote working, would help explain uneven ridership by mode during, and coming out of, the pandemic.

Drawing on data provided by the Canadian Urban Transit Association, Figure 6.3's top panel presents GTA transit ridership by mode and agency, relative to 2019. Although the TTC's bus ridership never dropped below 50 per cent of 2019 levels, it was outperformed by bus ridership in three suburban jurisdictions: Mississauga, Brampton, and Burlington. In contrast, GO rail ridership collapsed to less than 20 per cent of pre-pandemic levels, while TTC subway ridership bottomed out at 30 per cent. However, these agencies responded to the pandemic by rapidly adjusting their service levels, complicating raw ridership comparisons. To account for these adjustments, the bottom panel presents ridership per vehicle revenue hour relative to 2019. This data, also provided by CUTA, reveals a somewhat less apocalyptic picture for GO Transit and the subway. However, it further underscores the strength of suburban bus ridership. When we control for vehicle revenue hours, bus ridership in Brampton bottomed out at 70 per cent of pre-pandemic

Figure 6.5. Annual ridership by time of day relative to 2019 (from TTC 2022)

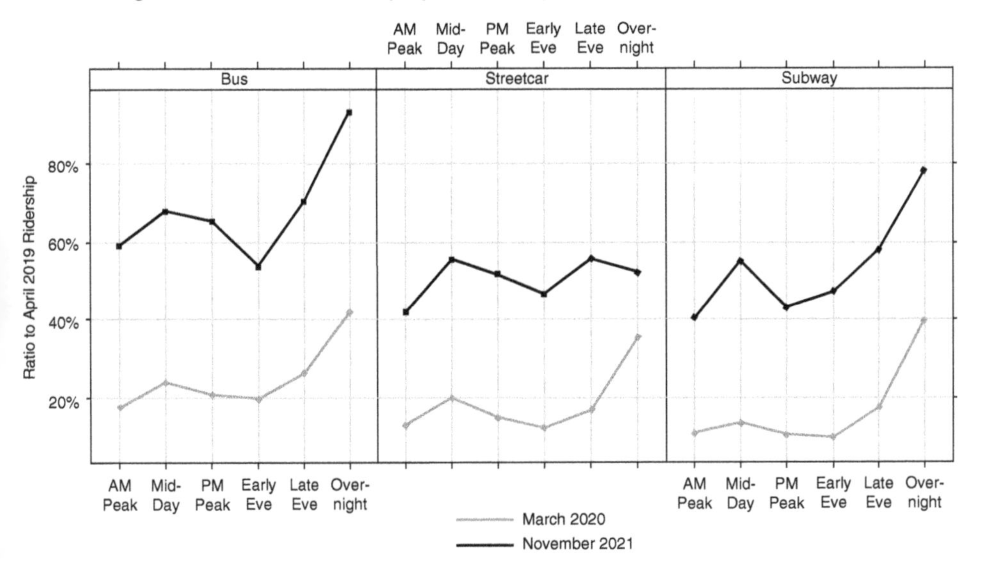

levels, with bus ridership in Burlington, Mississauga, and Durham Region performing almost as well.

To what extent will these trends continue? We sought to compare ridership across these agencies for summer 2022, but not all agencies made this data available or easy to access (see Vinodrai, chapter 4). In June 2022, TTC bus ridership returned to 64 per cent of pre-pandemic levels, versus 56 per cent for the streetcar and 49 per cent for the subway (TTC, 2022, p. 26). In the same month, however, Brampton bus ridership achieved 107 per cent of pre-pandemic levels (City of Brampton, 2022), while in July, ridership in Mississauga reached 86 per cent of the pre-pandemic norm (Cornwell, 2022). As of this writing, ridership is still inversely related to the income distribution of pre-pandemic ridership: in brief, modes serving people on lower incomes continue to outperform modes that serve higher-income travellers. Among bus services, the strongest levels of returning ridership are taking place in suburban areas, notably Brampton.

When Did Ridership Persist?

Before the pandemic, transit use in the region was also temporally polarized. Travellers from lower-income households made up a larger share of ridership in the midday (9 am–3 pm), early evening (7–10 pm), and late evening (10 pm–1 am) periods. Figure 6.4 above, drawing on TTS data, provides income distributions by time period. Travellers from low-income households being over-represented during off-peak periods is neither unique to transit nor to Toronto. Similar patterns have been identified in Salt Lake City (Farber et al., 2016), Montreal (Lachapelle & Boisjoly, 2022), and across Canada (Palm et al., 2023).

Although each mode experienced varying declines in ridership, they all showed similar patterns of decline by time of day, correlated with income, as illustrated in Figure 6.4. Time periods serving greater shares of people on low incomes experienced less ridership decline. The TTC provided us with boarding data by time of day at three data points: April 2019, March 2020, and November 2021. To illustrate ridership recovery, we've plotted the data from 2020 and 2021 as a proportion of the 2019 boardings (see Figure 6.5).

Overnight ridership recovered the fastest, particularly for bus travel. This rapid recovery may reflect the strength of Toronto's Blue Night Network, which is the largest evening bus network in North America (Palm, Shalaby, et al., 2020). Ridership also recovered faster during the late-evening and midday time periods. Notably, during these periods, the largest share of pre-pandemic ridership came from households earning less than forty thousand dollars annually (see Figure 6.4). Ultimately, although differences in ridership recovery by time period are less dramatic than those by mode, they still follow the same pattern of low-income ridership correlating positively with ridership recovery.

Lessons for an Equitable Twenty-First-Century Transit System

The pandemic altered transit ridership squarely along lines of socio-economic privilege and socio-spatial advantage. Decimated demand revealed who truly depended on transit: the essential riders, or those for whom transit is an essential service (He et al., 2022). Over-represented in this group were newcomers, people on low incomes, and people without vehicles, particularly those in neighbourhoods that made walking or cycling a poor alternative (Palm, Allen, et al., 2021; Palm, Sturrock, et al., 2021). Before the pandemic, regular riders also included older adults.

In the context of a gentrified region with a splintered transit infra-structure, these demographic patterns clarify the spatial, temporal, and modal dynamics of pandemic-era ridership. Ridership declined the most in the gentrified region of the GTA and in transit modes used by the highest income earners: GO Rail and TTC's LRT, streetcars, and to a lesser extent, subway lines. Ridership remained strongest where essential workers and newcomers live in less walkable communities or "in-between cities" with poorer suburb-to-suburb transit options, in regions like Brampton, Burlington, and Mississauga. Ridership declined the least during the times of day when transit serves shift workers and people who depend on transit for non-work trips like groceries.

As some have already argued, the people who need transit the most deserve better access to it (Martens et al., 2022). However, conventional strategies, services, and financing arrangements will not achieve this goal in the GTA. To ensure an equitable transit future in the region, solutions must respond to increased demand in suburban communities, including at off-peak periods, and meet the diverse travel needs of riders without vehicles, especially non-work trips. We offer two overarching recommendations for building towards an equitable, post-pandemic transit system in the GTA: shifting to a sustainable financing model and planning an everywhere-to-everywhere rapid transit network.

Creating Sustainable Revenue for Transit Operations

The riders who vanished during the pandemic – wealthier and less diverse than the riders who remained – had kept transit systems solvent before the crisis. That mass of commuters that crammed onto subways, GO Rail, and express buses to the downtown core during peak periods generated revenue that helped sustain the overall transit network, including less intensively used routes that served off-peak travellers and people in transport poverty. The continued appeal of remote work threatens the long-term viability of this financial approach. Nearly 80 per cent of Canada's new remote workers, or people who started remote work for the first time during the pandemic, want to continue working remotely for at least half of their work hours (Mehdi & Morissette, 2021). As these people did not work remotely at all before the pandemic, their preferences signal a dramatic shift in societal attitudes towards remote work. The old revenue model that leaned on these workers dutifully showing up to the office every day is untenable.

While our analysis does not point to what alternative financing model would best serve the GTA, global experience offers several examples that warrant investigation. The positive effect of transit on real estate values is well established (Mohammad et al., 2013), and multiple regions around the world use land value capture to pay for a wide array of public services. Tools such as tax increment financing, special property tax district levies, value-added real estate taxes, and betterment charges can all recapture value (Global Infrastructure Hub, 2020; OECD, 2022; Salon et al., 2019). Alternatively, Paris partially funds transit through payroll taxes applied to businesses within transit service areas (Salon et al., 2019), an approach that left the city's system less damaged by the pandemic (McArthur et al., 2020). Future research will need to carefully delineate the distributional impacts of different strategies along lines of income, race, and immigration. Potential solutions must not disproportionately burden residents who are underserved by the existing transit system.

Investing in Everywhere-to-Everywhere Transit When Riders Need It

As the pandemic has revealed, providing transportation equity in an auto-centric system is complex. It means providing high-quality transit that riders can use at all times of day, not only to reach a myriad of essential activities besides work but also to reach work sites outside major employment hubs. This recommendation is not new (Hertel et al., 2016), and the pandemic has made it harder to ignore. Several services or service improvements have the potential to help meet the region's diverse transit needs.

On-demand transit (ODT) offers one step towards everywhere-to-everywhere accessibility during off-peak periods. Our own review of pre-pandemic evidence suggests that ODT could advance transportation equity. Internationally, it has successfully served the same kinds of communities where transport poverty is growing in the GTA: lower-income, lower-density suburbs (Palm, Farber, et al., 2020). ODT has also effectively addressed service deficits during off-peak periods. One case study in Belleville, ON, demonstrated that ODT can provide evening and night-shift workers with direct access to their jobs, thereby reducing transport disadvantage for those participating in the evening economy (Zhang et al., 2021). These technologies are not new, but the rise of the smartphone and advanced AI could make both the riders' experience and operational management more efficient (Alsaleh et al., 2021; Klumpenhouwer et al., 2020; Sanaullah et al., 2021), as long as agencies can help riders overcome any digital or financial barriers (Palm, Farber, et al., 2020).

Interventions to speed up suburban bus service could also help us achieve robust everywhere-to-everywhere networks. Regional agencies can take several approaches to help buses operate more efficiently in mixed traffic, including transit priority signalling, bus-only lanes, express services, and bus rapid transit, among others (Lin et al., 2015; Sidloski & Diab, 2020). However, to ensure that such interventions build equitable systems, decision-makers must analyse accessibility impacts through an equity lens. Interventions deployed before the pandemic without rigorous equity evaluations have notably failed to improve equity in the Canadian context (DeWeese et al., 2022; Linovski et al., 2018). Despite such setbacks, these approaches have helped some international centres to move the needle on transportation equity (Pereira, 2019; Venter et al., 2018).

Conclusions

Ridership changes in the Greater Toronto Area during the pandemic mirrored the socio-demographic polarization of ridership that existed before the crisis. As our analysis of these changes reveals, policymakers need to provide service to suburban areas where poverty is rising, for trip purposes beyond just work and beyond traditional peak hours. We conclude that the pandemic has made the call by Hertel et al. (2016) for an "everywhere-to-everywhere" network more pertinent for transportation equity. This vision may seem idealistic, even utopian, especially under current financing and political arrangements. But a network that serves residents' increasingly diverse needs will not only improve equity but also advance climate goals by making public transit a more attractive alternative to the car. Policymakers should also consider Ontario's future population. If the province adds over four million newcomers by 2043 (Westoll, 2022), then not investing in this kind of transit network is simply impractical.

Acknowledgments

The authors would like to acknowledge staff at the Toronto Transit Commission (TTC) and the Canadian Urban Transit Association (CUTA) for making data available. Matthew Palm is supported in part by funding from the Social Sciences and Humanities Research Council's partnership grant Mobilizing Justice: Towards Evidence-Based Transportation Equity Policy.

REFERENCES

Addie, J.-P.D. (2013). Metropolitics in motion: The dynamics of transportation and state reterritorialization in the Chicago and Toronto city-regions. *Urban Geography*, *34*(2), 188–217. https://doi.org/10.1080/02723638.2013.778651.

Addie, J.-P.D., & Keil, R. (2015). Real existing regionalism: The region between talk, territory and technology: Debates & developments. *International Journal of Urban and Regional Research*, *39*(2), 407–17. https://doi.org/10.1111/1468-2427.12179.

Ades, J., Apparicio, P., & Seguin, A.-M. (2012). Are new patterns of low-income distribution emerging in Canadian metropolitan areas? *The Canadian Geographer/Le Géographe canadien*, *56*(3), 339–61. https://doi.org/10.1111/j.1541-0064.2012.00438.x.

Allen, J. (2022). *Transportation, poverty, and urban dynamics* [Dissertation]. Department of Geography and Planning, University of Toronto.

Allen, J., & Farber, S. (2019). Sizing up transport poverty: A national scale accounting of low-income households suffering from inaccessibility in Canada, and what to do about it. *Transport Policy*, *74*, 214–23. https://doi.org/10.1016/j.tranpol.2018.11.018.

Allen, J., & Farber, S. (2021). Suburbanization of transport poverty. *Annals of the American Association of Geographers*, *111*(6), 1833–50. https://doi.org/10.1080/24694452.2020.1859981.

Allen, J., Farber, S., Greaves, S., Clifton, G., Wu, H., Sarkar, S., & Levinson, D.M. (2021). Immigrant settlement patterns, transit accessibility, and transit use. *Journal of Transport Geography*, *96*, Article 103187. https://doi.org/10.1016/j.jtrangeo.2021.103187.

Allen, J., Palm, M., Tiznado-Aitken, I., & Farber, S. (2022). Inequalities of extreme commuting across Canada. *Travel Behaviour and Society*, *29*, 42–52. https://doi.org/10.1016/j.tbs.2022.05.005.

Alsaleh, N., Farooq, B., Zhang, Y., & Farber, S. (2021). On-demand transit user preference analysis using hybrid choice models. *ArXiv Preprint ArXiv:2102.08256*.

Carney, F., Long, A., & Kandt, J. (2022). Accessibility and essential travel: Public transport reliance among senior citizens during the COVID-19 pandemic. *Frontiers in Big Data*, *5*. https://doi.org/10.3389/fdata.2022.867085.

Chatman, D.G., & Noland, RB. (2014). Transit service, physical agglomeration and productivity in US metropolitan areas. *Urban Studies*, *51*(5), 917–37. https://doi.org/10.1177/0042098013494426.

City of Brampton. (2022, July). *Ridership*. https://www.brampton.ca/EN/residents/transit/About-Us/Pages/Ridership.aspx.

Cornwell, S. (2022, 16 July). Four things to know about Mississauga's next budget, which could lead to a 5.7% property tax hike. *Toronto Star*. https://

www.mississauga.com/news-story/10675183-four-things-to-know-about-mississauga-s-next-budget-which-could-lead-to-a-5-7-property-tax-hike/.

Cui, B., Boisjoly, G., Miranda-Moreno, L., & El-Geneidy, A. (2020). Accessibility matters: Exploring the determinants of public transport mode share across income groups in Canadian cities. *Transportation Research Part D: Transport and Environment*, *80*, Article 102276. https://doi.org/10.1016/j.trd.2020.102276.

CUTA-ACTU. (2021a). *Essential Public Transit: Submission to the National Infrastructure Assessment* (p. 9) [Submission]. Canadian Urban Transit Association. https://cutaactu.ca/wp-content/uploads/2021/06/Essential-public-transit-submission-to-the-National-Infrastructure-Assessment-2021.pdf.

CUTA-ACTU. (2021b). *The GHG Reduction Impact of Public Transit* (Issue Paper No. 52; Urban Mobility Issue Paper, p. 5). Canadian Urban Transit Association. https://cutaactu.ca/wp-content/uploads/2021/01/issue_paper_52_-_the_ghg_reduction_impact_of_public_transit_.pdf.

CUTA-ACTU. (2021c). *Why Public Transit Needs Extended Operating Support* (Issue Paper Vol. 53; p. 6). Canadian Urban Transit Association. https://cutaactu.ca/wp-content/uploads/2021/06/Issue-Paper-Why-public-transit-needs-extended-operating-support.pdf.

Deboosere, R., & El-Geneidy, A. (2018). Evaluating equity and accessibility to jobs by public transport across Canada. *Journal of Transport Geography*, *73*, 54–63. https://doi.org/10.1016/j.jtrangeo.2018.10.006.

DeWeese, J., Santana Palacios, M., Belikow, A., & El-Geneidy, A. (2022). Whose express access? Assessing the equity implications of bus express routes in Montreal, Canada. *Journal of Transport and Land Use*, *15*(1), 35–51. https://doi.org/10.5198/jtlu.2022.1879.

DMG. (2014). *TTS introduction*. Data Management Group. http://dmg.utoronto.ca/transportation-tomorrow-survey/tts-introduction.

Farber, S., Ritter, B., & Fu, L. (2016). Space–time mismatch between transit service and observed travel patterns in the Wasatch Front, Utah: A social equity perspective. *Travel Behaviour and Society*, *4*, 40–8. https://doi.org/10.1016/j.tbs.2016.01.001.

Global Infrastructure Hub. (2020, 5 November). *Leveraging property tax to finance transit infrastructure. CASE STUDIES*. https://www.gihub.org/innovative-funding-and-financing/case-studies/leveraging-property-tax-to-finance-transit-infrastructure/.

Grant, J., Walks, A., & Ramos, H. (2020). *Changing neighbourhoods: Social and spatial polarization in Canadian cities*. UBC Press.

Grube-Cavers, A., & Patterson, Z. (2015). Urban rapid rail transit and gentrification in Canadian urban centres: A survival analysis approach. *Urban Studies*, *52*(1), 178–94. https://doi.org/10.1177/0042098014524287.

He, Q., Rowangould, D., Karner, A., Palm, M., & LaRue, S. (2022). Covid-19 pandemic impacts on essential transit riders: Findings from a US survey. *Transportation Research Part D, 105*, Article 103217. https://doi.org/10.31235/osf.io/3km9y.

Hertel, S., Keil, R., & Collens, M. (2016). Next stop equity: Routes to fairer transit access in the Greater Toronto and Hamilton Area. City Institute at York University.

Hulchanski, J. (2010). *THE THREE CITIES WITHIN TORONTO Income polarization among Toronto's neighbourhoods, 1970–2005*. Cities Centre, University of Toronto. http://3cities.neighbourhoodchange.ca/wp-content/themes/3-Cities/pdfs/three-cities-in-toronto.pdf.

Kamga, C., & Eickemeyer, P. (2021). Slowing the spread of COVID-19: Review of "social distancing" interventions deployed by public transit in the United States and Canada. *Transport Policy, 106*, 25–36. https://doi.org/10.1016/j.tranpol.2021.03.014.

Kingdon, J.W. (1993). How do issues get on public policy agendas? *Sociology and the Public Agenda, 8*(1), 40–53. https://doi.org/10.4135/9781483325484.n3.

Klumpenhouwer, W., Shalaby, A., & Weissling, L. (2020). *The state of demand-responsive transit in Canada*. University of Toronto.

Kramer, A. (2018). The unaffordable city: Housing and transit in North American cities. *Cities, 83*, 1–10. https://doi.org/10.1016/j.cities.2018.05.013.

Lachapelle, U., & Boisjoly, G. (2022, March). The equity implications of highway development and expansion: Four indicators. *Findings*, Article 33180. https://doi.org/10.32866/001c.33180.

Lin, Y., Yang, X., Zou, N., & Franz, M. (2015). Transit signal priority control at signalized intersections: A comprehensive review. *Transportation Letters, 7*(3), 168–80. https://doi.org/10.1179/1942787514Y.0000000044.

Linovski, O., Baker, D.M., & Manaugh, K. (2018). Equity in practice? Evaluations of equity in planning for bus rapid transit. *Transportation Research Part A: Policy and Practice, 113*, 75–87. https://doi.org/10.1016/j.tra.2018.03.030.

Litman, T. (2002). Evaluating transportation equity guidance for incorporating distributional impacts in transportation planning. *World Transport Policy & Practice, 8*(2), 50–65.

Liu, L., Miller, H.J., & Scheff, J. (2020). The impacts of COVID-19 pandemic on public transit demand in the United States. *PLOS ONE, 15*(11), Article e0242476. https://doi.org/10.1371/journal.pone.0242476.

Lo, L., Shalaby, A., & Alshalalfah, B. (2011). Relationship between immigrant settlement patterns and transit use in the Greater Toronto Area. *Journal of Urban Planning and Development, 137*(4), 470–6.

Lucas, K. (2012). Transport and social exclusion: Where are we now? *Transport Policy, 20*, 105–13. https://doi.org/10.1016/j.tranpol.2012.01.013.

Martens, K., & Golub, A. (2021). A fair distribution of accessibility: Interpreting civil rights regulations for regional transportation plans. *Journal of Planning Education and Research, 41*(4), 425–44. https://doi.org/10.1177/0739456X18791014.

Martens, K., Singer, M.E., & Cohen-Zada, A.L. (2022). Equity in accessibility: Moving from disparity to insufficiency analyses. *Journal of the American Planning Association, 88*(4), 479–94. https://doi.org/10.1080/01944363.2021.2016476.

McArthur, J., Smeds, E., & Ray, R.S. (2020, 8 September). *Coronavirus showed the way cities fund public transport is broken – Here's how it needs to change.* The Conversation. https://theconversation.com/coronavirus-showed-the-way-cities-fund-public-transport-is-broken-heres-how-it-needs-to-change-145136.

Md. Mashrur, Sk., Wang, K., & Nurul Habib, K. (2022). Will COVID-19 be the end for the public transit? Investigating the impacts of public health crisis on transit mode choice. *Transportation Research Part A: Policy and Practice,* 164, 352–78. https://doi.org/10.1016/j.tra.2022.08.020.

Mehdi, T., & Morissette, R. (2021, 1 April). *Working from home: Productivity and preferences* (Catalogue no. 45280001; StatCan COVID-19: Data to Insights for a Better Canada, p. 9). Statistics Canada. https://www150.statcan.gc.ca/n1/en/pub/45-28-0001/2021001/article/00012-eng.pdf?st=CEMHTLWd.

Mohammad, S.I., Graham, D.J., Melo, P.C., & Anderson, R.J. (2013). A meta-analysis of the impact of rail projects on land and property values. *Transportation Research Part A: Policy and Practice, 50*, 158–70. https://doi.org/10.1016/j.tra.2013.01.013.

OECD. (2022). *Financing transportation infrastructure through land value capture: Concepts, tools and case studies* (OECD Regional Development Papers, p. 70). The Organization for Economic Cooperation and Development. https://www.oecd-ilibrary.org/docserver/8015065d-en.pdf?expires=1662401454&id=id&accname=guest&checksum=84CD158EC8D5578C98DD8255DFB8F4E4.

Palm, M., Allen, J., & Farber, S. (2023, 23 March). Shifted out: The well-being and justice implications of evening and night commuting. *Transportation Research Part D: Transport and Environment, Transportation Research Part D: Transport and Environment, 122*, Article 103875. https://doi.org/10.31235/osf.io/uy96s.

Palm, M., Allen, J., Liu, B., Zhang, Y., Widener, M., & Farber, S. (2021). Riders who avoided public transit during COVID-19: Personal burdens and implications for social equity. *Journal of the American Planning Association, 87*(4), 455–69. https://doi.org/10.1080/01944363.2021.1886974.

Palm, M., Allen, J., Widener, M., Zhang, Y., Farber, S., & Howell, N. (2020, 28 May). *Preliminary results from the public transit and COVID-19 survey* (p. 15). University of Toronto. https://drive.google.com/file/d/1xtbl9nNNcjQFB51-crAGOyC0vSLCHw3M/view.

Palm, M., Farber, S., Shalaby, A., & Young, M. (2020). Equity analysis and new mobility technologies: Toward meaningful interventions. *Journal of Planning Literature, 36*(1), 31–45. https://doi.org/10.1177/0885412220955197.

Palm, M., Shalaby, A., & Farber, S. (2020). Social equity and bus on-time performance in Canada's largest city. *Transportation Research Record: Journal of the Transportation Research Board*, 2674(11), 329–42. https://doi.org/10.1177/0361198120944923.

Palm, M., Sturrock, S.L., Howell, N.A., Farber, S., & Widener, M.J. (2021). The uneven impacts of avoiding public transit on riders' access to healthcare during COVID-19. *Journal of Transport & Health, 22*, Article 101112. https://doi.org/10.1016/j.jth.2021.101112.

Palm, M., Widener, M., & Farber, S. (2023). Getting groceries during the pandemic: How transit remained important despite the rise of e-delivery. *Journal of Transport & Health, 31*, Article 101623. https://doi.org/10.1016/j.jth.2023.101623.

Pereira, R.H.M. (2019). Future accessibility impacts of transport policy scenarios: Equity and sensitivity to travel time thresholds for Bus Rapid Transit expansion in Rio de Janeiro. *Journal of Transport Geography, 74*, 321–32. https://doi.org/10.1016/j.jtrangeo.2018.12.005.

Pereira, R.H.M., & Karner, A. (2021). Transportation equity. *International Encyclopedia of Transportation, 1*, 271–7. https://doi.org/10.1016/B978-0-08-102671-7.10053-3.

Pereira, R.H.M., Schwanen, T., & Banister, D. (2017). Distributive justice and equity in transportation. *Transport Reviews, 37*(2), 170–91. https://doi.org/10.1080/01441647.2016.1257660.

Salon, D., Sclar, E., & Barone, R. (2019). Can location value capture pay for transit? Organizational challenges of transforming theory into practice. *Urban Affairs Review, 55*(3), 743–71. https://doi.org/10.1177/1078087417715523.

Sanaullah, I., Alsaleh, N., Djavadian, S., & Farooq, B. (2021). Spatio-temporal analysis of on-demand transit: A case study of Belleville, Canada. *Transportation Research Part A: Policy and Practice, 145*, 284–301. https://doi.org/10.1016/j.tra.2021.01.020.

Sidloski, M., & Diab, E. (2020). Understanding the effectiveness of Bus Rapid Transit systems in small and medium-sized cities in North America. *Transportation Research Record: Journal of the Transportation Research Board, 2674*(10), 831–45. https://doi.org/10.1177/0361198120940993.

Spurr, B. (2020, 7 April). Who's still crowding into TTC buses amid the pandemic? Evidence suggests many are Toronto's working poor. *Toronto Star*. Retrieved 11 April 2020 from https://www.thestar.com/news/gta/who-s-still-crowding-into-ttc-buses-amid-the-pandemic-evidence-suggests-many-are-toronto/article_778c7434-8887-5b68-a2fb-194a4703b551.html#:~:text=amid%20the%20pandemic%3F-,Evidence%20 suggests%20many%20are%20Toronto's%20working%20poor,just%20 impossible%2C%E2%80%9D%20she%20said.

Statistics Canada. (2021, 4 August). *Working from home during the COVID-19 pandemic, April 2020 to June 2021*. The Daily. https://www150.statcan.gc.ca/n1/daily-quotidien/210804/dq210804b-eng.htm.

TransLink. (2020, April). *Coronavirus (COVID-19)*. Retrieved 14 June 2020 from https://new.translink.ca/rider-guide/coronavirus-precautions?bcgovtm=vancouver%20is%20awesome.

TTC. (2020). *Riding the TTC during the COVID-19 pandemic* [Toronto Transit Commission]. Retrieve 31 July 2020 from http://ttc.ca/Riding_the_TTC/Safety_and_Security/Riding_the_TTC_during_the_COVID-19_pandemic.jsp.

TTC. (2022). *CEO's Report*. Toronto Transit Commission. Toronto, Canada. Retrieved 1 August 2022 from https://cdn.ttc.ca/-/media/Project/TTC/DevProto/Documents/Home/Transparency-and-accountability/Reports/CEO-Reports/2022/CEOs-Report--August-2022.pdf?rev=aba96df3224347f28cf1b5ef90b2dfac&hash=832B76E179F63C1B90B849549F5740BD.

Venter, C., Jennings, G., Hidalgo, D., & Valderrama Pineda, A.F. (2018). The equity impacts of bus rapid transit: A review of the evidence and implications for sustainable transport. *International Journal of Sustainable Transportation, 12*(2), 140–52. https://doi.org/10.1080/15568318.2017.1340528.

Walker, J. (2020, 7 April). *In a pandemic, we're all "transit dependent."* CityLab. https://www.citylab.com/perspective/2020/04/coronavirus-public-transit-subway-bus-ridership-revenue/609556/.

Walks, R.A. (2001). The social ecology of the post-Fordist/global city? Economic restructuring and socio-spatial polarisation in the Toronto urban region. *Urban Studies, 38*(3), 407–441. https://doi.org/10.1080/00420980120027438.

Wang, K., Liu, Y., Reilly, B., & Habib, K.N. (2021a). *COVID-19 influenced households' interrupted travel schedules (COVHITS) survey: Fall 2020 cycle report* (p. 34). UTTRI. https://uttri.utoronto.ca/files/2021/04/UTTRI-Report-COVHITS-Survey-Fall-2020-Cycle-Report-Wang-2021.pdf.

Wang, K., Liu, Y., Reilly, B., & Habib, K.N. (2021b). *COVID-19 influenced households' interrupted travel schedules (COVHITS) survey: Fall 2021 cycle report* (p. 43). UTTRI. https://uttri.utoronto.ca/files/2022/01/UTTRI-Report-COVHITS-Survey-Fall-2021-Cycle-Report-Wang-2021.pdf.

Welch, T.F., Gehrke, S.R.,& Farber, S. (2018). Rail station access and housing market resilience: Case studies in Atlanta, Baltimore and Portland. *Urban Studies*, *55*(16), 3615–30.

Westoll, N. (2022, 24 August). *Ontario could see 4.2 million more residents by 2043: Statistics Canada projection*. CityNews. https://toronto.citynews.ca/2022/08/24/ontario-population-projection-statistics-canada/.

Zhang, Y., Farber, S., & Young, M. (2021). Eliminating barriers to nighttime activity participation: The case of on-demand transit in Belleville, Canada. *Transportation*, *49*, 1385–1408. https://doi.org/10.1007/s11116-021-10215-2.
Zhang, Y., Palm, M., Scheff, J., Farber, S., & Widener, M. (2020, 15 December). Travel survey recruitment through Facebook and transit app: Lessons from COVID-19. *Transport Findings*. https://doi.org/10.32866/001c.18066.

7 E-Scooter Sharing as Hope or Hype? Bridging Transportation Equity Divides in Calgary

VIVIAN KONG AND AGNIESZKA LESZCZYNSKI

Introduction

Since the mid-2010s, the popularity of micromobility sharing systems has risen rapidly in cities around the world. "Micromobility" refers to the use of small, lightweight vehicles such as bicycles and e-scooters intended for point-to-point short-distance trips, such as those to and from transit hubs (Yanocha & Allan, 2019). The "sharing" component of these mobility systems is contingent on their digital mediation. Riders frequently use a mobile app to locate, unlock (activate and/or gain access to), and pay for use of a shared vehicle, all through the convenience of a smartphone.

Most recently, cities have seen an increase in dockless micromobility sharing, which includes dockless bikes and electrified or "e"-scooters (see McNee and Miller, chapter 8). Because dockless systems do not rely on fixed stations, which anchor vehicles to predetermined trip origin and terminus locations, experts have celebrated this option as having the potential to "bridge the existing transportation divide" by closing spatial equity gaps in micromobility sharing (Kim et al., 2019, p. 263). Dockless systems in particular have been celebrated for their spatial equity potentials because they do not require an expensive expansion of infrastructure (e.g., installation of more bikesharing stations), and cities can use the funds saved to put more shared micromobility vehicles into circulation or to reduce the costs of ridership (McCarty Carino, 2018). Yet how well e-scooter sharing systems realize these potential equity gains remains understudied.

In the context of micromobility sharing systems, spatial equity (see Palm and Farber, chapter 6) refers to the evenness of spatial access to, and distribution of, the service and its components – such as docking infrastructure and shared vehicles – across a city, irrespective of differences in the comparative advantage and between different areas such as neighbourhoods (Hosford & Winters, 2018). Knowledge of a micromobility system's spatial equity profile is important because "[a]s more cities start piloting micromobility sharing programs as approaches to facilitating more sustainable, healthy, traffic congestion reducing modes of transportation, spatial equity is an important axis to consider in the planning and evaluation of these systems to ensure that their actual and potential benefits are evenly distributed across a city's spatial fabric" (Kong and Leszczynski, 2022, p. 98).

Using a publicly available open dataset representing three months' worth of e-scooter trip data collected as part of the City of Calgary's two-year Shared Mobility Pilot Project (2018–19), we analyse the spatial equity of e-scooter sharing in Calgary. This analysis of use considers the relative socio-economic deprivation of the locations at the beginning and end of e-scooter trips. Our approach improves upon previous studies of micromobility sharing, which have foregrounded access to vehicles at trip origins as a sole determinant of spatial equity. To take this different approach, we mobilize a gravity model, which allows us to identify patterns of e-scooter trip flows between July and September 2019. Specifically, we focus on differences in area-level deprivation metrics characteristic of the origin and destination locations of individual trips. If e-scooter sharing in Calgary during these three months was spatially equitable, then the model would indicate that differences in area-level deprivation at trip origin and destination locations had no influence on e-scooter trip volume.

This chapter begins with a review of the literature on micromobility sharing, focusing on the methods used in several key studies to evaluate spatial equity. This review leads to a discussion on the gravity model and its use in micromobility research, followed by a presentation and discussion of the analytical findings.

E-Scooters, Micromobility-Sharing, and Spatial Equity

E-scooter sharing – which involves the public shared use of a fleet of dockless electric scooters – initially gained momentum in 2017 with the emergence of two major scooter sharing companies, Lime and Bird (Kolodny, 2017). Since then, more than one hundred cities worldwide have adopted a scooter sharing system (Hawkins, 2018). Despite this growing popularity, e-scooter sharing remains an understudied mode of urban transportation. The limited research to date has overwhelmingly focused on e-scooter injuries, public health implications, and environmental impacts. In a few studies exploring geographical dimensions, preliminary spatio-temporal analyses of e-scooter data have investigated the spatial extent of e-scooter service areas, identified temporal patterns of use, and examined differences in regulatory frameworks, including in Canada (e.g., Frisbee et al., 2022; McKenzie, 2019; Younes et al., 2020). However, few scholars have addressed the spatial equity dimensions of e-scooter sharing.

Spatial equity is an indicator of whether micromobility infrastructures and vehicles are evenly distributed, shared, and used across neighbourhoods characterized by different levels of privilege, as measured by their socio-economic statuses in urban spatial hierarchies. Spatial equity is often determined on the basis of the spatial relationship between the deprivation profile of urban enclaves and the distribution of mobility assets (vehicles, infrastructures) across a sharing network (Babagoli et al., 2019; Bachand-Marleau et al., 2012; Médard de Chardon et al., 2017; Zhang et al., 2017). Measuring spatial equity can be used to identify differences in where mobility resources and infrastructures may be accessed, which often has a circumscribing effect on the demographics of mobility system use (often referred to as social equity).

The spatial equity profile of a micromobility network is key to understanding short-hop transport modes' potentials. The profile can reveal whether all members of an urban community, wherever they live, can enjoy the benefits of micromobilities, including increases in personal transportation options, flexibility, and efficiency (Howland et al., 2017; Johnston et al., 2020; McNeil et al., 2018). However, analytical research has only recently begun to assess whether dockless micromobility systems are equalizing spatial inequalities in urban transportation. Mooney et al. (2019), for instance, studied the spatial equity of dockless bikesharing in Seattle. Through analysing the locations to which bikes were rebalanced (redistributed) by system operators, the researchers found that areas with higher dockless bike availability were characterized by higher socio-economic factors, including higher median incomes and more college-educated residents.

In a direct comparison of dockless and docked micromobility sharing systems, Lazarus et al. (2020) investigated the JUMP dockless bike pilot program in San Francisco and found that dockless bikes had a far wider service area compared to GoBike, San Francisco's docked bikesharing system. Whereas docked bikesharing happened mainly in the central business districts (CBDs), dockless bikesharing took place mainly outside the CBDs. This comparison suggests that dockless bikesharing can offer more transportation options than docked counterparts, potentially improving the equity of spatial access to micromobility sharing.

More recently, research has begun to study e-scooters as a distinct mode of dockless micromobility. In an important analysis of the spatial dimensions of shared e-scooter use, Arnell (2019) found that riders not only initiated fewer trips in more marginalized neighbourhoods in Nashville and in Portland, OR, but also that operators frequently rebalanced e-scooters to CBDs and areas with high employment density. These findings suggest that scooter sharing programs focus on serving areas with higher population density and better employment opportunities, rather than providing alternative transportation modes to all communities.

Additional sources of e-scooter equity research come from investigative reports, often produced by organizations commissioned by policymakers and government officials. One such report, by Fedorowicz et al. (2020), examined new mobility equity (such as bikesharing and e-scooter sharing) in ten mid-sized American cities, reviewing the cities' transportation plans and drawing on data from interviews with transportation and equity representatives. This report identifies that equity in new mobility-sharing goes beyond a simple question of access; instead, historically neglected urban neighbourhoods are also less likely to have the pre-existing infrastructure – e.g., cycling lanes – needed to support the actual use of these new mobilities (Fedorowicz et al., 2020).

These findings are consistent with previous studies, which show that even though micromobilities should serve the urban fabric as a whole, they disproportionately serve city centres (e.g., Matthew et al., 2019; Médard de Chardon, 2019; Qian et al., 2020; Stehlin, 2015). In this chapter, we investigate how spatial equity played out "on the ground" in e-scooter sharing in Calgary during the pilot phase of the city's micromobility program. To do so, we adopt a gravity model to measure the equity of e-scooter use in Canada's fourth-largest city.

Gravity Models and Transportation Research

The gravity model is a technique for analysing origin-destination flows of transportation modes, such as e-scooters, and for establishing relationships between these flows and additional factors, such as distance and socio-economic indicators. It analyses spatial flows by calculating the probability of the interaction of flows between origin and destination points in space (Kincses & Tóth, 2014). The interaction is based on the inverse distance between the points, where points that are closer together have a higher probability of interaction than points that are farther apart. Researchers have applied gravity models to the analysis of bilateral transportation flows, including within micromobility sharing systems. For instance, Zhang et al. (2018) built a gravity model to predict the origin-destination distribution of bikesharing trips between bike-share docks in Ningbo, China. When tested against public bikesharing data, their model was shown to be an accurate predictor of bikesharing trip flows in Ningbo. The model also suggests that a higher number of attractions at destination locations had a positive influence on shared bike trip flows between two points (i.e., more trips), whereas increased travel time between two locations had a negative influence (i.e., fewer trips). Elsewhere, Li et al. (2021) produced a gravity model to infer the purposes of dockless shared bike trips in Shenzhen, China. They found that shorter-distance bike trips were more likely to be used to connect users directly between homes and workplaces rather than connecting them to transit hubs for commutes in the suburbs.

Although these studies reveal the utility of gravity models for analysing bikesharing, studies have not yet applied this model to e-scooter sharing. Compared to shared bikes, e-scooters have different patterns of use. They also require a separate analytic evaluation to determine whether the spatial inequities that characterize other modes of micromobility sharing also apply to them, or whether scooter sharing offers a more spatially equitable short-hop mode of transportation. Using a gravity model, we evaluated the spatial equity of e-scooter sharing in Calgary between July and September 2019. If the results of the model show that differences in area-level deprivation at trip start and end locations had no significant impact on trip flow patterns, then we can say that shared e-scooter use during was spatially equitable during that three-month period.

Data, Methods, and Analytic Rationale

Study Area

The City of Calgary, the largest city by population in the province of Alberta, is the fourth largest city in Canada. It is home to 1.2 million persons living within a roughly 825 km^2 area (Statistics Canada, 2018). The city features the largest bicycling path network in North America, which is also well suited for shared e-scooters, which the city allows on its surface cycling infrastructure (City of Calgary, n.d.; Frisbee et al., 2022). On 16 July 2018, Calgary implemented a two-year Shared Mobility Pilot Project (which we refer to as the "Pilot" for short) that started with a fleet of dockless shared bicycles provided by the commercial operator Lime (City of Calgary, 2019). In July 2019, fifteen hundred shared e-scooters were added to the Pilot. At the time, two commercial operators, Lime and Bird, owned and operated Calgary's scooter share services. More recently, however, Lime has ceased to operate in the city, and a new operator, Neuron, has joined the micromobility sharing mix (Krause, 2019; Lo, 2019; Smith, 2021).

Data

At the time of analysis (2020), Calgary was the only Canadian city that had publicly released trip data for both shared bikes and shared scooters under an open data licence. We accessed the trip data online from Calgary's Open Data portal (Calgary Open Data, 2019). The original dataset represents 482,021 shared dockless scooter and bike trips taken between 1 July 2019 and 30 September 2019 (City of Calgary, 2019). The published dataset excluded trips lasting less than thirty seconds or one hundred metres as well as trips where the geospatial data quality made analysis unsuitable (City of Calgary, 2019). The dataset provided the start and end locations of each trip taken, generalized to 30,000 m^2 (0.03 km^2) hexagonal bins (a hexbin, or hexagonal bin, is a hexagonal-shaped unit of consistent dimensions [in this case, 0.03 km^2] used to aggregate point data for spatial analysis) that tile the city. The latitude and longitude coordinates for the start and end points of the trips represent the centroid of the corresponding hexbin. Other information in the dataset included the trip start date, day of the week for the trip start, start hour, trip duration in seconds, and trip distance in metres. Importantly, the dataset did not include data on rebalancing locations to which e-scooters may have been redistributed by system operators. Additionally, the dataset did not include any unique e-scooter identification, so an e-scooter's general location could not be tracked through space and time. This means that while the general location of each taken trips' start and end are identified, there is no indication where the e-scooter was before the trip commenced, or where it was moved to after the trip.

After retrieving the data, we further filtered the information in advance of analysis to remove bike trips and exclude scooter trips over two hours in duration. As e-scooters have a limited battery life allowing for trips of up to two hours on a single charge, we could assume that these longer trips were not genuine (Marshall, 2018; McKenzie, 2019). To limit the analysis within the Calgary city limits, we also excluded trips starting or ending outside the city boundary. Another filter was average trip speed, calculated as the recorded trip distance divided by the trip time. The City of Calgary has upper speed limits of 20 km/h for the shared e-scooters (Potkins, 2019), so any e-scooter trips with an average speed higher than 20 km/h were likely not genuine or unfit for analysis. Trips that originated or ended in areas with no calculated Pampalon Deprivation Index (PDI) score (detailed below) were also excluded, as were trips with a calculated distance of 0, representing e-scooter trips that started and ended in the same hexbin. This is because the maximum distance that could be travelled within a single hexbin is 214.9 m (the long diagonal of a 30,000 m^2 hexbin), leading us to assume that these trips were not legitimate and likely represent either erroneous data or abandoned trips (where riders initiated a trip but changed their mind and left the e-scooter near the original location).

Measuring Deprivation

For this analysis, the PDI was used as the measure of relative socio-economic well-being along a spectrum from high deprivation (low well-being) to low deprivation (high well-being). Specifically, this analysis uses the material component of the Canada-specific PDI (Pampalon et al., 2012) as a proxy indicator of the relative socioeconomic advantage and disadvantage of areas in the City of Calgary. The PDI is an area-based measure of material deprivation in Canada determined by a principal component analysis of Canadian census variables (Pampalon et al., 2012). The index is calculated for Dissemination Areas (DAs), the smallest unit of census geography for which census data are disseminated (Statistics Canada, 2015). Three socio-economic indicators inform the calculation of a quintile score representing the level of deprivation for each DA ranging from the least deprived (PDI = 1) to the most deprived (PDI = 5) (Institut National de Santé Publique du Québec, 2016; Pampalon et al., 2012):

1 the proportion of persons without a high school diploma;
2 average personal income; and
3 the employment-population ratio.

The area-level PDI scores for 2016 (the most recent Canadian census for which data had been published at the time of analysis) were downloaded from the Institut National de Santé Publique du Québec website[1] and spatially joined to DA census geometries retrieved from the Statistics Canada Dissemination Area Boundary Files online catalogue.[2] This yielded a spatial data layer of DAs (clipped to the city of

1 https://www.inspq.qc.ca/en/expertise/information-management-and-analysis/deprivation-index/deprivation-index-canada-2016.
2 https://www150.statcan.gc.ca/n1/en/catalogue/92-169-X.

Figure 7.1. Calgary Dissemination Areas (DAs), shaded by their Pampalon Deprivation Index score (PDI). A PDI score of 1 represents membership in the least deprived (most advantaged) quintile while a PDI score of 5 represents membership in the most deprived (least advantaged) quintile. This map represents the distribution of PDI scores across the city of Calgary. *Spatial data sources: Statistics Canada (The City of Calgary Dissemination Area Boundary Files)*[3]

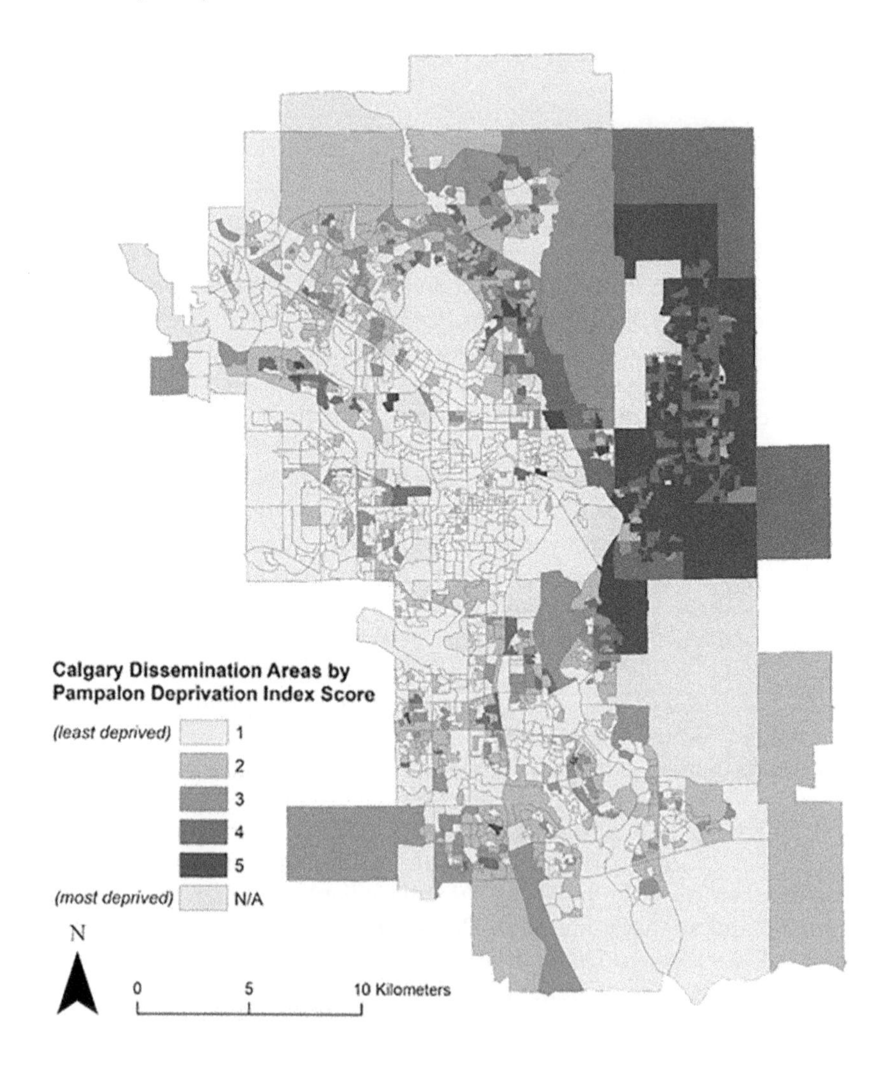

3 https://www150.statcan.gc.ca/n1/en/catalogue/92-169-X.

Calgary) populated with a PDI score attribute. Figure 7.1 shows the city of Calgary mapped by its DA-level geography of deprivation.

Analysis Methodology

To determine the deprivation score for the start and end locations of e-scooter trips taken during the three-month period (July to September 2019), we spatially joined the PDI scores for each DA to the hexbin centroid point locations approximating trip origins and terminuses. Hexbin centroids were considered to have the same deprivation score as the DA in which the point was spatially contained. This process allowed trip origins and terminuses to be associated with a measure of socio-economic value (the PDI score). Each hexbin in the dataset has an associated grid ID, which we used to generate an origin-destination table consisting of

i each unique origin and destination pair of e-scooter trips;
ii the deprivation quintile classification of the originating hexbin;
iii the deprivation quintile classification of the terminating hexbin;
iv the Euclidian distance between each unique pair's hexbin centroid in metres; and
v the number of trips made between each pair of hexbins.

After we filtered the data with the parameters described above, the final dataset consisted of 359,278 unique e-scooter trips to analyse. First, trip start and end locations were used to populate an Origin-Destination (O-D) matrix based on the PDI scores of the origin and destination hexbins. We used this matrix to describe the raw volume of trips between areas based on deprivation score. Next, to analyse the effects of PDI score on the distribution of e-scooter trips, we used the O-D matrix to calculate a gravity model to determine the effects of the deprivation profiles of origin and destination points on the flows of e-scooter trips. Using the matrix of 89,289 unique O-D pairs, we calculated a log-linearized gravity model using the open-source software *R*, generating output expressed as the percent change in the dependent variable resulting from changes in the independent variable. The 89,289 O-D pairs represent 89,289 unique sets of origin and destination hexbins in which trips began (origin hexbin) and ended (destination hexbin) across the dataset. There are fewer O-D pairs than trips as more than one trip could begin and end in the same hexbin pair. Table 7.1 presents a summary of O-D pairings by trip origin and destination PDI score.

Table 7.1. Summarized origin-destination matrix of the Pampalon Deprivation Index score (PDI) of the Discrimination Area (DA) in which each scooter trip began and ended. Data is for all trips taken within the city of Calgary during the Shared Mobility Pilot Project (1 July–30 September 2019)

	PDI of trip terminus DA[a]					
PDI of trip origin DA[a]	1	2	3	4	5	Total trips by PDI of origin DA
1	294,528 trips *(81.98% of all trips)*	15,398 *(4.29)*	1,924 *(0.54)*	10,214 *(2.84)*	1,055 *(0.29)*	323,126 *(89.94)*
2	15,359 *(4.27)*	1,347 *(0.37)*	236 *(0.07)*	427 *(0.12)*	73 *(0.02)*	17,443 *(4.85)*
3	1,626 *(0.45)*	253 *(0.07)*	392 *(0.11)*	119 *(0.03)*	34 *(0.01)*	2,424 *(0.67)*
4	11,223 *(3.12)*	530 *(0.15)*	202 *(0.06)*	2,561 *(0.71)*	281 *(0.08)*	14,797 *(4.12)*
5	821 *(0.23)*	64 *(0.02)*	27 *(0.01)*	196 *(0.05)*	388 *(0.11)*	1,496 *(0.42)*
Total trips by PDI of terminus DA	323,564 *(90.06)*	17,593 *(4.90)*	2,781 *(0.77)*	13,517 *(3.76)*	1,831 *(0.51)*	359,278[b] *(all trips)*

[a] A PDI score of 1 indicates the quintile of lowest deprivation, whereas a PDI score of 5 indicates the quintile of the highest deprivation.

[b] Number of e-scooter trips originating and/or terminating in dissemination areas with no calculated PDI score = 52,219 trips. Number of e-scooter trips originating and terminating in the same hexbin = 32,557 trips.

For the model, we defined e-scooter trips between origin and destination points as the dependent variable, while distance, trip origin PDI score (Origin PDI), and trip terminus PDI score (Terminus PDI) comprised independent variables. Origin PDI represents the PDI score of the DA spatially containing the hexbin in which a trip was initiated (started), while the Terminus PDI represents the PDI score of the DA spatially containing the hexbin in which a trip ended. The classical approach of the gravity model using distance decay was used. The model for this analysis is described by Equation 1:

$$log X_{od} = b_1 PDI_0 + b_2 PDI_d + b_3 log \tau_{od} + e_{od}$$

where X_{od} represents the volume of e-scooter flows between the origin and destination, PDI_o is the PDI score of the origin location, PDI_d is the

PDI score of the destination location, τ_{od} is the distance between the two locations, b represents the estimated coefficients for the three terms of the model, and e_{od} is the random error term.

For the purposes of this analysis, one of the model results is of particular significance: how changes in the PDI score of the start and end locations of e-scooter trips impact the volume of trips. In this case, if e-scooter sharing in Calgary was spatially equitable during the three-month pilot period, then increases in deprivation score (from less to more deprived) for the origin and terminus locations would *not* have a significant impact on the volume of trip flows. In other words, if e-scooter sharing was spatially equitable, then the gravity model coefficients for the explanatory variables of Origin PDI and Terminus PDI would be identified as non-significant.

Results

According to our results, e-scooter sharing in Calgary was *not* spatially equitable during the pilot phase. The analysis shows that areas characterized by higher levels of deprivation (lower advantage) were associated with lower volumes of e-scooter sharing trips. Had e-scooter sharing been relatively spatially equitable, the results of the model would have shown no statistically significant effect of the PDI score on e-scooter sharing trip volume. In other words, in a spatially equitable scenario, the volume of trips between areas with different deprivation profiles would have been comparable, unaffected by the PDI score of different neighbourhoods. As detailed below, our model shows that PDI score *did* affect differences in trip volumes, demonstrating e-scooter sharing to have been spatially inequitable during the three-month period for which data were available.

Figure 7.2 shows trip flows visually, identifying where trips occurred across the entire city. As per the summary of O-D pairings provided in Table 7.1, the majority of trips taken flowed between the most socio-economically advantaged areas in the city. Indeed, 90.91 per cent of all trips originated and ended in hexbins in the two most advantaged quintiles (hexbins with a DA-level PDI score of 1 or 2). By contrast, only 0.95 per cent of all e-scooter trips began and ended in DAs within the two most deprived quintiles (PDI score of 4 or 5). Moreover, 3.52 per cent of trips starting in DAs within the two most deprived quintiles ended in DAs within the two least deprived quintiles. While 3.52 per cent represents a small share of trips, it is still 3.7 times more trips than those that began and ended in the two least deprived quintiles (0.95% of trips). This difference means that trips originating in areas of highest

Figure 7.2. E-scooter trip flows across Calgary, Alberta. Calgary Dissemination Areas (DAs) are shaded by Pampalon Deprivation Index (PDI) score. E-scooter trip flows are summarized and represented by line thickness based on volume of flows. *Spatial data sources: City of Calgary (Calgary DAs), OpenStreetMap (base map)*

Table 7.2. Gravity model analysis results for number of shared e-scooter trips

Variable	E-scooter trip volume gravity model		
	Estimated coefficient	t-value	sig
Distance[a]	−0.382	−71.24	<0.001
Origin PDI[b]	−0.054	−14.68	<0.001
Terminus PDI[b]	−0.061	−17.04	<0.001

Multiple R-squared = 0.097
Adjusted R-squared = 0.097
[a] Distance is modelled as log (distance) to provide a more accurate estimate of the impact of the change in rate of distance on a percent change in trip frequency. As a result, the results of the effects of distance are reported as a percent change rate of growth.
[b] Origin PDI and Terminus PDI are not continuous variables, unlike distance, so these factors cannot be modelled as a per cent change rate of growth. Instead, the table reports them by a percent change of the number of e-scooter trips with an increase in one PDI score.

deprivation were more than twice as likely to end in areas of lowest deprivation than in comparatively deprived areas. This result shows that rather than taking place equitably across all areas regardless of deprivation score, e-scooter trips were concentrated within the most advantaged (least deprived) areas of the city during the three-month pilot period.

Moving beyond summary description, Table 7.2 provides the output of the gravity model. The most significant components here are the estimated coefficients, which represent the strength and direction of the relationship between independent and dependent variables. Distance is included because the model necessarily accounts for it, even though it was not a focus of our analysis. What *is* important is that the results of the model show that PDI score *did* significantly influence e-scooter trip volumes in Calgary at both trip start and end locations between July and September 2019. At e-scooter trip origin locations, the PDI score was negatively associated with e-scooter sharing trip flows. Specifically, each one-level increase in PDI score from one quintile to the next most deprived quintile (e.g., from quintile 2 to 3, moving from more to less advantaged) was significantly associated with a 5.4% decrease in e-scooter trips ($t = -14.68$, $p < 0.001$; see Table 7.2. Similarly, for terminus locations of e-scooter trips, a single-quintile increase in PDI score was significantly associated with a 6.1% decrease in e-scooter trip volume ($t = -17.04$, $p < 0.001$; see Table 7.2). In brief, the PDI scores of both trip origin and terminus locations have statistically significant coefficients, indicating that in this model the deprivation profiles of start and end locations had a significant impact on the volume and patterns of e-scooter sharing.

Discussion

As we consider the implications of these results, several factors require attention. To begin, the R-squared value in the resulting gravity model is significant because it suggests that the model explains less than 10 per cent of the observed variation in the data. This low percentage means that other variables beyond deprivation could also account for the differences in trip volumes between different areas. Models often have trouble producing a high level of fit in the social sciences due to the complex nature of social phenomena, and in our analysis, we tested for the impact of only one factor – PDI score – on patterns of e-scooter use. For instance, our model did not analytically account for other variables that are known to influence micromobility sharing use, such as proximity to the downtown core (i.e., where the areas of high economic activities are; Arnell, 2019), and the location of transit hubs (such as bus stops and train stations; Espinoza et al., 2019). Including such additional variables could improve the R-squared value and provide a more nuanced, complex understanding of e-scooter sharing usage in Calgary.

Beyond analysing whether differences in area-level deprivation had statistically significant effect on the volume of e-scooter trips, our model does not account for reasons for *why* these trip volumes were different between areas of low and high deprivation. For instance, in the absence of access to rebalancing and idleness data (which would indicate how long an e-scooter sat unused before being unlocked by a rider or redistributed by a system operator) or real-time e-scooter availability, we are unable to comment on whether differences in trip frequency between deprived and non-deprived areas were due to differences in demand for and/or supply of e-scooters.

Similarly, our model did not account for the spatial density and proximity of dedicated surface infrastructure, and how these factors may have affected trip volume. Research has shown that dedicated bike lanes, cycleways, and recreational networks of paths, for example, have a strong positive influence on micromobility use, increasing the likelihood that people will ride shared bikes and e-scooters (Caspi et al., 2020; Smith et al., 2015; Xu et al. 2019; Zhang et al. 2021). According to research from the United States, racialized and ethnic-minoritized neighbourhoods have much less cycling infrastructure than whiter neighbourhoods (Hoffman, 2016; Stehlin & Payne, 2022). Similarly, docked systems have a disproportionate concentration in whiter, less diverse neighbourhoods compared to more diverse and/or ethnic minority-majority enclaves (Meng & Brown, 2021). This latter point links to considerations of *social* equity.

Rather than examining which *areas* transportation services, modes, and infrastructures underserve, social equity analyses examine which *people* – demographics, socio-economic cohorts, and communities – may be underserved and why. For example, social equity may focus on the "[s]ocial, cultural, safety, and language" barriers that may inform these exclusions (Shaheen et al. 2017, p. 23). Based on a survey of >1,256 e-scooter riders in Tempe, Arizona, Sanders et al. (2020) found that racial difference affected use. While the survey revealed no significant racial differences in e-scooter ridership frequency, respondents of colour were more likely than white respondents to identify the complexity of e-scooter rental – which requires app download, installation, and configuration – as a barrier to use.

As our gravity model constitutes a spatial-centric method, it does not analytically account for important dimensions of social equity alongside spatial equity. Nevertheless, our model's finding that e-scooter sharing in Calgary during the three-month period was spatially inequitable has implications for our understanding of mobility transitions. First, our analysis suggests that cities cannot solve long-standing geographical disparities in urban transportation simply by adding digital conveniences – namely, the ability to locate, initiate/unlock, and pay for the use of a shared vehicle via a mobile app – to the urban modal mix. Cities must evaluate each new mode added on the basis of its equity dimensions; indeed, cities need to *plan* for the equity of micromobility (see also Palm and Farber, chapter 6). Whether to ride or not ride an e-scooter is ultimately an individual choice, and neither the City of Calgary nor the e-scooter operators are responsible for the spatially inequitable patterns we observed in our analysis, and we neither assume nor ascribe any intentionality to these or any other actors as relates to these observed patterns. Rather, we suggest that cities and operators may find spatial equity analysis to be a meaningful tool for evaluating new mobility systems at the early stages of their introduction into an urban area. As Meng and Brown (2021) write, "[p]olicymakers should use program design and performance metrics" to actively address "the mismatch between" desired "and actual service geographies" so as to "ensure that micromobility services benefit marginalized communities" (p. 1).

This idea of intentionally planning for equity in micromobility sharing informs the second implication of our results. Our findings suggest that we need to understand the spatial equity profile of a micromobility sharing system before we can understand where the benefits of short-hop transportation are likely to flow and to whom. As we know, communities are always grounded in and have strong ties to place. Thus, as cities look to micromobilities as solutions for reducing urban CO_2 emissions, they must make equity a priority. Micromobilities have the potential to substitute for single-occupancy motor vehicle trips and thereby help improve air quality, but these trips will likely take place locally, reducing air pollution where these new mobility modes are used (see Rojas and Miller, chapter 5). In this way, the spatial equity of micromobility sharing is an important consideration towards achieving urban climate justice.

Conclusion

In analysing spatial equity through the geographies of e-scooter use, which considers the deprivation profiles of both trip beginning and end locations, this study improves upon previous methodologies that rely solely on trip origins – determined by dock or rebalancing locations – as an analytic entry point, without considering where trips end. Our study is also one of the first to analyse available open data on scooter sharing trips in Canada, as scooter sharing is a recently available service and trip data has been limited.

Based on a spatial gravity model, the results indicate that e-scooter use in Calgary was spatially inequitable during the three-month period, with e-scooter trip volume decreasing with increases in area-level deprivation of both trip origin and destination locations. This conclusion parallels and confirms findings of spatial inequity identified within earlier studies of both docked and dockless micromobility sharing systems. It also dispels conjectures about the potential for dockless scooter sharing to redress some of the spatial inequalities in pre-existing micromobility sharing services, such as docked and dockless bikesharing (see also Médard de Chardon, 2019). As per Federowicz et al. (2020), cities tend to plan the introduction of new mobilities before deciding whether or not to add any new modes permanently to the mix of transportation options. As a result, plans to address equity often occur *after* a testing/piloting phase of the program (Federowicz et al., 2020). Thus, inequities in the micromobility system become "designed-in" rather than pre-empted through equity-first approaches to planning, roll-out, and testing. In other words, digitally-mediated, app-centred urban mobility solutions – such as micromobilities that may be rented via a digital screen – cannot, in and of themselves, fix pre-existing equity gaps that plague urban transportation systems.

Our findings may benefit scooter sharing stakeholders – such as service operators, urban and transportation planners, and policymakers – by helping them to understand where these systems may be spatially inequitable and how to remedy the inequities in those areas. Specifically, stakeholders need to focus not only on how they can redistribute (rebalance) dockless vehicles, but also on where people actually use these newer mobilities and where associated benefits, such as reductions in CO_2 emissions, are likely to flow. To make decisions that benefit the entire geography of a city equally, stakeholders need data to help them understand patterns of e-scooter sharing across individual micromobility sharing systems operating in different cities. Without this knowledge, decision makers may have difficulty improving these systems.

REFERENCES

Arnell, B.M. (2019). Shared electric scooters and transportation equity: A multivariate spatial regression analysis of environmental factors on revealed travel behavior and mode shift potential. *MIT Theses, 53*(9), 1689–99. https://doi.org/10.1017/CBO9781107415324.004.

Babagoli, M.A., Kaufman, T.K., Noyes, P., & Sheffield, P.E. (2019). Exploring the health and spatial equity implications of the New York City bike share system. *Journal of Transport and Health, 13*, 200–9. https://doi.org/10.1016/j.jth.2019.04.003.

Bachand-Marleau, J., Lee, B., & El-Geneidy, A. (2012). Better understanding of factors influencing likelihood of using shared bicycle systems and frequency of use. *Transportation Research Record, 2314*(1), 66–71. https://doi.org/10.3141/2314-09.

Calgary Open Data. (2019). *Shared mobility pilot trips.* https://data.calgary.ca/Transportation-Transit/Shared-Mobility-Pilot-Trips/jicz-mxiz.

City of Calgary. (n.d.a). *Pathways and trails.* Retrieved 19 April 2021 from https://www.calgary.ca/csps/parks/pathways/pathways-in-calgary.html.

City of Calgary. (2019). *Shared electric scooter pilot*. Retrieved 2 April 2020 from https://www.calgary.ca/Transportation/TP/Pages/Cycling/Cycling-Strategy/Shared-electric-scooter-pilot.aspx#.

Caspi, O., Smart, M., & Noland, R. (2020, September). Spatial associations of dockless shared e-scooter usage. *Transportation Research Part D: Transport and Environment 86*, Article 102396. https://doi.org/10.1016/j.trd.2020.102396.

Espinoza, W., Howard, M., Lane, J., & Van Hentenrych, P. (2019). *Shared e-scooters: Business, pleasure, or transit?* ArXiv. Retrieved from https://arxiv.org/abs/1910.05807.

Fedorowicz, M., Bramhall, E., Treskon, M., & Ezike, R. (2020). *New mobility and equity: Insights for medium-sized cities*. Urban Institute. https://www.urban.org/sites/default/files/publication/102529/new-mobility-and-equity-insight-in-medium-cities_2.pdf.

Frisbee, N.C., Dyce, B., Wray, A., Bradford, N., Lee, J., & Gilliland, J. (2022). *Muddling through e-scooters, an impending wave of e-bikes: Examining Policy approaches to light electric vehicles in Canada's transportation system*.

Hawkins, A.J. (2018). *The electric scooter craze is officially one year old – What's next?* The Verge. https://www.theverge.com/2018/9/20/17878676/electric-scooter-bird-lime-uber-lyft.

Hoffman, M.L. 2016. *Bike lanes are white lanes: Bicycle advocacy and urban planning*. University of Nebraska Press.

Hosford, K., & Winters, M. (2018). Who are public bicycle share programs serving? An evaluation of the equity of spatial access to bicycle share service areas in Canadian cities. *Transportation Research Record*, *2672*(36), 42–50. https://doi.org/10.1177/0361198118783107.

Howland, S., McNeil, N., Broach, J., Rankins, K., MacArthur, J., & Dill, J. (2017). Breaking barriers to bike share: Insights on equity from a survey of bike share system owners and operators. *TREC Final Reports*. https://doi.org/10.15760/trec.173.

Institut National de Santé Publique du Québec. (2016). Deprivation index, Canada, 2016. https://www.inspq.qc.ca/en/expertise/information-management-and-analysis/deprivation-index/deprivation-index-canada-2016.

Isard, W. (1954). Location theory and trade theory: Short-run analysis. *Quarterly Journal of Economics*, *68*(2), 305–20. https://doi.org/10.2307/1884452.

Johnston, K., Oakley, D.A., Durham, A., Bass, C., & Kershner, S. (2020). Regulating micromobility: Examining transportation equity and access. *Journal of Comparative Urban Law and Policy*, *4*(1), 682–720. https://readingroom.law.gsu.edu/jculp/vol4/iss1/35.

Kim, AJ., Brown, A., Nelson, M., Ehrenfeucht, R., Holman, N., Gurran, N., Sadowski, J., Ferreri, M., Sanyal, R., Bastos, M., & Kresse, K. (2019). Planning and the so-called 'sharing' economy … *Planning Theory and Practice*, *20*(2), 261–87. https://doi.org/10.1080/14649357.2019.1599612.

Kincses, Á., & Tóth, G. (2014). The application of gravity model in the investigation of spatial structure. *Acta Polytechnica Hungarica, 11*(2), 5–19. https://doi.org/10.12700/APH.11.02.2014.02.1.

Kolodny, L. (2017). *LimeBike raises $12 million to roll out bike sharing without kiosks in the US*. Techcrunch. https://techcrunch.com/2017/03/15/limebike-raises-12-million-to-roll-out-bike-sharing-without-kiosks-in-the-us/.

Kong, V., & Leszczynski, A. (2022) Dockless micromobility sharing in Calgary: A spatial equity comparison of e-bikes and e-scooters. *Canadian Journal of Urban Research, 31*(1), 97–110. https://cjur-rhel8.uwlib.ca/index.php/cjur/article/view/390.

Krause, D. (2019). *Calgary bike share supply, demand dwindling*. LiveWire Calgary. https://livewirecalgary.com/2019/09/23/calgary-bike-share-supply-demand-dwindling/.

Lazarus, J., Pourquier, J.C., Feng, F., Hammel, H., & Shaheen, S. (2020). Micromobility evolution and expansion: Understanding how docked and dockless bike-sharing models complement and compete – A case study of San Francisco. *Journal of Transport Geography, 84*, Article 102620. https://doi.org/10.1016/j.jtrangeo.2019.102620.

Li, S., Zhuang, C., Tan, Z., Gao, F., Lai, Z., & Wu, Z. (2021). Inferring the trip purposes and uncovering spatio-temporal activity patterns from dockless shared bike dataset in Shenzhen, China. *Journal of Transport Geography, 91*, Article 102974. https://doi.org/10.1016/j.jtrangeo.2021.102974.

Lo, T. (2019). Calgary e-scooters will disappear for the next 4 ½ months. *CBC News*. https://www.cbc.ca/news/canada/calgary/lime-bird-escooter-hiatus-winter-calgary-1.5340786.

Marshall, A. (2018). Lime's new scooter is hardier, heavier, and built for life on the streets. *Wired*. https://www.wired.com/story/lime-scooter-gen3-design/.

Matthew, J., Liu, M., Li, H., Seeder, S., & Bullock, D. (2019). Analysis of e-scooter trips and their temporal usage patterns. *ITE Journal, 89*(6), 44–9. Retrieved from http://www.nxtbook.com/ygsreprints/ITE/G107225_ITE_June2019/index.php#/44.

McCarty Carino, M. [host]. (2018, 5 December). Scooters could improve mobility in low-income areas, but have an image problem. *Marketplace* [podcast]. Retrieved 31 August 2020 from https://www.marketplace.org/2018/12/05/scooters-could-improve-mobility-low-income-areas-they-have-image-problem/.

McKenzie, G. (2019). Spatiotemporal comparative analysis of scooter-share and bike-share usage patterns in Washington, D.C. *Journal of Transport Geography, 78*, 19–28. https://doi.org/10.1016/j.jtrangeo.2019.05.007.

McKenzie, G. (2020). Urban mobility in the sharing economy: A spatiotemporal comparison of shared mobility services. *Computers, Environment and Urban Systems, 79*. https://doi.org/10.1016/j.compenvurbsys.2019.101418.

McNeil, N., Broach, J., & Dill, J. (2018). Breaking barriers to bike share: Lessons on bike share equity. *ITE Journal (Institute of Transportation Engineers), 88*(2), 31–5. https://doi.org/10.15760/trec.191.

Médard de Chardon, C. (2019). The contradictions of bike-share benefits, purposes and outcomes. *Transportation Research Part A: Policy and Practice, 121*, 401–19. https://doi.org/10.1016/j.tra.2019.01.031.

Médard de Chardon, C., Caruso, G., & Thomas, I. (2017). Bicycle sharing system 'success' determinants. *Transportation Research Part A: Policy and Practice, 100*, 202–14. https://doi.org/10.1016/j.tra.2017.04.020.

Meng, S., & Brown, A. (2021). Docked vs. dockless equity: Comparing three micromobility service geographies. *Journal of Transport Geography, 96*, Article 103185, 1–11. https://doi.org/10.1016/j.jtrangeo.2021.103185.

Mooney, S.J., Hosford, K., Howe, B., Yan, A., Winters, M., Bassok, A., & Hirsch, J. A. (2019). Freedom from the station: Spatial equity in access to dockless bike share. *Journal of Transport Geography, 74*, 91–6. https://doi.org/10.1016/j.jtrangeo.2018.11.009.

Moreau, H., de Meux, L. de J., Zeller, V., D'Ans, P., Ruwet, C., & Achten, W.M.J. (2020). Dockless e-scooter: A green solution for mobility? Comparative case study between dockless e-scooters, displaced transport, and personal e-scooters. *Sustainability, 12*(5), 1803. https://doi.org/10.3390/su12051803.

Pampalon, R., Hamel, D., Gamache, P., Philibert, M.D., Raymond, G., & Simpson, A. (2012). An area-based material and social deprivation index for public health in Quebec and Canada. *Canadian Journal of Public Health, 103*(Suppl 2), S17–S22. https://doi.org/10.1007/BF03403824.

Potkins, M. (2019). Lime riders report new limits on speed, parking. *Calgary Herald*. Retrieved 10 April 2021 from https://calgaryherald.com/news/local-news/lime-users-report-new-speed-parking-restrictions-on-e-scooters.

Qian, X., Jaller, M., & Niemeier, D. (2020). Enhancing equitable service level: Which can address better, dockless or dock-based bikeshare systems? *Journal of Transport Geography, 86*, Article 102784. https://doi.org/10.1016/j.jtrangeo.2020.102784.

Sanders, R.L., Branion-Calles, M., and Nelson, T.A. (2020). To scoot or not to scoot: Findings from a recent survey about the benefits and barriers of using E-scooters for riders and non-riders. *Transportation Research Part A*, 139, 217–27. https://doi.org/10.1016/j.tra.2020.07.009.

Shaheen, S., Bell, C., Cohen, A., & Yelcuru, B. (2017). *Travel behavior: Shared mobility and transportation equity*. US Department of Transportation Report

No. PL-18-007. https://www.fhwa.dot.gov/policy/otps/shared_use_mobility_equity_final.pdf.

Shepherd, B., Doytchinova, H.S., & Kravchenko, A. (2019, 24 June). *The gravity model of international trade: A user guide* [R version]. United Nations ESCAP. Retrieved from https://www.unescap.org/resources/gravity-model-international-trade-user-guide-r-version.

Smith, C.S., Oh, J.S., & Lei, C. (2015, 31 August). *Exploring the equity dimensions of US bicycle sharing systems*. Transportation Research Center for Liveable Communities (TRCLC) Report No. 14-01. https://rosap.ntl.bts.gov/view/dot/30675.

Smith, M. (2021). Calgary's summer e-scooter starts Friday. *Calgary Herald*. https://calgaryherald.com/news/local-news/calgarys-summer-e-scooter-comeback-starts-friday.

Statistics Canada. (2015). Dissemination area. Retrieved 31 August 2020 from https://www12.statcan.gc.ca/census-recensement/2011/ref/dict/geo021-eng.cfm.

Statistics Canada. (2018). Population and dwelling count highlight tables, 2016 census [data table]. Retrieved 19 April 2021 from https://www12.statcan.gc.ca/census-recensement/2016/dp-pd/hlt-fst/pd-pl/Table.cfm?Lang=Eng&T=302&SR=1&S=86&O=A&RPP=9999&PR=48.

Stehlin, J. (2015). Cycles of investment: Bicycle infrastructure, gentrification, and the restructuring of the San Francisco Bay Area. *Environment and Planning A*, *47*(1), 121–37. https://doi.org/10.1068/a130098p.

Stehlin, J.G., & Payne, W.B. (2022). Mesoscale infrastructures and uneven development: Bicycle sharing systems in the United States as "already splintered" urbanism. *Annals of the American Association of Geographers* *112*(4), 1065–83. https://doi.org/10.1080/24694452.2021.1956874.

Xu, Y., D. Chen, X. Zhang, W. Tu, Y. Chen, Y. Shen, and C. Ratti. (2019). Unravel the landscape and pulses of cycling activities from a dockless bike-sharing system. *Computers, Environment and Urban Systems*, *75*, 184–203. https://doi.org/10.1016/j.compenvurbsys.2019.02.002.

Yanocha, D., & Allan, M. (2019). The electric assist: Leveraging e-bikes and e-scooters for more livable cities. *Institute for Transportation & Development Policy*. https://www.itdp.org/wp-content/uploads/2019/12/ITDP_The-Electric-Assist_-Leveraging-E-bikes-and-E-scooters-for-More-Livable-Cities.pdf.

Younes, H., Zou, Z., Wu, J., & Baiocchi, G. (2020). Comparing the temporal determinants of dockless scooter-share and station-based bike-share in Washington, D.C. *Transportation Research Part A*, *134*, 308–20. https://doi.org/10.1016/j.tra.2020.02.021.

Zhang, S., Ji, Y., Sheng, D., & Zhou, J. (2018). Origin-destination distribution prediction model for public bicycles based on rental characteristics. In

W. Wang, B. Klaus, & X. Jiang (Eds.), *Green intelligent transportation systems* (pp. 293–303). Springer. https://doi.org/10.1007/978-981-10-3551-7.

Zhang, W., Buehler, R., Broaddus, A., & Sweeney, T. (2021). What type of infrastructures do e-scooter riders prefer? A route choice model. *Transportation Research Part D: Transport and Environment 94*, Article 102761. https://doi.org/10.1016/j.trd.2021.102761.

Zhang, Y., Thomas, T., Brussel, M., & van Maarseveen, M. (2017). Exploring the impact of built environment factors on the use of public bikes at bike stations: Case study in Zhongshan, China. *Journal of Transport Geography, 58*, 59–70. https://doi.org/10.1016/j.jtrangeo.2016.11.014.

8 Bike Share Toronto: Building a Network to Meet the Demands of COVID and Beyond

SPENCER McNEE AND ERIC J. MILLER

Introduction

Urbanization trends around the world continue to drive strong growth in city populations. In Canada, 73.7 per cent of the population now lives in a large urban area, defined as an area with one hundred thousand or more people, and most of Canada's population growth from 2016 to 2021 has occurred in these urban areas (Statistics Canada, 2022). For cities facing congestion and grappling with decarbonization, improvements to cycling infrastructure and encouragement of bicycle use offer compelling solutions. Cycling provides users with an active, inexpensive, and zero-emissions method of transportation. In the city of Toronto, 46 per cent of trips take place by car, compared with 13 per cent made by foot or bicycle. However, many car trips within cities are short. As the median trip length by car is 5.5 kilometres, people could complete these trips by bicycle instead (Ashby, 2018). To maximize cycling uptake, cities can encourage bicycle trips by providing safe cycling infrastructure, which increases people's willingness to cycle (El-Assi et al., 2017), and by offering bike-share systems. These systems lower the barriers to taking up cycling by removing storage requirements and mitigating the risk of stolen bicycles. Bike-share systems also facilitate both one-way trips by bicycle and trips paired with public transit, providing a convenient last-mile solution for transit journeys (Kim & Cho, 2021).

Bike-share systems represent a growing and flexible mobility service (see Kong and Leszczynski, chapter 7). The City of Toronto's docked system, Bike Share Toronto, began in 2011 with 80 stations and 1,000 bicycles (Taylor, 2011). It has since grown to 625 stations and over 6,000 bicycles in 2022, and it has plans to expand to over 1,000 stations by 2025 (Toronto Parking Authority, 2022).

In this chapter, we look at ridership trends on the system as it has grown and analyse how the system adapted to the COVID-19 pandemic. Bike Share Toronto has shown itself to be flexible and resilient, rising to meet the challenges of urban transportation. While the number of stations remained unchanged during the pandemic, ridership continued to grow despite travel restrictions and remote work. As pandemic restrictions lift, Bike Share Toronto has the opportunity to build on this growth. We examine some of the factors for planning and operations that the system can apply to maximize its potential.

Our analysis uses data from the City of Toronto's Open Data Portal, which offers two types of information about the Bike Share Toronto network. The first type is trip data, which includes a log of every bicycle trip made on the network with details such as start time, end time, start station, and end station. The second type of information is station data, which includes the location, unique identifiers, names, and capacities of each station in the network. In addition to these data provided by Bike Share Toronto, the analysis uses the City of Toronto's road network map and census data from Statistics Canada to understand the demographics of users within the network.

Literature Review

History of Bike-Share Systems

Bike-share systems have exploded in popularity in recent years, and many cities around the world now have extensive networks (see Rojas and Miller, chapter 5). The first system began in 1965, when Amsterdam made a number of "white bikes" available for shared use. Ultimately, the initiative lacked effective security, and the bicycles were regularly vandalized and stolen (Fishman, 2020).

Researchers have identified four waves of bike-share systems. The first wave, such as the Amsterdam initiative, and a similar arrangement in 1976 in La Rochelle, France, included smaller systems that offered the bicycles free of charge. Due to a lack of security measures, these programs had little success (Fishman, 2020).

To address these issues, the second wave of bike-share systems introduced basic payment requirements and locks. One example of the second wave took place in Copenhagen in the 1990s. While these measures somewhat improved security, the coin-operated locks, similar to those on coin-operated grocery carts, still did not prevent widespread theft. The deposits paid were too small relative to the bicycles' value, and so vandalism continued (Fishman, 2020).

The third wave of bike-share systems began in the mid-2000s. These systems used docking stations for bicycles and featured payment methods tied to memberships and credit cards (Fishman, 2020). Because these changes improved security, the systems grew rapidly. Many third-wave systems, including Bike Share Toronto, continue to operate successfully in cities around the world today.

The fourth bike-share wave features dockless systems (see Kong and Leszczynski, chapter 7). Riders rely on GPS and smartphones to rent and lock these bicycles, which do not need docking stations. Fourth-wave systems have proliferated in many regions, especially in China. Dockless systems exhibit slightly different travel patterns when compared with docked systems, notably showing an increase in bicycles and ridership in residential and industrial areas and fewer trips in the central business district (Guo et al., 2022). While these dockless systems share some similarities with docked systems, the challenges around network expansion and operations are quite different. As our analysis focuses on a third-wave docked system, this chapter will not consider fourth-wave systems further.

Bike Share Toronto, our focus, uses the bicycle and docking system developed in Montreal for the Bixi bike-share system (see Hashemi et al., chapter 9). Many bike-share systems use the Bixi platform, including those in New York City and London (*Smart bike-sharing systems for cities*, n.d.). Bike Share Toronto launched in 2011 under the name Bixi with 80 stations and 1,000 bicycles (Taylor, 2011). Originally, the system was owned by the Public Bike System Company (PBSC), which created Bixi in Montreal. However, after PBSC filed for bankruptcy in 2014, the system was taken over by the Toronto Parking Authority, a municipally owned corporation, and renamed Bike Share Toronto (Gerster, 2014). Under the Toronto Parking Authority, which subcontracts the system's operation to Shift Transit, a mobility operations company, Bike Share Toronto has grown to over 625 stations and over 6,000 bicycles.

Establishing and Growing a Bike-Share Network

Several studies have looked at the logistical and infrastructure-related challenges of building a bike-share network. When designing a bike-share system, operators must first decide on the type of network to build. With current technologies, both fixed-station (docked) and free-floating (dockless) systems exist (Guo et al., 2022).

After that initial choice, designers must determine where to place stations within the host city. When evaluating station placement, researchers can use geographic information system (GIS) tools to evaluate the characteristics of potential sites along with decision-making criteria to propose locations (Kabak et al., 2018). Another strategy to determine station placement is to use tools that predict cycling demand within a city and place stations to maximize demand while minimizing operational costs (Frade & Ribeiro, 2015). A study in Washington, DC, sought to improve on these strategies by applying a maximal covering location problem to expand network coverage, with the goal of optimizing bike-share network expansion (Salih-Elamin & Al-Deek, 2021). A study of the bike-share network in Baltimore used trip GPS data to evaluate candidate sites for new stations expanding the twenty-one-station network (Banerjee et al., 2020). All these studies focus on establishing systems and providing stations to maximize coverage of a city. While they offer good insights, they examine cities with either no bike-share system or small bike-share systems. In Toronto's case, the system is already extensive, with over 625 stations and an effective catchment area of 59 km^2 in the city.

Once designers choose a possible location for a bike-share station, they consider how to predict ridership trends at that spot. In particular, they consider how neighbouring stations will influence demand at the new location. To answer this question, researchers have built gravity models that can predict ridership at new stations without historical ridership information (Liu & Oshan, 2022). As systems continue to grow beyond an initial coverage network, operators must also decide whether to continue expanding the network or to add density and redundancy within it. A study in London looked at the network effects of the stations in London's bike-share system. Specifically, the study examined how the usefulness of a given station is characterized not by the station itself but by the larger network and the parts of the city to which that station provides access (He et al., 2021).

Beyond system coverage, the density of stations also impacts bike-share ridership in two important ways. First, ridership increases as the distance to a given station decreases. Riders clearly prefer a short walk to a station, so as the distance between stations increases, ridership drops (Kabra et al., 2020). Second, density of available stations can help reduce delays caused by full or empty stations by reducing the travel time to the next available station with a bike if the station is empty or the next non-full station if the station is full.

Considerations about station placement also align with the concept of accessibility in transportation, which refers to the measurement of the catchment area available to a user who is using a given mode of transportation and who is starting from a given point (Handy & Niemeier, 1997). Accessibility, which is mode specific, quantifies the level of attractiveness of a set of destinations (Xi & Miller, 2019). For the analysis in this chapter, we computed an accessibility score for each Bike Share Toronto station by quantifying the areas surrounding possible destination stations within a given travel time. This method of determining accessibility differs from the London study's method because it examines the areas within a short walking distance to destination stations, as opposed to looking at features along bicycle routes.

Establishing a bike-share network can lead to many challenges. Some move beyond a given network's coverage or density and rest on intangibles, such as the users' experience purchasing a pass or the users' knowledge of the network (Leister et al., 2018). While these are important factors for bike-share operators to consider and may play a role in a system's growth, they are difficult to capture in a model.

Finally, operators must also consider equity when they plan networks (see Kong and Leszczynski, chapter 7). Bike-share systems can provide access to users at a low cost. The bike-share system in Hamilton, Ontario, has added "equity" stations in underserved areas of the city in an effort to provide better service (Desjardins et al., 2022). Bike Share Toronto already provides good coverage in the city's core and hopes to expand to all City of Toronto wards in the coming years (Toronto Parking Authority, 2022).

COVID-19 and Its Effects on Bike Share Toronto

The COVID-19 pandemic has radically affected travel within cities. Travel changed dramatically as governments around the world imposed lockdowns and as businesses switched to remote work models. Studies have assessed the impacts of lockdowns on travel in Toronto (see Palm and Farber, chapter 6) and in other jurisdictions, such as Singapore.

Ontario declared a state of emergency on 17 March 2020. Businesses, recreational spaces, and establishments deemed non-essential were closed, and the province strictly limited gathering sizes to reduce the spread of the virus. The Ontario government first issued a road map for reopening on 27 April 2020, although capacity numbers and services were still largely restricted (*Report on Ontario's Provincial Emergency*, n.d.). The reopening process occurred slowly over several years, with the province declaring additional restrictions and emergency measures on 12 January 2021 and 7 April 2021, respectively, in response to viral spread trends (*Report on Amendments and Extensions of Orders*, n.d.). All provincial restrictions were removed on 27 April 2022 (*Report on Amendments, Extensions, and Revocations of Orders*, n.d.). These restrictions had wide-ranging impacts on travel behaviour within the city of Toronto (see Palm and Farber, chapter 6; Brail and Donald, chapter 10). With businesses closed or shifted to remote work, and with social gatherings limited, shopping and socialization patterns changed. Some changes peaked in 2020 when restrictions were the strictest, while others have persisted beyond the end of pandemic-related measures (see Cavalli and McGahan, chapter 2).

A Toronto study surveyed participants throughout and following the lockdowns in order to assess some of the temporary and long-term impacts of the pandemic on travel behaviour within the city. The study looked at changes to trips of varying types such as work, grocery shopping, other shopping, and food preparation and eating. The results showed an increase in home meal preparation and a decrease in shopping trips during the pandemic. Online socialization also remained above pre-pandemic levels even as in-person social gatherings returned after lockdowns ended. Finally, the study found that a long-term trend towards remote work was difficult to assess. While results showed an explosion in working from home during the lockdowns, the researchers found it premature to quantify long-term remote work trends. However, the analysis did reveal that access to a suitable home office was the strongest driver of a preference for remote work (Dianat et al., 2022).

These changes in travel behaviour also significantly impacted bike-share ridership levels. A study of bike-share traffic conducted in Singapore found that this activity increased during lockdown. The study also found that areas with high residential densities showed the highest ridership increases, indicating that people were using shared bikes for different types of trips, not just for commuting. Bike-share trips also seemed to replace transit trips, suggesting that riders were switching travel mode to avoid enclosed spaces (Song et al., 2022).

Demand Patterns

To understand overall ridership behaviour and how it changed after the outbreak of COVID-19 pandemic, we must first examine historical trip data about Toronto's bike-share system from January 2017 through June 2022.

Trip Length and Speed

In Toronto, the length, speed, and cost of a bike-share trip are closely interconnected. Bike Share Toronto charges a base cost for a trip of thirty minutes or less and applies additional charges if the trip exceeds thirty minutes. Because of this cost factor, coupled with the relative heaviness of a bicycle, trip lengths generally fall below the thirty-minute mark. Available data reflect this pattern, with 93 per cent of all trips from 2017 through June 2022 lasting thirty minutes or less.

Trip Counts

Ridership has grown on the Bike Share Toronto network each year since 2017, the first year considered in this analysis. The number of stations also grew between 2017 and 2020 and then remained fairly stable. The system first established a strong core network downtown and then worked on expanding the network's boundaries. Before 2017 and through 2018, Bike Share Toronto focused on providing a density of stations through the central part of Toronto, but thereafter it increased the extent of the network, which now includes isolated pilot areas to the north and east of the core. Although Bike Share Toronto does intend to add more stations in the coming years, that plan is outside our analysis period (Toronto Parking Authority, 2022).

In addition to the yearly ridership increases shown in Table 8.1, monthly ridership from 2017 to 2022 increased in trip count each year across all months. In 2021, the period of high ridership extends for longer throughout the year beyond the observed peak in the summer months. This extension may result from ridership increases that occurred during the pandemic. Specifically, users who have integrated bike-share into their regular commuting routines might be less likely to switch transportation modes when the weather is cooler. This theory is reinforced when we examine the fraction of yearly trips made each month. The curve for 2021 flattens over the period from May to October, with previous years showing a much narrower peak in July and August.

COVID-19 Impacts on Travel Behaviour

The COVID-19 pandemic radically changed travel behaviour across the world. In 2020 and 2021, Torontonians experienced several lockdowns and indoor capacity restrictions. These restrictions precipitated widespread changes to travel patterns as people began working remotely and avoided modes of travel with a higher risk of viral transmission. In this section, we explore how travel behaviour on the Bike Share Toronto network changed through 2020, 2021, and into 2022.

As cycling provides a flexible mode of transportation with a low risk of viral transmission, Bike Share Toronto saw ridership grow during this period, despite a lower volume of overall travel. The number of

Table 8.1. Summary of yearly ridership and station count on the bike share network

Year	Number of trips	Number of unique stations
2017	1,492,369	200
2018	1,922,955	359
2019	2,439,517	469
2020	2,911,308	612
2021	3,575,182	629
2022*	1,632,204	637

*2022 includes data only up to the end of June 2022.

stations in the network did not significantly change during these years, and we can therefore attribute changes to changing demand patterns instead of expanded coverage.

In a first notable change in ridership patterns, as shown in Figure 8.1, the length of trips increased during this growth phase. The years 2017, 2018, and 2019 all have a similar distribution of trip lengths, with a median trip length around eleven minutes and an eighty-fifth-percentile trip length around twenty-two minutes. In 2020, there is a noticeable increase in trip length, with the median length around fourteen minutes and the eighty-fifth percentile around twenty-five minutes. These data might indicate a change in the type of trip that users took this year. This change did appear to be temporary. Mean trip length decreased in 2021, a year that featured fewer restrictions. Although we could not collect a full year of data for 2022, there is again a trend towards the shorter pre-COVID trip length distribution.

The data also show a significant upward trend in the percentage of Bike Share Toronto trips taken by casual users, defined as users without an annual membership. From 2020 onwards, the percentage of trips taken by casual users has risen significantly, suggesting that the bike share system is reaching new customers.

Another significant change in travel patterns relates to where trips started and ended. In 2020, the number of trips starting and ending at the same station increased from around 3 per cent to 5 per cent. With the mean duration of these trips over double that of trips made between different stations, this increase affects the overall distribution of trip length. An increase in the number of trips starting and ending at the same station also indicates an overall rise in leisure trips, a change that helps explain the increase in trip length observed in 2020. Table 8.2

Figure 8.1. Distribution of trip duration from 2017 to 2022

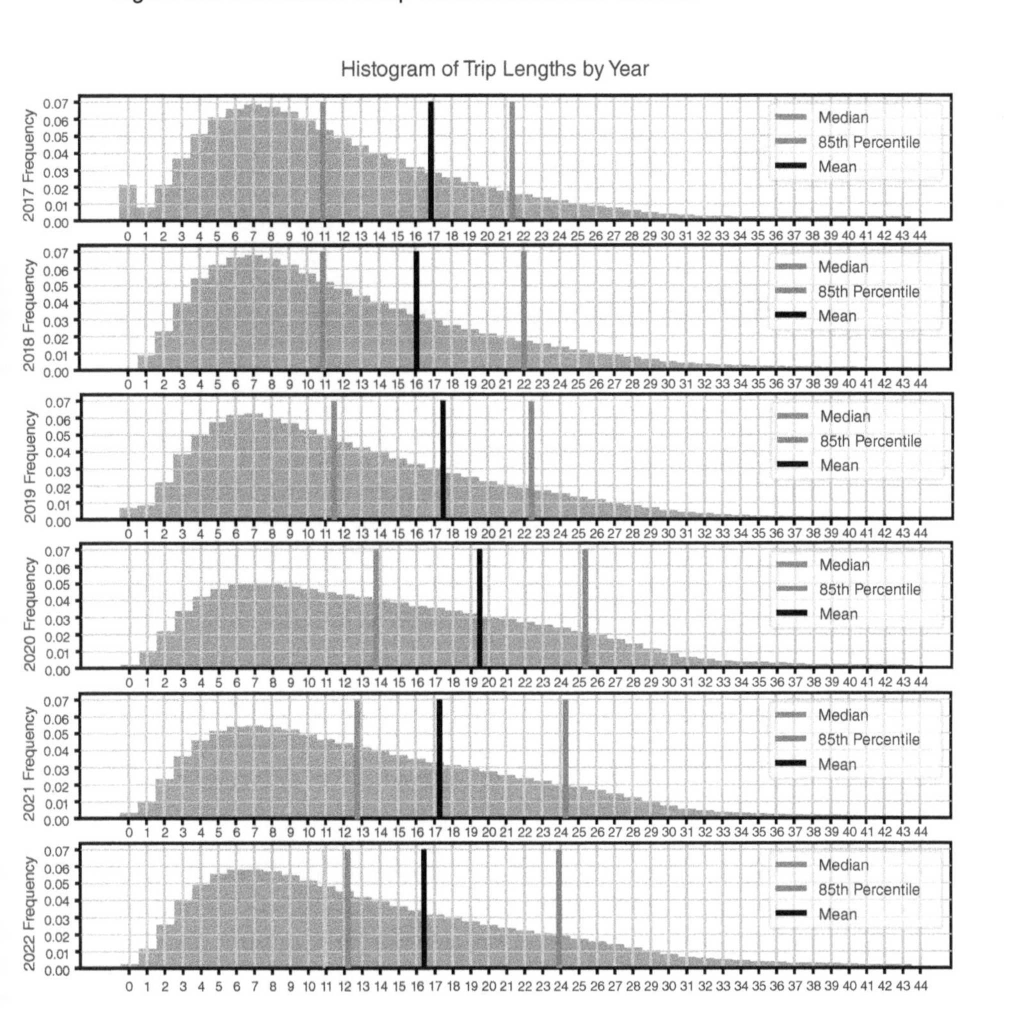

Table 8.2. Percentage of trips to and from the same station by year

Year	Trips to/from same station (%)	Mean duration of trips to/from same station(s)	Mean duration of trips to/from different stations(s)
2017	2.5	3022	818
2018	2.9	2701	850
2019	3.3	1824	841
2020	5.9	2196	957
2021	5.1	1922	897
2022*	3.8	1622	877

*2022 includes data only to the end of June 2022.

shows the percentage of all trips that started and ended at the same station as well as the mean duration of the two trip types.

In general, the length of trips along the same route did not increase in this period. An analysis of the distribution of the change in trip duration among all routes present in both 2019 and 2020 shows that routes changed by only thirty-five seconds on average. As well, the distribution of the change in trip lengths was near to being symmetrical around zero indicating that there is no underlying change in length, only natural variation.

Following the outbreak of the COVID-19 pandemic, the number of weekend trips taken relative to weekday trips rose, indicating an increase in recreational trips made on the system. In 2020 especially, trips taken on the weekend increased as compared with trips taken on weekdays. During the weekdays, ridership is highest on Wednesdays, likely because Bike Share Toronto periodically runs a "Free Ride Wednesdays" promotion. As with other noted trends, the data suggest a return to previous ridership patterns in 2021 and again in 2022. Weekend ridership declines slightly as compared with weekday travel, indicating that riders are once again using Bike Share Toronto for other trip types. However, the number of leisure trips taken remains significant.

Some of the changes in trip distribution across days of the week may result from new routes made possible by new stations added in 2020. In 2020, 495,587 of 2,911,308 trips took place along routes that did not exist in 2019. Although trips taken in 2021 and 2022 to date show a reduction in trip length trending back towards pre-pandemic numbers, the increase in leisure or recreational trips likely accounts for the increase in trip duration relative to the pre-pandemic period.

Station Patterns

In this section, we analyse the distribution of demand across various stations in the system and investigate whether certain stations exhibit distinct characteristics. This understanding is essential because a bike share system can reach its potential only if it can balance supply and demand.

Systematic Imbalances

To understand the challenge of balancing available bicycles to demand, we must first quantify the presence of systematic imbalances at stations. We define stations with a systematic imbalance as those that experience persistent imbalances over a sustained period. For example, a station that gains one bicycle per day on average has a systematic imbalance and would need regular rebalancing. Stations can also experience imbalances within a day. Such stations may show no regular loss or gain over a full day, but they may have a within-day imbalance. For example, a downtown station with significant commuter ridership might experience significant bicycle arrivals in the morning and a reversing trend in the afternoon. These stations pose a larger rebalancing challenge because an operator must support the station within a narrow time frame.

Our analysis of the net gains of Bike Share Toronto stations shows a net migration of bicycles from north to south. As trips from north to south in Toronto generally run downhill, according to the city's geography, this pattern makes sense. Stations east and west of the core are more balanced.

We can also observe this phenomenon by looking at within-day patterns for specific stations. For instance, station 7015 is a downtown station located at King Street West and Bay Street that exhibits a significant commuter trend, as revealed by an analysis of the station's weekly profile during the week of 23 September 2019. In the morning, there is a large peak of arrivals and few departures, causing the station to fill, while in the afternoon, there is a large peak of departures and few arrivals, causing the station to empty. The result is a within-day imbalance. In addition, because the morning peak of arrivals is larger than the afternoon peak of departures, the station has a systematic imbalance. In brief, the station presents rebalancing challenges on both a within-day and a longer-term basis, as the station has a net tendency towards being full. Finally, the relative amount of station activity from Monday to Friday compared to Saturday and Sunday also shows a strong commuter influence. There is almost no activity at this station on the weekend as compared with weekdays. Usage by commuters, like the pattern observed at this station, are also reflected in the general redistribution of bicycles from stations uptown to stations downtown.

Station 7015 also reveals the radical impacts of the COVID-19 pandemic on ridership patterns. While the Bike Share Toronto ridership grew during the pandemic, certain bicycle distribution patterns caused by commuting changed. In September 2020, one year after the original profile, the expansion of remote work altered commuting patterns. Downtown stations such as station 7015 experienced less activity, and the distinct commuter pattern of riders arriving in the morning and departing in the evening collapsed. Another year later, in September 2021, the commuter pattern has started to recover but is less distinct than in 2019.

Commuter travel does not affect every station, however. For example, station 7239 at the intersection of Bloor Street and Manning Avenue in Toronto's west end exhibits greater balance. Here, arrivals and departures stay in relative balance with each other throughout the day, and by the end of the day, this station experiences a net gain of a single bicycle, a small systematic imbalance.

Accessibility

The growth of a bike share system affects the entire network, including accessibility. To measure accessibility, we measure the potential catchment area that a person has access to during a given time using a given mode of transportation. We can apply an accessibility score not only to the entire Bike Share network but also to each station within the network. These individual scores offer insight into the options available to users.

To compute an individual accessibility score, we first compute the travel time between each pair of stations. For each station, the upper quartile travel time is used to capture the point at which 75 per cent of all trips would be below this travel time. Next, we apply a maximum travel time of 20 minutes as the cutoff for distance. Any station where the upper quartile travel time is below 20 minutes we count as a "nearby station." Next, using QGIS, we draw a 300 m walking buffer around each nearby station with the overlaps dissolved so that we do not count each area twice. Finally, the buffered area is intersected with population data from the 2016 and 2021 censuses. We can summarize this calculation as follows:

A_{ic}: *Area within 300m of station i in census tract c*
A_{ic} *cannot overlap with any other* A_{jc} *for all stations j*

Figure 8.2. Plot of station arrivals vs station accessibility in 2021

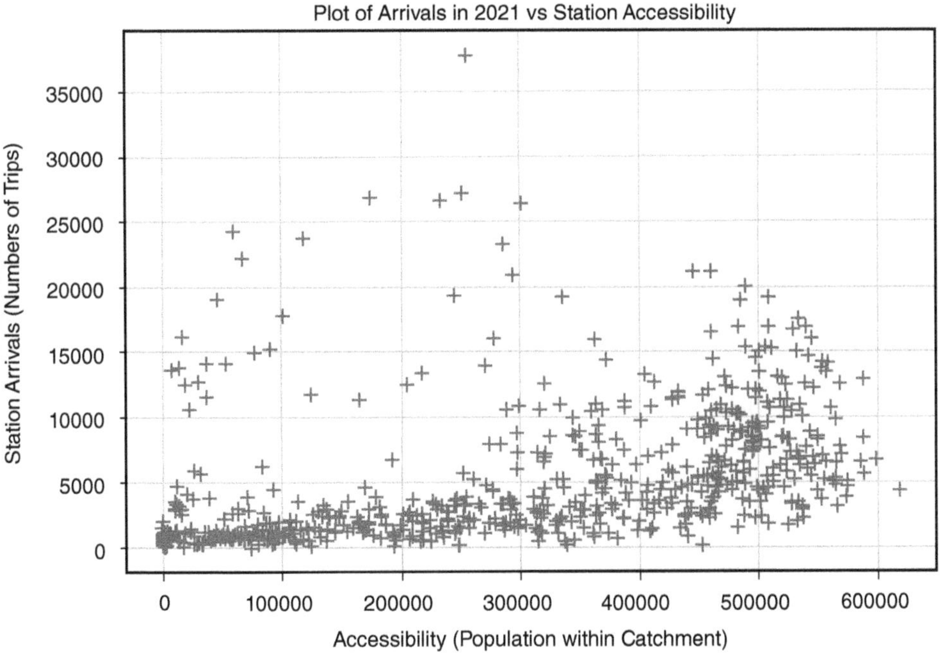

P_c: Population density in census tract c
T_{ij}: Upper quartile travel time from station i to station j in minutes

$$Accessibility_i = \sum_j \sum_c A_{jc} p_c \text{ for all } c \text{ and all } j \text{ where } T_{ij} \leq 20$$

These accessibility scores have important implications. For example, they provide predictive insight into stations likely to have high ridership. Stations with high accessibility generally have a much higher ridership, and the ridership at a station increases dramatically when the accessibility score exceeds a noticeable threshold of around four hundred thousand as shown in Figure 8.2. When we look at station departures and total trips, similar patterns emerge (see Figure 8.2).

However, there are also some stations with high ridership despite a low accessibility score. The characteristics of these particular stations offer insight into the network's different use cases and opportunities. One example is station 7242, located at Lake Shore Boulevard West and Ontario Street. This station sits along the waterfront on a linear park and the Waterfront Trail. While it does connect to many population

Table 8.3. Ridership and accessibility at station 7091 by year

Year	Departures	Arrivals	Accessibility
2017	389	410	87,497
2018	2023	1898	156,998
2019	2266	2052	199,213
2020	2796	2555	218,164
2021	2287	1874	218,164

centres, station 7242 is also in a prime location to offer users recreational cycling along Toronto's waterfront. Several stations with low accessibility scores and strong ridership are located along the Waterfront Trail in Toronto's west end and at Cherry Beach in the east end.

It is more difficult to find explanations for why some stations with high accessibility scores have low ridership. However, these stations tend to be placed within smaller residential pockets of the city and likely do not experience significant foot traffic. For example, station 7141 at Bridgeman Ave and Kendall Ave had very low ridership despite its high accessibility, likely due to the fact that it is on a small street that is not a common route for pedestrians.

Accessibility scores also reveal that stations may reach a ridership saturation point, when ridership and accessibility increase to a certain point and then remain level. Table 8.3 shows an example of station 7091 at Donlands subway station in Toronto's east end. Before 2018, all Bike Share stations on the Danforth were located near the subway stations. In 2018, the system expanded and added many stations to the surrounding neighbourhoods. These stations experienced a large jump in ridership as accessibility grew from 2017 to 2018. However, ridership has remained fairly consistent following this jump, even as accessibility has continued to grow with more stations added. This observation suggests that the station has reached a saturation point due to either demand or capacity limits.

As Bike Share Toronto has grown, the expansion has created a network effect, with each subsequent addition of new stations increasing the accessibility of nearby stations. This network effect indicates that the value of a new addition is greater than the value of previous ones because the new addition enhances the usefulness of all other nodes in the network. Thus, the growth in the number of stations has dramatically increased the number of stations with high accessibility. This growth has also led to higher ridership each year.

Conclusions

Bike-share systems offer an attractive, low-cost, and zero-emissions transportation option within cities. Technological improvements in the early twenty-first century have allowed such systems to proliferate. In this chapter, we have examined the Bike Share Toronto system, opened in 2011, looking at trip and station patterns through Bike Share Toronto's growth from 2017 and after the major shock of the COVID-19 pandemic in 2020. The system adapted well to the changes imposed by the pandemic, and it looks set to help Toronto tackle the transportation challenges of the coming years.

Understanding station and ridership patterns is essential to enhancing the system. This chapter has examined several factors affecting these patterns. First, understanding systematic and within-day imbalances can help operators to address the crucial challenge of rebalancing supply and demand at stations. Second, understanding typical trip lengths and travel patterns can help inform potential changes to the membership and pricing model. Finally, understanding the network effects of expansion, as revealed by accessibility scores, can help inform station placement as the system grows towards its goal of having stations in every Toronto ward. With an understanding of these factors, the Toronto Parking Authority can make operational and capital improvements to help the Bike Share Toronto network, already a flexible and adaptable system, reach its full potential.

REFERENCES

Ashby, B. (2018). *Transportation Tomorrow Survey 2016*. Malatest.

Banerjee, S., Kabir, Md.M., Khadem, N.K., & Chavis, C. (2020). Optimal locations for bikeshare stations: A new GIS based spatial approach. *Transportation Research Interdisciplinary Perspectives, 4*, Article 100101. https://doi.org/10.1016/j.trip.2020.100101.

Desjardins, E., Higgins, C.D., & Páez, A. (2022). Examining equity in accessibility to bike share: A balanced floating catchment area approach. *Transportation Research Part D: Transport and Environment, 102*, Article 103091. https://doi.org/10.1016/j.trd.2021.103091.

Dianat, A., Hawkins, J., & Habib, K.N. (2022). Assessing the impacts of COVID-19 on activity-travel scheduling: A survey in the greater Toronto area. *Transportation Research. Part A, Policy and Practice, 162*, 296–314. https://doi.org/10.1016/j.tra.2022.06.008.

El-Assi, W., Salah Mahmoud, M., & Nurul Habib, K. (2017). Effects of built environment and weather on bike sharing demand: A station level analysis of commercial bike sharing in Toronto. *Transportation (Dordrecht), 44*(3), 589–613. https://doi.org/10.1007/s11116-015-9669-z.

Fishman, E. (2020). *Bike share*. Routledge.

Frade, I., & Ribeiro, A. (2015). Bike-sharing stations: A maximal covering location approach. *Transportation Research Part A: Policy and Practice, 82*, 216–27. https://doi.org/10.1016/j.tra.2015.09.014.

Gerster, J. (2014, 30 March). New name, look and prices for Toronto's Bixi. *Toronto Star*. https://www.thestar.com/news/gta/2014/03/30/new_name_look_and_prices_for_torontos_bixi.html.

Guo, Y., Yang, L., & Chen, Y. (2022). Bike share usage and the built environment: A review. *Frontiers in Public Health, 10*, Article 848169. https://doi.org/10.3389/fpubh.2022.848169.

Handy, S.L., & Niemeier, D.A. (1997). Measuring accessibility: An exploration of issues and alternatives. *Environment and Planning A: Economy and Space, 29*(7), 1175–194. https://doi.org/10.1068/a291175.

He, P., Zheng, F., Belavina, E., & Girotra, K. (2021). Customer preference and station network in the London bike-share system. *Management Science, 67*(3), 1392–412. https://doi.org/10.1287/mnsc.2020.3620.

Kabak, M., Erbaş, M., Çetinkaya, C., & Özceylan, E. (2018). A GIS-based MCDM approach for the evaluation of bike-share stations. *Journal of Cleaner Production, 201*, 49–60. https://doi.org/10.1016/j.jclepro.2018.08.033.

Kabra, A., Belavina, E., & Girotra, K. (2020). Bike-share systems: Accessibility and availability. *Management Science, 66*(9), 3803–24. https://doi.org/10.1287/mnsc.2019.3407.

Kim, M., & Cho, G.-H. (2021). Analysis on bike-share ridership for origin-destination pairs: Effects of public transit route characteristics and land-use patterns. *Journal of Transport Geography, 93*, Article 103047. https://doi.org/10.1016/j.jtrangeo.2021.103047.

Leister, E.H., Vairo, N., Sims, D., & Bopp, M. (2018). Understanding bike share reach, use, access and function: An exploratory study. *Sustainable Cities and Society, 43*, 191–6. https://doi.org/10.1016/j.scs.2018.08.031.

Liu, Z., & Oshan, T. (2022). Comparing spatial interaction models and flow interpolation techniques for predicting "cold start" bike-share trip demand. *Transactions in GIS, 26*(4), 81–98. https://doi.org/10.1111/tgis.12933.

Report on Amendments and Extensions of Orders under the Reopening Ontario (A Flexible Response to COVID-19) Act, 2020 from July 24, 2020, to July 24, 2021 | Ontario.ca. (n.d.). Retrieved 10 August 2022, from https://www.ontario.ca/document/report-amendments-and-extensions-orders-under-reopening-ontario-flexible-response-covid-19-act-2020.

Report on Amendments, Extensions, and Revocations of Orders under the Reopening Ontario (A Flexible Response to COVID-19) Act, 2020 from December 2, 2021, to March 28, 2022. (n.d.). Ontario.ca. Retrieved 10 August 2022, from http://www.ontario.ca/document/report-amendments-extensions-and-revocations-orders-under-reopening-ontario-flexible-3.

Report on Ontario's Provincial Emergency from March 17, 2020, to July 24, 2020 | Ontario.ca. (n.d.). Retrieved 10 August 2022, from https://www.ontario.ca/document/report-ontarios-provincial-emergency-march-17-2020-july-24-2020.

Salih-Elamin, R., & Al-Deek, H. (2021). A new method for determining optimal locations of bike stations to maximize coverage in a bike share system network. *Canadian Journal of Civil Engineering, 48*(5), 540–53. https://doi.org/10.1139/cjce-2020-0143.

Smart bike-sharing systems for cities. (n.d.). PBSC Urban Solutions. Retrieved 26 July 2022, from https://www.pbsc.com/.

Song, J., Zhang, L., Qin, Z., & Ramli, M.A. (2022). Spatiotemporal evolving patterns of bike-share mobility networks and their associations with land-use conditions before and after the COVID-19 outbreak. *Physica A: Statistical Mechanics and Its Applications, 592*, Article 126819. https://doi.org/10.1016/j.physa.2021.

Statistics Canada. (2022). *Canada's large urban centres continue to grow and spread.* https://www150.statcan.gc.ca/n1/daily-quotidien/220209/dq220209b-eng.htm.

Taylor, L.C. (2011, 26 April). Need to borrow a bicycle? Bixi launches in May. *Toronto Star.* https://www.thestar.com/news/gta/need-to-borrow-a-bicycle-bixi-launches-in-may/article_88a11b6b-6d06-5006-b043-2117454b9f59.html.

Toronto Parking Authority. (2022). *BIKE SHARE TORONTO FIRST QUARTER (Q1) 2022 UPDATE.* https://www.toronto.ca/legdocs/mmis/2022/pa/bgrd/backgroundfile-199512.pdf.

Xi, Y. (Luna), & Miller, E.J. (2019). Accessibility: Definitions, measurement & implications for transportation planning analysis. *TRANSP RES PROC, 41*, 159–61. https://doi.org/10.1016/j.trpro.2019.09.029.

9 Montreal's Digital Mobility Ecosystem: A Place-Based Story of a City Undergoing a Mobility Transition

MOE HASHEMI, HAMED MOTAGHI, AND
DIANE-GABRIELLE TREMBLAY

Introduction

Montreal, Canada's second most populous metropolitan area after Toronto (Statistics Canada, 2021), has a vibrant mobility ecosystem, with an impressive set of businesses, non-profits, multi-level government actors, students and immigrant populations playing a role in Montreal's mobility future (Berger et al., 2019; Darchen & Tremblay, 2010; de Lorimier & El-Geneidy, 2012; Radio-Canada, 2019b; Rotaris & Danielis, 2018; Vultur et al., 2022).

What makes Montreal a compelling case for studying digital mobility lies in examples of continuous experimentation (success and failures) in technological innovation, business practices, policy and social justice with the aim of identifying opportunities to improve how people and goods move throughout the city. This chapter summarizes the research conducted between August 2020 and August 2022[1] to explore Montreal's digital mobility platform economy. It starts by reviewing the extant literature on the digital mobility sector in the Montreal area and follows the research process, results, and conclusions.

Montreal's Digital Mobility Ecosystem: State of the Art of Research and Practice

The platform economy has recently received much attention from policymakers, businesses, the media, and the public. Enabled by technology and social trends, digital platforms transform most industries

1 We thank the Social Sciences and Humanities Research Council of Canada (SSHRC) for the funding to conduct this research.

(de Reuver et al., 2018) and change how people work and receive services (Kenney & Zysman, 2016). The platform economy has spread over different sectors, with mobility and transportation among the most critical (Shaheen & Chan, 2016). Various business models were developed as rivals to traditional mobility solutions and new services. Platforms such as ride-sharing (Zwick & Spicer, 2018), car-sharing (Borgato, 2018), micromobility (i.e., bike and scooter), grocery (Wygonik & Goodchild, 2012), and food delivery (Vasista, 2020) have rapidly entered Canadian cities (Shaheen et al., 2018) (see also Rojas and Miller, chapter 5).

With different approaches, many governments and taxi associations worldwide have been putting obstacles in the way of shared mobility services such as Uber since their inception (Thelen, 2018). However, because of positive impacts such as reducing gas emissions (Martin & Shaheen, 2011), curbing traffic congestion (Yu et al., 2017), reducing private car ownership, and improving customer satisfaction (Liao et al., 2020), many governments around the world have become more aware of these platforms' importance (Smorto, 2020). Digital mobility solutions are now considered an integral part of economic reality. Most governments attempt to maximize opportunities while mitigating threats by enforcing specific rules and regulations (Shaheen et al., 2018). However, Canadian cities face ongoing challenges in creating policies and regulating digital mobility services (see Brail and Donald, chapter 10, and Zwick, Young, and Spicer, chapter 13). These problems include the safety and security of passengers (Thompson, 2017), the degradation of taxi licences, the reduction in municipality revenues, social issues (e.g., gender equality and discrimination), and insurance issues (Ashraf & Habib, 2020). Despite all these challenges, government acts such as Bill 17 in Quebec (Ministère des Transports du Québec, 2019) put technology-based mobility modes under an integrated umbrella of passenger transportation.

Chronologically ordered, this section traces the evolution of the digital mobility sector in the Montreal region over the last decade.

Evolution of Digital Mobility in Montreal

Driving without Owning a Car: The Evolution of Car-Sharing Platforms in Montreal

Research and practice have shown a meaningful correlation between traffic congestion and private car ownership (de Lorimier & El-Geneidy, 2012; Klincevicius et al., 2014). One of the first ideas to reduce private car ownership was "car-sharing," an innovation based on digital technology that lets people find, reserve, and drive cars without owning them (see Rojas and Miller, chapter 5) (Kashani & Trépanier, 2018). This innovation differs from traditional car rental services, which are concentrated in a few retail locations and require hefty customer paperwork for insurance and other matters. In brief, car-sharing platforms offer more flexible and usually more affordable services (Bieszczat & Schwieterman, 2012).

The province of Quebec has been a North American pioneer, introducing and welcoming this innovation when "Communauto" launched in Quebec City in 1994 (Sioui et al., 2012). One year after its services began in Quebec City, Communauto entered Montreal with its initial model based on "fixed stations." In this model, using Communauto's website or mobile app, passengers reserve a car parked in a specific location and return it to the same parking spot without the hassle of interacting directly with car rental agents (Communauto, 2022). As of October 2022, Communauto operates in seventeen Canadian cities and Paris with two models: "round-trip vehicles by reservation" and "one-way FLEX without reservation."

In 2013, Car2Go, a German car-sharing company as a subsidiary of Daimler AG, saw the Quebec government's policy on promoting electric transportation as an opportunity. It launched its network of fuel-efficient "smart cars" with a new operating model: "free-floating" cars (Lesteven & Godillon, 2020). In this model, passengers did not need to return the vehicles to the same station. Instead, they could park the "Fortwo" mini cars in any regular parking spots within the city as well as many other "permitted" municipal parking places. Car2Go managers negotiated these parking spots with City of Montreal officials, bringing Car2Go members more convenience and flexibility (Rotaris & Danielis, 2018).

Just a few months after Car2Go's entry into Montreal, Communauto also introduced its first free-floating model, "Auto-mobile," with more than 270 cars floating on Montreal streets, not parked in designated stations (Wielinski et al., 2017). As of 2022, Communauto has more than 50,000 cars and more than 450 stations in Montreal (Communauto, 2022). Among these 50,000 cars, 65 per cent are hybrid vehicles and 35 per cent are gas cars, a distribution that shows the company's attempt to move towards electrification and create an eco-friendly fleet.

Due to competitive pressures, profitability challenges, and issues with municipalities in many Canadian cities and North American cities overall, Car2Go abandoned Montreal in 2020 as one of its last locations in North America (CBC News, 2019; Wurst, 2019). This decision gave its local competitor, Communauto, more room to offer new products and collaborate with public transportation networks.

Business-to-consumers (B2C) car-sharing platforms such as Communauto and Car2Go have not been the only car-sharing initiatives in Montreal. A consumer-to-consumer (C2C) or peer-to-peer (P2P) model has also gained popularity recently, especially in more dispersed geographical locations. In this model, people or small businesses can advertise and rent their cars to others for short periods in exchange for an agreed fee (Schwieterman & Smith, 2020).

As the largest P2P car-sharing platform (Castellanos et al., 2022) in North America, Turo launched its services in Montreal in 2016 and, as of July 2019, Montreal ranked as the ninth-largest international market for Turo in terms of the number of trips. Domestic users from Montreal, Toronto, Quebec City, and Vancouver as well as international visitors from France, the US, Belgium, and Brazil make up the most frequent users of Turo in Montreal. The average trip using a Turo car lasts three days, which shows that Turo car-sharing better suits longer travels than other car-sharing platforms (Siu, 2019). As of November 2017, Turo-listed cars in the province of Quebec constituted 46 per cent of all Turo-listed vehicles in Canada. Turo attracts Montrealers, especially the city's young and student-centred population, with its flexible and competitive pricing options. As well, many international travellers to Montreal and Quebec City from European countries such as France choose Turo, perhaps because they are already familiar with P2P car-sharing platforms. Attracting international users is an essential penetration criterion.

An Alternative to "Solo Driving": Ride-Sharing through Carpooling Platforms

Solo driving has always been a major factor in traffic congestion, especially in densely populated areas (Shaheen & Chan, 2015). One way to reduce solo driving is carpooling: a "casual grouping of travellers" into the same private cars using online networks (Dinesh et al., 2021). In other words, carpooling connects people living or working in the same areas and commuting to the same destinations so that they can share a ride (Shaheen et al., 2018).

As a carpooling platform with a social approach, Netlift, a Montreal-based company, launched its services in 2012. It helped people to find carpooling opportunities using a website and later, in 2013, through the platform's mobile app (Netlift, 2022). With the ultimate goal of reducing gas emissions, Netlift bases its value on two critical factors: reducing the number of vehicles on the road and making more efficient use of parking lots (Ferraris, 2015). In 2018, Netlift shifted its business model towards a business-to-business (B2B) approach to help large organizations struggling with parking shortages for their employees (Newswire, 2019). Among these organizational clients, hospitals are Netlift's main target focus (Netlift, 2022).

With the arrival of Uber in Montreal, some taxi agencies thought about creating a similar platform for classic taxis to compete against the ride-hailing giant (Morency et al., 2015). In this context, Téo Taxi was born in 2015 with an emphasis on environmental issues and drivers' working conditions, two concerns that have gained Uber many critics (Lesteven & Godillon, 2020; Radio Canada, 2019a; Shingler, 2019). Téo Taxi's fleet was fully electric, and its five hundred drivers were paid by the hour, not by fare. They also received benefits, including paid vacation days (Shingler, 2019). Téo Taxi's parent, Taxelco, received $9.5 million in government grants, and the provincial government authorized a $4 million loan. The Caisse de dépôt et placement du Québec, the province's public pension fund, and Investissement Québec, a provincial business development fund, also injected cash in 2018 to help expand the fleet (Shingler, 2019).

Despite these local supports, Téo Taxi failed to compete with Uber and went bankrupt in 2019 (Radio-Canada, 2019b). Three factors seemed to play a significant role: problems with electric cars (notably the need to recharge them several times a day, especially in winter), the high cost of the fleet, and a socially responsible business model, which also increased labour costs (Shingler, 2019). After its bankruptcy, Téo Taxi was bought by billionaire Quebecor president and CEO Pierre Karl Péladeau. In 2021, Téo Taxi relaunched its services with more efficient electric vehicles that do not need frequent daily recharging (Kirkwood, 2020).

The traditional taxi sector in the Montreal area still exists and includes companies such as Taxi Atlas, Taxi Coop, Taxi Diamond, and Taxi Hochelaga (Bureau du taxi de Montréal, 2021). Due to the competition from the above-mentioned revolutionary digital ride-hailing services, all of these more traditional companies have modernized over time, improving vehicle cleanliness and aesthetics and adding an online app to follow the taxi's approach to the client. In addition, mobile applications now allow passengers to access taxi services more easily (Bureau du taxi de Montréal, 2021).

Moreover, Bill 17 has opened the door to new entrants, including local startups, to innovate in the ride-hailing sector. Eva, a startup founded by two young entrepreneurs with software engineering and political science educational backgrounds, launched a new ride-hailing app in 2019 based on a "cooperative" model where drivers are shareholders who participate in the company's management (Eva, 2022). According to its co-founders, Eva takes only 15 per cent of each ride's fee as a commission, while Uber takes 25 per cent, and at the end of the year, Eva shares any profit with all members. In addition, Eva says it uses blockchain technologies that do not store passengers' data on its servers, and therefore the company cannot sell this data to any third party (Spector, 2019).

Impacts of Digital Mobility Solutions in Montreal

In studies of car-sharing, previous research has proven that the number of vehicles on the streets has gone down (Klincevicius et al., 2014) and the need for parking spaces has decreased (Morency et al., 2015). In densely populated cities such as Montreal, car-sharing supports sustainable transportation (Sioui et al., 2012; Wielinski et al., 2017) and reduces both air pollution and traffic congestion. Car-sharing works best in areas where people can walk, cycle, share their journeys, and use public transport to access shared cars (Shaheen & Chan, 2015).

However, the impacts of ride-hailing platforms such as Uber in Montreal have seemed contradictory. Lesteven and Godillon (2020), through a discourse analysis of 354 articles published in local newspapers between 2014 and 2016, discovered that users perceive these services as more affordable, more accessible, more reliable, easier to use, and, in a word, more "modern" than traditional taxis, although the latter has evolved in the last years to face this new competition. This positive perception originates from using digital tools, electric payment solutions, and driver evaluations. However, from a social perspective, Uber is criticized for creating unfair competition against traditional taxi drivers and job insecurity for gig workers (Lesteven & Godillon, 2020). Furthermore, research done in the US showed the negative impacts of increased road congestion and a decline in transit ridership due to ride-hailing companies such as Uber (Diao et al., 2021). Overall, the research on the impacts of digital mobility on Montreal is thus inconclusive, and many questions remain.

Research Method

Due to a lack of adequate empirical research focused on Montreal's digital mobility ecosystem, we undertook an exploratory qualitative study. We selected interview participants in two steps. First, we identified individuals relevant to our research questions via LinkedIn, the professional networking platform, and then invited them to participate in interviews via the same platform or by e-mail. Finally, we used a snowball sampling technique to recruit the remaining participants.[2] This process allowed us to choose fifteen participants representing thirteen key players in the passenger transport sector in the Montreal area. Based on their role in the digital mobility ecosystem, these participants fall into two main categories: digital mobility service providers and digital mobility ecosystem administrators.

The service provider category, which includes actors offering ride-hailing, car-sharing, carpooling, and bike-sharing services in the Montreal area, consists of eight participants. The administrator category, which includes people who shape this mobility ecosystem in one way or another, consists of seven participants. More specifically, these actors are the government, represented by its two levels (the provincial government and the City of Montreal), and various associations and non-profit organizations (NPOs) whose mission is to influence the digital mobility sector.

In this study, we drew upon Strauss & Corbin's paradigm model (2008), a qualitative research framework known for its grounded theory approach. Grounded theory involves developing theories directly from the data rather than relying on pre-existing ones. The model consists of three stages: Open Coding, Axial Coding, and Selective Coding. Open Coding entails analysing data to identify concepts and themes without preconceptions. Axial Coding involves organizing codes into categories and exploring relationships. Lastly, Selective Coding is the process of developing a unified theory by refining core categories.

We conducted semi-structured interviews from January 2020 to October 2020. All interviews were conducted online due to the COVID-19 situation. After transcribing the interview data, we initiated the coding process using the Atlas-ti software. This process commenced with Open Coding, wherein we dissected the qualitative data into discrete parts and assigned codes for labelling. In total, 1257 open codes were meticulously recorded. Subsequently, we progressed to the integration of these open codes into axial codes, resulting in thirty-seven codes. These codes were further categorized into sixteen distinct groups, providing valuable insights into the primary subjects of our study: the impacts of digital mobility on Montreal, the interaction among various actors within the digital mobility ecosystem, and the contextual conditions and intervening factors.

---

2 For a researcher, snowball sampling is the process of asking the original participants to point out other people they think would be helpful and to link the two parties if possible.

This systematic approach allowed us to comprehend complex phenomena like digital mobility ecosystems effectively. By applying this research process, we created a comprehensive framework that can serve as a valuable resource for future scholars investigating digital mobility ecosystems.

Results and Discussion

Table 9.1 presents the comprehensive results of our rigorous qualitative data analysis. The first column displays the frequency of open codes, representing the number of interview quotations coded for each axial code. Moving to the second column, we provide three selected sources for each code, indicating also the interviewee's affiliation: GM (Government/Municipal entities), B (Businesses), and UA (Unions/Associations). For instance, GM1:15 indicates that the interviewee belonged to the first Government/Municipality authority interviewed, and the corresponding reference quotation number is 15.

In the third column, we introduce the axial codes, which are central themes that emerged from the data analysis. Concerning the impacts of digital mobility, we utilized the symbols +, -, and M to signify our observations based on the interview data. Specifically, we used "+" to denote subjects with positive impacts, "-", for negative impacts, and "M" when the arguments presented contradictory evidence, leading to "mixed" observations.

In the subsequent phase, in order to gain a deeper understanding of our findings, we organized all the identified categories and axial codes into a graphical framework (see Figure 9.1). This framework revolves around Montreal's digital mobility ecosystem. Through meticulous coding and thematic analysis, we uncovered four primary themes: 1) stakeholders of digital mobility platforms in Montreal, 2) impacts of digital mobility platforms on the stakeholders, 3) government policies, and 4) contextual conditions such as COVID-19. We elaborate on these findings below.

Stakeholders of Digital Mobility Platforms in Montreal

At the heart of any ecosystem are its actors or stakeholders. In a dynamic environment, actors interact with each other, bring input into the system, undertake processes, and collectively generate the ecosystem's output. In other words, the stakeholders' expectations, actions, and reactions together build an ecosystem. Identifying and exploring the specific features of any critical stakeholder is the first step towards understanding the whole ecosystem.

Table 9.1. Qualitative data analysis

Open codes (no.)	(Selected) Sources	Axial codes	Categories
5	B5:72; B8:21; UA3:15	Utilizing personal assets	Stakeholders: People: Gig workers
5	GM1:68; GM4:15; UA4:11	Work supplier	
19	GM1:30; B1:66; GM2:20	Demand services	Stakeholders: People: Passengers/users
13	GM1:51; GM3:26; UA3-28	Protest\Resistance	
51	UA4:36; B8:32; GM2:25	Compete	Stakeholders: People: Traditional taxi drivers
37	UA3:65; B7:52; GM2:18	Reform	
78	B1:10; GM2:20; B4:10	Business innovations	Stakeholders: Businesses: Ride-hailing
36	GM1:27; B3:16; UA1:28	Platform management	
7	B4:14; GM1:29; UA3:14	Job creation	Stakeholders: Businesses: Carpooling
13	B1:53; GM2:12; B4:36	Corporate social responsibilities (CSR)	Stakeholders: Businesses: Carsharing stakeholders: Businesses: Food/ grocery delivery Stakeholders: Businesses: Bike-sharing
111	GM1:20; B1:36; GM2:18	Regulating	
17	GM1:11; B1:51; B8:25	Building infrastructure	Stakeholders: Municipal entities and government
27	UA1:45; B1:26; GM1:26	Incentives and financial support	
32	B6:17; GM3:22; GM1:42	Planning	
26	UA4:36; UA3:24; GM4:21	Representation, lobbying	Stakeholders: Unions/ associations
64	B2:22; GM3:32; B4:32	Better services/ behaviours in the taxi/ car rental industries (+)	
39	GM1:28; B5:30; B7:36	Welfare (traditional taxi drivers, gig workers) (M)	

(Continued)

Table 9.1. *Continued*

Open codes (no.)	(Selected) Sources	Axial codes	Categories
13	B4:37; B7:50; UA2:15	Losing the sense of justice (taxi drivers) (-)	
17	B1:40; B3:56; B6:22	Job creation for the underprivileged (immigrants, students) (+)	
23	B1:23; GM3:58; B5:80	Public health, safety, and security (M)	Impacts: Social
17	UA1:39; B5:35; B7:42	Work-life balance (-)	
19	GM1:53; B4:10; B7:60	Labour Voice and working conditions (-)	
52	UA3:33; B8:71; UA2:37	Citizen satisfaction (+)	
32	B1:53; GM2:13; GM:24	Municipality revenue creation/loss (M)	
15	B1:13; GM2:9; B4:15	Job creation/loss (M)	
26	B3:49; B4:31; UA1:37	Private car ownership (M)	Impacts: Economic
73	GM1:14; B3:64; B4:31	Public transport usage (M)	
47	GM2:8; B7:24; B4:16	Competition and innovation in the taxi industry, insurance, etc. (+)	
15	GM2:15; B3:44; B4:20	Privacy and information security challenges (-)	
60	GM1:18; B1:45; B2:25	Innovations in areas such as artificial intelligence, big data, blockchain and cryptocurrencies (+)	Impacts: Technological
45	B1:21; B3:64; UA1:44	Investments in Canadian talents, startups or incubators (+)	
31	B2:29; B4:31; B5:28	Traffic congestion (M)	
56	GM2:24; B4:31; UA1:28	Gas emissions (M)	Impacts: Environmental
18	UA2:37; B5:72; B7:27	Parking/land issues (M)	
52	GM1:68; B2:29; UA1:34	COVID-19	Contextual conditions
5	GM4:40; B8:49; UA3:63	Diversity and demographic trends	
61	GM1:20; B1:25; GM2:7	Government policies (e.g., Bill 17)	Intervening Factors

Figure 9.1. The digital mobility ecosystem in Montreal: Stakeholders, impacts, context, and intervening conditions

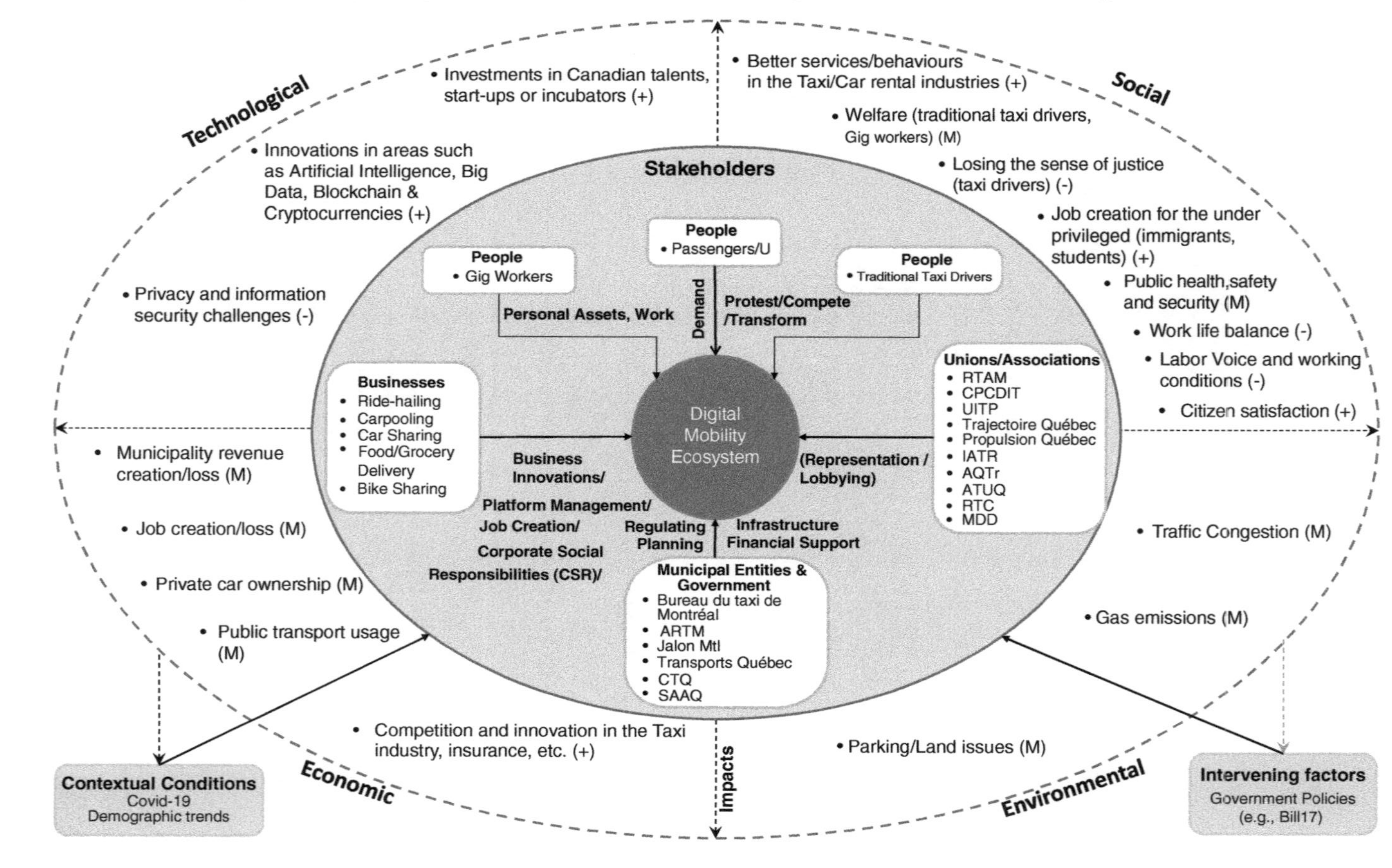

Our analysis explores four primary stakeholders: businesses, government and municipal entities, associations and unions, and people (passengers-users). We will first introduce each stakeholder and the leading players in each category. We will later explain their actions and interactions leading to impacts on the digital mobility ecosystem in Montreal.

BUSINESSES

In our research, we classified digital mobility platforms into five groups:

1. Ride-hailing (or ride-sharing) companies that connect passengers to drivers of privately owned cars (e.g., Uber, Eva) or formal taxi drivers (e.g., Téo Taxi)
2. Carpooling services, which are primarily non-profit, that connect people who live and work in close areas to each other to save commute and parking costs (e.g., Netlift)
3. Car-sharing platforms that connect people who need flexible rental cars to private car owners (e.g., Turo) or companies (e.g., Communauto, Car2Go)
4. Bike-sharing platforms (e.g., Bixi, Jump, Lime)
5. Food/Grocery online delivery (e.g., Uber Eats, DoorDash, SkipTheDishes)

In this chapter, we discuss only the first three categories.

Another way of classifying and studying digital mobility businesses is by these groups:

1. International players (e.g., Uber, Car2Go)
2. Local players (Communauto, Bixi, Téo Taxi, Netlift)
3. Traditional businesses that transformed into digital (e.g., Diamond Taxi)

This alternative classification helps us understand the local-global dynamics of digital mobility in Montreal.

The arrows in Figure 9.1 demonstrate some of the stakeholders' significant interactions, contributions, and influences on the ecosystem. Digital mobility companies offer four main contributions to the ecosystem: business innovations, platform management, job creation, and corporate social responsibility (CSR).

GOVERNMENT AND MUNICIPAL ENTITIES

Many digital mobility solutions were launched in Montreal before governing policies or regulations were in place. However, public

Table 9.2. Government and municipal entities in Quebec involved in Montreal's digital Mobility ecosystem

Row	Stakeholder	Role
1	Transport Québec (Ministry of Transportation)	Policymaking and regulating the transportation rules.
2	Commission des transports du Québec (Commission of transportation-operating under the ministry)	Authorizing and monitoring the transport system operators. Setting passenger fares unless the fare is calculated by technological means (e.g., mobile apps) and the customer has pre-admitted the maximum fare Inspecting and investigating drivers and cars.
3	Société de l'assurance automobile du Québec (Quebec Auto Insurance Company-operating under the ministry)	Authorizing official taxi drivers to run passenger cars. Regarding the drivers who work for transport system providers (i.e., ride-hailing companies), these companies would evaluate the drivers' eligibility for providing services to passengers according to criteria set by SAAQ. Authorizing all general driving licences is under the jurisdiction of SAAQ.
4	Bureau du taxi de Montréal (Montreal taxi office) – under Ville de Montréal	Previously, all responsibilities and provisions related to the Commission des transports du Québec and Société de l'assurance automobile du Québec were exclusively managed for Montreal city. However, an exception existed, whereby the regulation process was delegated to Ville de Montréal (Montreal municipality). A significant change occurred on 31 December 2022, when the Ville de Montréal ratified a decision to terminate the surveillance, control, and investigation activities of the Bureau du taxi de Montréal. This marked a notable shift in the authority and oversight of taxi-related matters within the city. Additionally, the MTQ (Ministère des Transports du Québec) further delegated the responsibilities and administrative activities of oversight and control of the TRPA (Taxi, Limousine, and other Passenger Transport) industry in Montreal to the Société de l'assurance automobile du Québec and Contrôle routier Québec, effective from 1 January 2023.

(Continued)

Table 9.2. *Continued*

Row	Stakeholder	Role
5	Autorité régionale de transport métropolitain (ARTM) (Regional metropolitan transport authority)	Planning, organizing, funding, and promoting public transit services for the metropolitan region of Montreal. Promoting the integration of services between the different modes of transport.
6	Jalon Mtl	Carrying out a wide range of activities that support manufacturers of various backgrounds working to expedite innovative processes into practice. Jalon Mtl is a non-profit organization founded in 2017 by the Montreal municipality. Formerly known as L'institut des transports électriques et intelligents, Jalon has since been supported by the provincial and federal governments.

organizations recognized their regulatory role in the digital mobility ecosystem when threats increased to specific stakeholders, such as traditional taxi drivers. Table 9.2 lists essential government entities in Montreal's digital mobility sector and their functions.

UNIONS/ASSOCIATIONS

Our research shows that many unions and associations are involved in the digital mobility ecosystem, representing some stakeholder or interest groups. These actors lobby with other stakeholders, mostly government organizations, to maximize their stakeholders' benefits and minimize their risks. The interactions of these organizations (which in many cases are not for profit) have made Montreal's digital mobility ecosystem diverse and dynamic. Table 9.3 lists some crucial unions and associations related to our research.

PEOPLE

Our research identified three groups of people as essential stakeholders in Montreal's digital mobility ecosystem: traditional taxi drivers, gig workers, and passengers/end users.

For many months, traditional taxi drivers protested against ride-sharing services, as they found their businesses at risk and lost the value of their taxi licences. When they realized that new platforms were becoming a reality in passenger-paid services, and when the government started to regulate this new sector, taxi drivers began competing

Table 9.3. Unions and associations related to digital mobility

Row	Union/association	Role
1	Regroupement des Travailleur Autonomes métallos (RTAM)	Represents taxi drivers from all over Quebec. It is the most important association of representation in transporting people and goods.
2	Comité provincial de concertation et de développement de l'industrie du taxi (CPCDIT)	Represents the interests of the taxi industry in general. It ensures consultation and development and enhances the industry's image.
3	International Association of Public Transport (UITP)	A non-profit advocacy organization that represents public transport authorities and operators, policy decision-makers, scientific institutes, and the public transport supply and service industry.
4	Trajectoire Québec	Represents citizens and public transit users, promoting their interests in sustainable mobility throughout Quebec.
5	Propulsion Québec	Conducts joint projects aimed at positioning Quebec as a global leader in developing and implementing smart and electric modes of ground transportation. It is financed by companies and city of Montreal.
6	International Association of Transportation Regulators (IATR)	A non-profit, professional association that represents government transportation officials.
7	Association québécoise des transports (AQTr)	Conducts tables of expertise, composed of volunteer specialists. These experts have a mandate to develop activities to disseminate expertise and to advise the association on the need for continuing education in transportation.
8	Association du transport urbain du Québec (ATUQ)	Represents the ten Quebec public transportation organizations serving the province's major urban centres. ATUQ considers itself a key player in integrated urban mobility in Quebec.
9	Maison du développement durable (MDD)	A group of eight socially and environmentally minded organizations that has created the Centre for Sustainable Development, which offers space for reflection, innovation, education, and the meeting of minds on sustainable development.

with digital mobility companies and even transforming their business models. While some traditional taxi drivers kept their prices fixed for some time, others realized that the taxi industry needed something more to compete. Some of these agencies launched mobile apps with similar features to Uber (e.g., Diamond Taxi, Taxi Coop). While these companies lack the resources of a giant company like Uber (e.g., customer base, international brand reputation, routing algorithms, Artificial Intelligence, and financial reserves), the modernization has made it possible for them to keep part of their local customer base.

Gig workers are essential stakeholders of digital mobility platforms (see Brail and Donald, chapter 10). They use their assets (cars) to provide services in this industry. Two contradictory viewpoints have emerged regarding the impact of digital mobility platforms on gig workers. The first perspective considers these platforms great job opportunities for people with underprivileged working conditions. This group includes students who have not yet completed their studies, new immigrants who may lack the necessary language skills to enter more professional job markets, and anyone needing temporary or part-time work to improve their financial situation (Tremblay et al., 2023). However, the second perspective criticizes these platforms for their lack of employee benefits and support. These critics see platform employment as especially problematic because the gig workers use their personal assets for work while the company does not share any profits or values with them, considering these workers as owners or partners. According to this perspective, platform providers exploit gig workers.

For the final group of stakeholders – passengers or end users – platforms provide a reliable and economical new service that generally satisfies their expectations. Although passenger surveys still reveal concerns about data privacy and security, most users express satisfaction with digital mobility. This general satisfaction has helped to push back opposing stakeholders and convince them of the advantages of digital mobility platforms.

Impacts of Digital Mobility Platforms on Stakeholders

We identified positive and negative effects of digital mobility platforms on Montreal's stakeholders and classified them into four types of impacts: economic, social, technological, and environmental. Of course, many of these effects overlap.

Figure 9.1 shows these significant impacts. We used "+" for positive effects, "-" for negative effects, and "M" for "mixed" results. We interpret mixed results in two ways. First, during our interviews, we realized that stakeholders view some implications differently. This disagreement shows that Montreal's digital mobility ecosystem has multiple stakeholders, each with different expectations and views on the results of the system. Second, for some effects, we found contradictory results in our research and even in the literature. To clarify these contradictions, future research should analyse these impacts in more depth.

ECONOMIC IMPACTS

Most stakeholders still consider the economic impacts of the digital mobility ecosystem to be most significant. Our research revealed concerns on various economic subjects: the economics of this ecosystem, specifically job creation or losses, the revenues of public agencies such as municipalities, the effects on public transportation use, the impacts on car ownership, and the competition among new companies and traditional players.

Analysing the digital mobility ecosystem's impact on labour, we observed two different effects. On one side, the ecosystem has created new jobs for some people; on the other, simultaneously, it has led to job losses for traditional service providers. However, some digital mobility companies offer specific services to established taxi drivers and car rental companies. Moreover, many platform mobility employees work in administrative jobs, call centres, and newly formed "data science" groups. Considering the launch of many startups over the past years, we conclude that job creation helps to outweigh job loss, although the changing economic context and development of working from home in the post-pandemic context have had impacts on the sector.

Analysing the mutual effects of this ecosystem on the public sector, government, and municipalities, we must address the economic impacts created due to the new mobility ecosystem. Our research shows that while the municipality lost some of its revenues from traditional models such as licences, it has gained some revenues from new models such as parking facilities for car-sharing companies and taxes from digital mobility companies. Understanding this shift in revenues for municipalities is critical to understanding related policies in this ecosystem. Public agencies also benefit from issuing permits for car-sharing companies. So, we can consider this impact on the public sector a change in revenue models, not necessarily a loss.

As we investigated the impact of digital mobility on the use of public transportation, we discovered a debate in the literature, and our research also yielded contradictory results. The Bixi bike-sharing platform's partnership with Montreal's public transportation network, STM, positively affected public transportation use. We observed no significant correlation between car-sharing and the use of public transportation. However, ride-sharing and carpooling appear to hurt the demand for public transportation. Yet, this decrease in demand for public transportation would also decrease the city's long-term capital investments. Overall, we don't know yet whether the cost saving from shifts towards carpooling/ride-sharing would outweigh the costs incurred by the increase in the amount of carbon emissions. Future studies should evaluate the carbon emission impact of decrease in demand for public transportation due to use of ride-sharing and carpooling services, but also due to telework, in order to provide a clearer picture for policymakers

Although public transportation seeks to provide sustainable, affordable, and environmentally friendly access to all areas of a city, private cars appear to still be essential for many in Montreal. Based on the Canadian lifestyle and landscapes, many individuals and families still find a car necessary, especially in the suburbs. Like many previous studies, our research showed that digital mobility platforms decreased private car ownership, especially in metropolitan and densely populated areas. In a different light, car-sharing has helped many low-income populations in Montreal to meet their mobility needs. For those without a car who live in urban cores, the opportunity to hire a car enhances the quality of life, as this would give them extra access to other areas of the city for leisure and shopping purposes.

In another meaningful economic impact, the new ecosystem is creating competition and innovation in traditional sectors (e.g., the taxi industry, insurance, and car rental). By encouraging traditional businesses to innovate and compete, digital mobility companies have positively affected urban transport, especially for end users. Although Uber dominates the ride-sharing industry in Montreal, new companies can compete with it and provide better services. This competition is especially visible in food delivery services, where companies such as DoorDash have a considerable share, taking more of a foothold during COVID (see Brail and Donald, chapter 10).

SOCIAL IMPACTS

The digital mobility ecosystem has changed businesses' and individuals' norms, values, preferences, and lifestyles. Therefore, it should be considered an essential driver of social and institutional change. Several social impacts have been explored in our research and are depicted in Figure 9.1.

Better services/behaviours in the taxi/car rental industries: As new digital platforms have arrived in Montreal, competition has risen. Thus, drivers and companies have been encouraged to provide better services. Otherwise, the intense competition would eliminate weaker players from the market.

Welfare (traditional taxi drivers, gig workers): The new digital mobility platforms have also provided new opportunities for specific groups (such as immigrants, young individuals, and students) who before could not as easily participate. This is because these platforms offer flexible work schedules, leading to job satisfaction for many gig workers. As a part-time activity, these jobs can also provide low-income employees with more revenue streams (Tremblay et al., 2023). On the other hand, traditional taxi drivers now find their licences less valuable, a change that has significantly affected their income and quality of life. Every day, a growing number of customers choose digital platforms because they are more convenient, affordable, and user-friendly. Taxi drivers now perceive what they consider unjust competition and less justice due to these disruptive innovations. In Montreal, traditional taxi drivers are no longer essential players in the new mobility ecosystem.

In addition, big companies that have invested in this ecosystem have essential resources, and they can impact the working conditions of gig workers because of the high labour supply. In contrast, drivers do not have a strong position. As a result, companies can exploit the drivers. An officer of a government institution explains this problem:

> Those are substantial multinational companies; corporations only aim to squeeze the last drop of every lemon. Their objective is essentially to develop their business. I think Uber spends two out of three of its annual budget on developing autonomous vehicles because they want to get rid of its higher costs right now. However, they pay less to drivers.

Digital platforms also raise safety concerns. For example, ride-sharing drivers receive less supervision than traditional taxi drivers, who can have their licence revoked for behavioural or legal violations. In addition, regarding public health, ride-sharing was perceived as a way of spreading the virus during the pandemic.

Another safety concern is work-life balance. Since the digital platform is more flexible in terms of working hours and locations, some low-income groups may participate in these platforms overwhelmingly, which can have a negative impact on their family life, although some workers see this flexibility as positive for work-life balance (Tremblay et al., 2023). This issue might also increase drivers' and passengers' health risks.

Ultimately, passengers experience the most positive social impacts from digital mobility platforms. End users are benefiting from reliable and low-cost mobility solutions that were not accessible before. As well, the ongoing competition between traditional and new riding players has led to innovations, better-quality services, and greater customer satisfaction. A government official acknowledges the platforms' satisfactory performance: "Every year, we have 600 examiners who get into cars and do an evaluation of the services, and this year companies were evaluated at close to 95 per cent satisfaction, which is an amazing score compared to what we had a couple of years ago."

TECHNOLOGICAL IMPACTS

The digital mobility ecosystem, emerging from digital transformation and tech revolution, has vast technological impacts. We have categorized these impacts into three areas: investments in Canadian talents, startups, or incubators; innovations in Artificial Intelligence, big data, blockchain, and cryptocurrencies; and privacy and information security challenges.

New platforms are all tech-based and require investment in different areas. This investment will lead to a foundation for new startups and an opportunity for talent to work in this ecosystem. Many companies offer internships, job possibilities, and advanced training opportunities that significantly impact Montreal's human capital. Eventually, more skilled labour can boost economic performance for the region.

Due to high competitive pressures in this ecosystem and the rising demand for innovation, businesses should invest in new technologies that will bring positive benefits to other economic ecosystems, such as the insurance and banking sectors. On the one hand, Montreal has provided many facilities for tech companies and presents itself as an AI hub in Canada. On the other hand, technological advances have caused many concerns. For example, platforms collect a vast amount of data from drivers and customers; they can analyse this big data and use the results in an abusive way. Companies may sell this information to other companies for business purposes or even trace the citizens and spy on them for governments. This exploitation may harm the privacy of individuals and their freedom.

ENVIRONMENTAL IMPACTS

The digital mobility ecosystem has environmental impacts in Montreal, and we have explored them in three categories: traffic congestion, gas emissions, and parking/land issues. However, evaluating environmental impacts is complicated because of the concurrent positive and negative impacts as well as the difficulty in finding reliable data and respondents on these issues, at least for the moment.

Some digital mobility modes (e.g., car-sharing) proved to have positive impacts on traffic congestion, while others (e.g., ride-hailing) showed adverse effects. For instance, as an officer of a carsharing company states:

> We had a study done by Barclay university, and Montreal city was included, and the takeaway was that one carsharing was able to withdraw ten cars from the street. From my experience, it was more than ten after the study. It was in 2013. I am a good example; I purchased a car when [name of a carsharing company] shut down.

To add to environmental effects, our findings also emphasize gas emissions as one of the critical impacts of digital mobility platforms. However, we do not have results on this, and quantitative research is needed on this issue. We can conclude that the variety of available transportation options can help avert the climate crisis, but it is not clear to what extent. More transportation options can mean fewer cars, which is positive for the environment, but it can also reduce public transportation (bus, metro), which would be negative. In addition, the new ecosystem can employ environmentally friendly vehicles like electric cars. In terms of emissions and carbon footprint, because many exogenous conditions affect these environmental measurements (production of electric batteries. for example), we cannot estimate the ecosystem's individual impact.

Furthermore, one of the intentions of all digital mobility platforms has been to reduce the need for parking in downtown areas. However, our research had mixed findings on this issue. First, car-sharing platforms must still negotiate defined parking spaces with the City of Montreal. As well, as our interviews revealed, there are still many real-estate and land-use issues related to digital mobility platforms in Montreal. Nonetheless, ride-hailing and carpooling services can save people time and money as passengers will spend less on parking.

Conclusion

Despite protests, obstacles, and challenges, Quebec has been a pioneer in embracing many digital mobility innovations in Canada. Communauto, one of the first car-sharing platforms in North America, appeared in Montreal in 1995. It now helps people in Quebec and some other Canadian provinces to use its fleet of hundreds of cars parked on city streets for flexible car rentals in order to reduce car ownership and traffic congestion. Uber, the giant American ride-hailing company, launched its services in Montreal in 2014. Despite severe opposition from taxi unions and many political leaders, it found its way and is still a significant mobility player in Montreal.

Furthermore, many smaller local players have attempted to enter the digital mobility arena in Montreal. For example, Téo Taxi launched its electric-cab services, driven by professional taxi drivers, in 2015, went bankrupt in 2019, and returned to the streets in 2021 with new ownership and a new operational model. Another local player, Netlift, began a carpooling service in 2012 to connect commuters from the same neighbourhoods to downtown or other nearby working areas. Eva introduced its democratized ride-hailing platform using blockchain technologies in 2019, intending to share value with the original service providers – the drivers who use their cars to transfer passengers.

As a public bike-sharing platform, Bixi was first established in 2009, and after it filed for bankruptcy in 2014, the City of Montreal took it over and transformed it. It is one of the best examples of active mobility modes in North America, offering many flexible programs to connect people to the public transport network.

Since 2020, and with COVID-19, lockdowns, and social-distancing policies, some other companies such as DoorDash, SkipTheDishes, and Uber Eats have developed their online food and grocery ordering/delivery services faster than ever before (see Brail and Donald, chapter 10).

Many other players shape Montreal's digital mobility ecosystem and are affected by its evolution. With various demographic and socio-economic backgrounds and needs, "gig workers," mainly students and immigrants, contribute their assets and time while they lack average employment benefits. Traditional taxi drivers have lost the value of their taxi licences and a considerable portion of their revenues. End users (passengers) have benefited from better-quality services but have sometimes experienced a lack of safety or loss of personal data.

With protests and other concerns rising, the provincial government and the City of Montreal have tried to play a more active role in regulating the newly emerged industry (see Zwick, Young, and Spicer, chapter 13). The Quebec Ministry of Transportation, which launched its first version of a bill governing the activities of "online paid passenger transport" in 2019, is still trying to evaluate the bill's impacts and improve it. Many unions and associations also represent their stakeholders and contribute to the digital mobility ecosystem.

In our research, we tried to answer three primary questions:

1 Who are the stakeholders who are contributing to the evolution of Montreal's digital mobility ecosystem?
2 What is their contribution to the ecosystem?
3 How is this disruptive evolution affecting them?

We see our work as documenting the diverse stakeholders involved in Montreal's digital mobility ecosystem. This documentation is crucial because it reveals the dynamics of stakeholders now engaged in trying to "fix" transportation infrastructures in Montreal. This research shows that some approaches have worked, and others have failed. There does not seem to be a silver bullet for fixing transportation woes in the city. Governments have invested significant money into failed companies, drivers and unions have resisted new ideas and technologies, and companies have come and gone – some local and some global. Nevertheless, no group has come forward with a comprehensive plan to address systemic mobility challenges in Montreal. This is the first step in recommending policy changes for a better mobility future.

Previous research on the impacts of digital mobility platforms on cities has been contradictory and inconclusive. Our exploratory study sheds some light on the reasons behind these contradictory perspectives. Measuring the effects of digital mobility platforms is difficult because these platforms have multiple stakeholders with different and even opposing expectations. Sometimes, what makes customers happy (lower prices, for example) is precisely what makes workers unhappy (less income and thus lower quality of life). Each stakeholder plays a specific role in the ecosystem, and the ecosystem generates different impacts that we have categorized as economic, social, technological, and environmental. More research is needed to further examine our research findings, especially considering multiple stakeholders and different units of analysis (e.g., individuals and organizations).

Our research revealed that two factors significantly impact Montreal's digital mobility stakeholders: the interactions between stakeholders and the ecosystem's effects on them. First, government policies such as Bill 17 intervene in many of the dynamic processes in the system, and second, contextual conditions such as COVID-19 have created numerous opportunities and threats. Although we could not go into detail here on these two essential factors, we will address them in future research publications.

In a nutshell, our research shows why Montreal has improved its mobility system – and continues to do so – because there is a continuous effort on the part of the ecosystem (e.g., governments [local, provincial], Quebec-based businesses, global actors, labour unions, etc.) to make the transition work for as many people as possible who live in, work in, and visit the city. While we analyse digital mobility, we also need to consider the constant evolution of transportation, with the new public initiative of the region's Réseau express métropolitain (REM). The REM and its future extensions will function twenty hours a day, seven days a week, and will therefore clearly have impacts on Montreal's urban mobility systems.

REFERENCES

Ashraf, A., & Habib, M.A. (2020). A review of regulations and media discourse on technology-enabled shared mobility in Canada. *Transportation Research Procedia, 48*, 2757–70. 1. https://doi.org/10.1016/j.trpro.2020.08.241.

Berger, T., Frey, C.B., Levin, G., & Danda, S.R. (2019, July). Uber happy? Work and wellbeing in the 'Gig Economy.' *Economic Policy, 34*(99), 429–77. Retrieved from https://doi.org/10.1093/epolic/eiz007.

Bieszczat, A., & Schwieterman, J. (2012). Carsharing – Review of its public benefits and level of taxation. *Transportation Research Record: Journal of the Transportation Research Board, 2319*(1), 105–12. Retrieved from https://doi.org/10.3141/2319-12.

Borgato, S. (2018). *The role of carsharing in urban mobility: Relationship with human, spatial, and modal features in Metro Vancouver* [Master's thesis, Simon Fraser University]. Retrieved from https://doi.org/10.13140/RG.2.2.23026.50886.

Bureau du taxi de Montréal. (2021). *Programme clients-mystères: L'industrie du taxi à Montréal se surpasse cette année encore!* ISSUU. Retrieved 12 October 2022 from https://issuu.com/bureaudutaxi/docs/jlt_2021_printemps_vol40_no1_site_web/s/11881741.

Castellanos, S., Grant-Muller, S., & Wright, K. (2022). Technology, transport, and the sharing economy: Towards a working taxonomy for shared mobility. *Transport Reviews, 42*(3), 318–36. Retrieved from https://doi.org/10.1080/01441647.2021.1968976.

CBC News. (2019, 18 December). *Car2Go to shut down in Montreal – and across North America*. CBC News. Retrieved 12 October 2022 from https://www.cbc.ca/news/canada/montreal/car2go-montreal-north-america-1.5401130.

Communauto. (2022). *Who are we?* Retrieved 12 October 2022, from https://communauto.com/qui/index.html.

Darchen, S., & Tremblay, D.G. (2010, August). What attracts and retains knowledge workers/students: The quality of place or career opportunities? The cases of Montreal and Ottawa. *Cities, 27*(4), 225–33. https://doi.org/10.1016/j.cities.2009.12.009.

de Lorimier, A., & El-Geneidy, A.M. (2012). Understanding the factors affecting vehicle usage and availability in carsharing networks: A case study of Communauto carsharing system from Montreal, Canada. *International Journal of Sustainable Transportation, 7*(1), 35–51. Retrieved from https://doi.org/10.1080/15568318.2012.660104.

de Reuver, M., Sørensen, C., & Basole, R.C. (2018). The digital platform: A research agenda. *Journal of Information Technology*, *33*(2), 124–35. Retrieved from https://doi.org/10.1057/s41265-016-0033-3.

Diao, M., Kong, H., & Zhao, J. (2021). Impacts of transportation network companies on urban mobility. *Nature Sustainability*, *4*(6), 494–500. Retrieved from https://doi.org/10.1038/s41893-020-00678-z.

Dinesh, S., Rejikumar, G., & Sisodia, G.S. (2021). An empirical investigation into carpooling behaviour for sustainability. *Transportation Research Part F: Traffic Psychology and Behaviour*, *77*, 181–96. https://doi.org/10.1016/j.trf.2021.01.005.

Eva. (2022). *Comment avoir Eva dans sa ville?* Retrieved from https://eva.coop/#/faq.

Ferraris, F.S.G. (2015, 7 December). Au temps du covoiturage 2.0. *Le devoir*. Retrieved 12 October 2022 from https://www.ledevoir.com/societe/transports-urbanisme/457237/au-temps-du-covoiturage-2-0.

Kashani, H.B., & Trépanier, M. (2018). A typology of carsharing customers in Montreal based on large-scale behavioural dataset. *Technical Report, CIRRELT-2018-16*. Retrieved from https://publications.polymtl.ca/39751/.

Kenney, M., & Zysman, J. (2016, Spring). The rise of the platform economy. *Issues in Science and Technology*, 32(3). Retrieved from https://issues.org/rise-platform-economy-big-data-work.

Kirkwood, I. (2020, 21 October). *Téo Taxi returns to Quebec more than a year after bankruptcy*. BetaKit. Retrieved 4 March 2024 from https://betakit.com/teo-taxi-returns-to-quebec-more-than-a-year-after-bankruptcy/.

Klincevicius, M.G.Y., Morency, C., & Trépanier, M. (2014). Assessing impact of carsharing on household car ownership in Montreal, Quebec, Canada. *Transportation Research Record*, *2416*(1). Retrieved from https://doi.org/10.3141/2416-06.

Lesteven, G., & Godillon, S. (2020). Fuelling the controversy on Uber's arrival: A comparative media analysis of Paris and Montreal. *Cities*, *106*, Article 102864. Retrieved from https://doi.org/10.1016/j.cities.2020.102864.

Liao, F., Molin, E., Timmermans, H., & van Wee, B. (2020). Carsharing: The impact of system characteristics on its potential to replace private car trips and reduce car ownership. *Transportation*, *47*(2), 935–70. Retrieved from https://doi.org/10.1007/s11116-018-9929-9.

Martin, EW., & Shaheen, S.A. (2011). Greenhouse gas emission impacts of carsharing in North America. *IEEE Transactions on Intelligent Transportation Systems*, *12*(4). Retrieved from https://doi.org/10.1109/TITS.2011.2158539.

Ministère des Transports du Québec (2019). *Modernisation de l'industrie du taxi – Le ministre François Bonnardel souligne l'adoption du projet de loi no 17*. Retrieved from https://www.quebec.ca/nouvelles/actualites/details/r-e-p-r-i-s-e-modernisation-de-lindustrie-du-taxi-le-ministre-francois-bonnardel-souligne-ladoption-du-projet-de-loi-no-17.

Morency, C., Verreault, H., & Demers, M. (2015). Identification of the minimum size of the shared-car fleet required to satisfy car-driving trips in Montreal. *Transportation*, *42*(3), 435–47. Retrieved from https://doi.org/10.1007/s11116-015-9605-2.

Netlift. (2022). *Better use of vehicles for a better mobility*. Retrieved 12 October 2022, from https://www.netlift.me/en/about.

Newswire. (2019, 22 November). *A world first: A technology project developed in Montreal that combines carpooling and street parking*. Retrieved 12 October 2022 from https://www.newswire.ca/news-releases/a-world-first-a-technology-project-developed-in-montreal-that-combines-carpooling-and-street-parking-892850953.html.

Radio-Canada. (2019a, January). *Téo Taxi disparaît de la circulation, tous les chauffeurs licenciés*. Retrieved 12 October 2022 from https://ici.radio-canada.ca/nouvelle/1149599/teo-taxi-fermeture.

Radio-Canada. (2019b, June). *The eternal problem of parking*. Retrieved 12 October 2022 from https://ici.radio-canada.ca/ohdio/premiere/emissions/le-15-18/segments/chronique/123400/mobilite-transport-urbanisme.

Rotaris, L., & Danielis, R. (2018, September). The role for carsharing in medium to small-sized towns and in less-densely populated rural areas. *Transportation Research Part A: Policy and Practice*, *115*, 49–62. Retrieved from https://doi.org/10.1016/j.tra.2017.07.006.

Schwieterman, J.P., & Smith, C.S. (2020). Estimating the earnings from peer-to-peer carsharing for vehicle owners on the Turo platform using anonymized data. *Transportation Research Record*, *2674*(9), 256–65. https://doi.org/10.1177/0361198120928341.

Shaheen, S.A., & Chan, N.D. (2015). Evolution of e-mobility in carsharing business models. *Part of the Lecture Notes in Mobility book series (LNMOB)*. Retrieved from https://doi.org/10.1007/978-3-319-12244-1_10.

Shaheen, S., & Chan, N. (2016). Mobility and the sharing economy: Potential to facilitate the first- and last-mile public transit connections. *Built Environment*, *42*(4), 573–88. Retrieved from https://doi.org/10.2148/benv.42.4.573.

Shaheen, S., Cohen, A., & Bayen, A. (2018). *The benefits of carpooling*. Institute of Transportation Studies. https://doi.org/10.7922/G2DZ06GF.

Shaheen, S., Martin, E., & Bansal, A. (2018). *Peer-to-peer (P2P) carsharing: Understanding early markets, social dynamics, and behavioral impacts*. Institute of Transportation Studies. https://doi.org/10.7922/G2FN14BD.

Shingler, B. (2019, 30 January). Why Téo Taxi's socially conscious business model may have been its undoing. *CBC News*. Montreal. Retrieved 12 October 2022 from https://www.cbc.ca/news/canada/montreal/teo-taxi-quebec-1.4997440.

Sioui, L., Morency, C., & Trépanier, M. (2012). How carsharing affects the travel behavior of households: A case study of Montréal, Canada. *International Journal of Sustainable Transportation*, *7*(1), 52–69. Retrieved from https://doi.org/10.1080/15568318.2012.660109.

Siu, Rachel. (2019, 22 July). *2019 Turo market guide: Montreal*. Retrieved 12 October 2022 from https://turo.com/blog/community/turo-market-guide-montreal-2.

Smorto, G. (2020). Regulating and Deregulating Sharing Mobility in Europe. In G. Smorto, G. & I. Vinci (Eds.), The role of sharing mobility in contemporary cities (pp. 13–33). UNIPA Springer Series. Springer. Retrieved from https://doi.org/10.1007/978-3-030-57725-4_2.

Spector, D. (2019, May 24). New Montreal-made Uber competitor now accepting clients and drivers. *Global News*. Montreal. Retrieved 12 October 2022 from https://globalnews.ca/news/5311885/new-montreal-made-uber-competitor-now-accepting-clients-and-drivers.

Statistics Canada. (2021). *Montreal (Census Metropolitan Area), Quebec*. Retrieved 12 October 2022 from https://www150.statcan.gc.ca/n1/fr/geo?geocode=S0503462&geotext=Montr%C3%A9al%20%5BR%C3%A9gion%20m%C3%A9tropolitaine%20de%20recensement%5D,%20Qu%C3%A9bec&trier=releasedate.

Strauss, A., & Corbin, J. (2008). *Basics of qualitative research: Techniques and procedures for developing grounded theory*, 3rd ed. Sage Publications Inc. https://doi.org/10.4135/9781452230153.

Thelen, K. (2018). Regulating uber: The politics of the platform economy in Europe and the United States. *Perspectives on Politics*, *16*(4), 938–53. Retrieved from https://doi.org/10.1017/S1537592718001081.

Thompson, A. (2017). *Evaluating for-profit ridesharing regulations in Canada* [Master's thesis]. University of Calgary. 2. Retrieved from http://hdl.handle.net/1880/106819.

Tremblay, D., Yagoubi, A., St-Hilaire, J., & Motaghi, H. (2023, forthcoming). Les enjeux de la conciliation famille-travail-études: quels soutiens pour les parents étudiants?'. In M. Longo & et al. (Eds.), *Le travail des jeunes au XXIe siècle. État de la situation et nouveaux enjeux au Québec et au Canada*. Presse de l'Université Laval.

Vasista, B. (2020). *A study of sharing economy: Impact of sharing economy in mobility, food delivery, co-working and short-term rental, on built environment* [Master's thesis, Politecnico di Milano]. https://doi.org/10.13140/RG.2.2.23026.50886.

Vultur, M., Enel, L., Barette, L.-P., & Viviers, S. (2022). *Les travailleurs des plateformes numériques de transport de personnes et de livraison de repas au Québec: profil et motivations*. Retrieved from https://www.cirano.qc.ca/fr/sommaires/2022s-15.

Wielinski, G., Trépanier, M., & Morency, C. (2017). Electric and hybrid car use in a free-floating carsharing system. *International Journal of Sustainable Transportation*, *11*(3), 161–9. Retrieved from https://doi.org/10.1080/15568318.2016.1220653.

Wurst, B. (2019). Car2Go shutting down operations. *Montreal Times*. Retrieved 12 October 2022 from https://mtltimes.ca/montreal/car2go-shutting-down-operations/.

Wygonik, E., & Goodchild, A. (2012). Evaluating the efficacy of shared-use vehicles for reducing greenhouse gas emissions: A US case study of grocery delivery. *Journal of the Transportation Research Forum, 51*(2), 115–26. Retrieved from https://doi.org/10.5399/osu/jtrf.51.2.2926.

Yu, B., Ma, Y., Xue, M., Tang, B., Wang, B., Yan, J., & Wei, Y.-M. (2017). Environmental benefits from ridesharing: A case of Beijing. *Applied Energy, 191*, 141–52. https://doi.org/10.1016/j.apenergy.2017.01.052.

Zwick, A., & Spicer, Z. (2018). Good or bad? Ridesharing's impact on Canadian cities. *The Canadian Geographer/Le Géographe Canadien, 62*(4), 430–6 Retrieved from https://www.researchgate.net/publication/326132759_Good_or_bad_Ridesharing%27s_impact_on_Canadian_cities.

10 Examining Pandemic Pivots: From Ride-Hailing to Food-Hailing

SHAUNA BRAIL AND BETSY DONALD

Introduction: Mobility, COVID-19, and Digital Platforms

When COVID-19 shutdowns began in spring 2020, mobility changed dramatically. In response to these changing mobility patterns, some observers debated the future of digital platform firms, questioning whether these firms – premised on concentration, mobility, and exchange – could survive the pandemic (Goldstein, 2020). In practice, however, digital platform firms, including those that provided personal mobility services, demonstrated an ability to pivot and adapt. This chapter highlights the ways in which digital platforms adjusted to sudden and protracted urban mobility shifts precipitated by COVID-19, with an emphasis on the pivot from ride-hailing to food-hailing.

Changes in mobility became stark indicators of cities in lockdown. As city after city enacted physical distancing interventions to reduce the transmission of COVID-19 and protect the health care system, a range of metrics pointed to changes in mobility. Apple Mobility reports, which measure mobility based on cell-phone searches for directions, indicated that driving, walking, and transit all experienced massive decline in cities like London, New York, Tokyo, and Toronto in March and April 2020. In New York City, ride-hailing trips fell dramatically in the earliest stages of COVID-19 (Chang & Miranda-Moreno, 2022). Similarly, a City of Toronto analysis of ride-hailing trip data demonstrated that in March 2020, daily ride-hailing trips dropped from 199,000 to 31,000 (City of Toronto, 2021). With drops like these, it makes sense that observers questioned ride-hailing's prospects in cities (for policy responses to ride-hailing regulations during COVID-19, see Zwick, Young, and Spicer, chapter 13). Could ride-hailing survive?

At the same time, cities experienced a noticeable uptick in food delivery. As restrictions limited personal mobility and restaurants closed to indoor and outdoor dining, people wanted to order in. Platforms that provided digital menu, payment, and delivery services connecting restaurants with both delivery drivers and customers grew substantially. While restaurants struggled to adapt to pandemic-related restrictions, on-demand food delivery (ODFD) platforms operating in Canadian cities reported their highest-ever volumes of customer orders. Because the slowdown in ride-hailing relates to the increase in ODFD, we need to understand the characteristics of ODFD and the implications for cities, restaurants, delivery workers, and digital platform firms.

In the remainder of this chapter, we examine the pivot from ride-hailing to food-hailing during the pandemic, and at the same time, we raise new questions about the role of digital platforms in cities. In the next section, we review scholarship and debate on the urban impacts of digital platforms with respect to ride-hailing and ODFD. Through a case study of digital platforms and ODFD, we propose three lenses through which to interpret the implications for twenty-first-century cities: governance, pivots, and hacks. We conclude by discussing how these findings influence our understanding of cities, mobility, and urban innovation.

Urban Impacts of Digital Platforms: Ride-Hailing and On-Demand Food Delivery

Ride-Hailing

Before the pandemic, a rich body of scholarship on ride-hailing as a digital platform economy activity took shape. Urban economic geographers and other scholars studied the urban implications of ride-hailing as a form of mobility, including the need for governance and ride-hailing's interactions with urban form (Brail, 2018; Flores & Rayle, 2017). They also investigated impacts and repercussions related to labour, race and gender, power, political economy, and surveillance technologies (Attoh et al., 2019; Kim et al., 2019). Finally, some scholars examined the interrelationships between ride-hailing and public transit (Hall et al., 2018).

Early studies of ride-hailing regulations focused mainly on whether or not cities should permit this form of transportation in jurisdictions around the world (Ash et al., 2018; Flores & Rayle, 2017; Ong, 2017; Pelzer et al., 2019; Ranchordas, 2015; Spicer et al., 2019). The first qualitative studies on ride-hailing regulations in North America emerged out of transportation research. These studies, primarily descriptive, provided details about the number of ride-hailing vehicles operating in a municipality, the number of background checks on drivers, and the number of ride-hailing trips within a city (Beer et al., 2017). Many studies emphasized modelling scenarios and travel behaviour, seeking to inform ride-hailing regulation at the urban level by identifying challenges and opportunities for future policies. For example, some researchers promoted pooled ride-hailing, analysed efforts to reduce vehicle miles travelled, and highlighted connections between ride-hailing and transit ridership (Clewlow & Mishra, 2017; Tirachini & del Río, 2019; Young et al., 2020).

While some scholars have examined the mobility implications of ride-hailing, others have considered the activity's political economy, situating it within the broader digital platform economy and concerns about the excessive power of digital platform firms (Attoh et al., 2019; Graham, 2020; Kenney & Zysman, 2016). Another stream of research emphasized policy-oriented outcomes associated with political decision-making and policy development (Brail, 2018; Flores & Rayle, 2017; Spicer et al., 2019). Graham (2020) argued, and we concur, that further research is needed to understand how platform firms, unconstrained by geography, are subject to local regulations.

Last, urban planning and policy scholars have focused on the development and deployment of mobility experiments, pilot projects, and new governance processes. In particular, these studies investigate how such initiatives might connect cities to the innovation economy and help them to govern emerging mobility technologies (Aoyama & Alvarez León, 2021; Dowling & McGuirk, 2020; Dudley et al., 2017). Pilot programs and other forms of policy experimentation have an allure that attracts "like" places grappling with similar challenges. As Temenos and McCann (2013) suggest, a city that creates a popular (i.e., transferable) policy model earns a reputation for policy leadership in that particular field. Where policy development is concerned, however, policy mobility debates encourage exercising caution around the notion that municipal governments can succeed in their political endeavours by mimicking one another's policies (Temenos & McCann, 2012). While city governments may be under pressure to deal with digital platform services in an urgent manner, developing responsive, place-based policies requires exercising caution while exploring the opportunity to learn from, experiment with, or adapt another city's approach.

Food-Hailing

Until recently, the delivery of prepared foods consisted mainly of fast-food items such as pizza provided by restaurants that coordinated their own delivery (Allen et al., 2021). In the years leading up to the pandemic, however, consumer behaviour shifted substantially. People began to consume and spend more on prepared food, boosting restaurant sales and delivery orders (CBRE Research, 2019a, 2019b).

In 2016, the traditional model of food delivery made up approximately 90 per cent of the market. A customer placed an order directly with a restaurant by phone or online and then relied on the restaurant to deliver the food (Hirschberg et al., 2016). Furthermore, while the restaurant sector was growing by about 3 to 4 per cent annually in the United States before 2020, delivery sales were increasing by 7 to 8 per cent each year (Ahuja et al., 2021). In Canada, current estimates suggest that most customers still order and receive prepared meals directly from restaurants. However, as people began to use food delivery apps, the proportion of revenues from ODFD platforms grew appreciably between 2017 and 2022 (Statista Market Forecast, 2022).

ODFD platforms arose in the early 2000s. The earliest food delivery platforms served a two-sided market, enabling consumers to connect with restaurants via a website (Ecker & Strüver, 2022). With the widespread adoption of cell phones and GPS, the potential for digital platform firms increased, especially as large sums of venture capital were invested to seed digital platforms serving multiple places at a time. These factors helped to develop the ODFD model, in which digital platforms serve a three-sided market connecting restaurants, delivery workers, and consumers. By 2021, a small number of large multinational firms dominated ODFD markets (Ecker & Strüver, 2022).

Debate on the impact of food delivery via digital platforms overlaps with debate on the emergence and implications of ride-hailing. This scholarly overlap includes questions about the challenges and opportunities that on-demand food platforms generate in cities. As we discuss below, this includes building an understanding of the interconnections between ODFD and long-standing industries such as the restaurant industry, use of publicly provided urban infrastructure, labour, governance, and power.

Restaurants not only contribute to a city's cultural economy and built environment but also add to a neighbourhood's vibrancy (Liang & Andris, 2021). Because digital platforms allow customers to enjoy a delivered restaurant meal anywhere, they can disrupt the traditional restaurant business model, which focuses on in-person dining. Although empirical evidence is lacking, experts frequently suggest that commission costs associated with ODFD, around 30 per cent, challenge restaurant profitability (Allen et al., 2021; Bissell, 2020; Ecker & Strüver, 2022; Ferreira et al., 2022). At the same time, restaurateurs lament the no-win situation that they find themselves in. If they choose not to participate in food delivery using digital platform firms, they may lose customers, who expect flexible options, especially after COVID-19 restrictions on in-person dining. Digitization has also accelerated a move away from place-based restaurants towards ghost kitchens. In these businesses, a kitchen provides a location for preparing delivered meals and an interface for ordering, but it has no discernible entrance, seating space, or customer-focused access (Shapiro, 2022). While ghost kitchens may offer value for investors and platforms, their prevalence threatens the vibrancy of cities, neighbourhoods, and streetscapes.

Within cities, early evidence suggests that ODFD activity is changing the location patterns of restaurants by encouraging dispersal. Through their business models, ODFD platforms take advantage of existing urban infrastructure, particularly by coordinating activities within urban networks (Richardson, 2020b). According to Richardson (2020b, 460), platforms imply "a reorganization of urban operations … not through new physical infrastructures, but instead through novel technologies of coordination that can reterritorialize those already existing." Indeed, Talamini et al. (2022), in a study of restaurants in Nanjing, China, demonstrate that ODFD platforms are leading to the geographic diffusion of restaurant location.

Significantly, on-demand food delivery relies on an on-demand labour force that is precarious, under-supported, and undervalued. (Gregory & Maldonado, 2020; Richardson, 2020b; van Doorn, 2020). In ODFD, where delivery workers form one side of a three-sided market, food delivery platforms algorithmically manage interactions between restaurants, drivers, and consumers. According to Shapiro (2020), the deliberate obfuscation of information, and sharing data with delivery workers on a need-to-know basis only, is an intentional feature of the technology underlying ODFD platforms. ODFD delivery workers may face a range of challenging working conditions, including conflicts over wages, delays (and wage penalties) as a result of long waiting times, and long hours of work (Popan, 2021). In several instances, drivers in food delivery have organized. Sometimes these efforts have improved working conditions, and sometimes they have resulted in delivery platforms leaving these markets (Popan, 2021; Richardson, 2020a; van Doorn, 2020). However, Gregory & Maldonado (2020) demonstrate that, in Edinburgh, ODFD bicycle delivery workers are not without agency. Sometimes delivery workers tinker with strategies – referred to as urban hacking – within the constraints of the platform and algorithm, in an effort to manage their own work, street safety, or data (Gregory & Maldonado, 2020; van Doorn & Badger, 2020).

ODFD platforms also challenge governments in their role as regulators. Governance concerns related to ODFD span issues related to labour, unfair business practices, data, and congestion management. From a labour perspective, delivery workers in gig economy positions usually lack employment protection, which regulation could provide (Graham, 2020). For example, governments can deploy commission caps, which limit the percentage of each sale that a platform can charge restaurants. High commissions were especially an issue during COVID-19 lockdowns, when restaurants relied on delivery (Griswold, 2022a). While some scholars tend to agree that governments have the power to benefit workers or restaurants, they disagree on what form those benefits ought to take.

Given the growth, evolution, and adaptation of digital platforms, the chapter now examines pandemic-related shifts, interventions, and responses on the part of governments, platforms, restaurants, and delivery workers.

Pandemic: Platform Governance, Platform Pivots, Platform Hacks

In March 2020, federal, provincial, and municipal governments across Canada declared states of emergency, marking the onset of the COVID-19 pandemic. Financial reports released by Uber revealed that in the first quarter (January–March) 2020, global ride-hailing activity contributed more than 60 per cent of the firm's total revenues. However, in the subsequent reporting period (April–June 2020), Uber ride-hailing revenues dropped by 75 per cent. At the same time, Uber Eats revenues increased by 106 per cent. By the end of September 2020, Uber financial reports indicated that Uber Eats had become the major revenue stream for the firm, out-earning revenues from ride-hailing. Figure 10.1 demonstrates that at the start of the pandemic, ride-hailing activity dropped while food-hailing (that is, food delivery) activity rose, and that food-hailing contributed the dominant share of total revenues for several quarters. This increase is especially notable given that at the end of 2019, food delivery made up only 11 per cent of Uber's total revenues. At the peak of the pandemic, in the first quarter of 2021, food delivery contributed 60 per cent of all revenues while ride-hailing contributed 29 per cent. By the second quarter of 2022, ride hailing revenues began to surpass food delivery revenues once again.

Below, we examine three interconnected shifts associated with the pivot from ride-hailing to food-hailing. First, we look at impacts and changes related to the governance of platforms. Next, we identify platform pivots in response to external factors, including both COVID-19

Figure 10.1. Uber: From ride-hailing to food-hailing

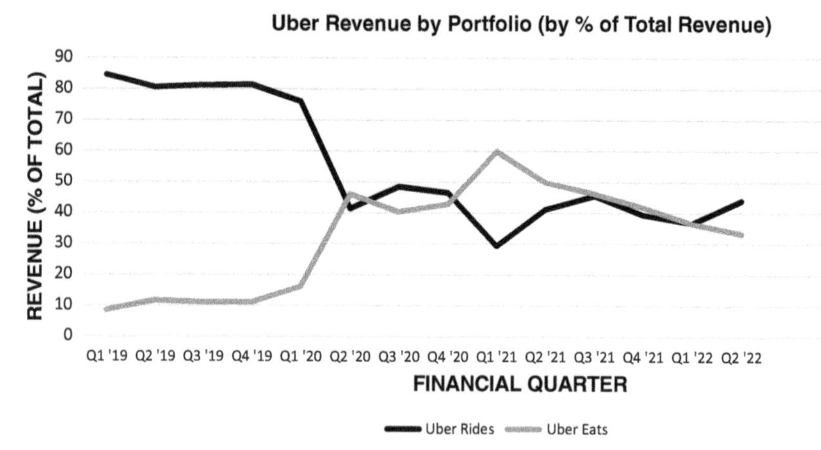

Source: Uber 2019, 2020, 2021, 2022 Quarterly Press Releases, authors' calculations

and government regulation. Finally, we analyse how restaurants and delivery workers used platform hacking to benefit from the growing popularity of these services.

Platform Governance

All around the world, before the pandemic, governments spent a great deal of effort developing ride-hailing regulations, especially in North America. In Canada, these regulations mainly occurred at the municipal level, though in both Quebec and British Columbia, provincial governments held primary responsibility for legalizing and regulating the service (Tabascio & Brail, 2021).

Setting off an early debate, ride-hailing firms insisted that their businesses should be considered technology companies and not transportation firms (Brail, 2017). Ultimately, regulations recognized ride-hailing activities (and firms that developed and operated ride-hailing platforms) as providers of a transportation service. However, the evolution of digital platform firms demonstrates that these businesses have the potential to use their platforms and networks to offer services well beyond mobility. This finding challenges the argument that ride-hailing companies should be classified as transportation firms.

Uber provides a well-known example, pivoting its network of drivers and customers as well as its bookings and revenues away from moving people towards moving food and other goods. As this evolution demonstrates, even though governments regulate ride-hailing as a ground transportation service, the platform's other activities defy simple classification. How then should digital platforms be governed, and are existing regulations sufficient?

As people began to rely more on digital platforms at the onset of the pandemic, observers suggested that the industry needed greater regulation (Lockhart et al., 2021). In the food delivery sector, these calls for government oversight often focused on commissions charged by digital platforms as well as on labour rights and protections for gig economy workers.

During long periods of government mandated COVID-19 closures, when restaurants could not offer indoor and outdoor dining, many relied on ODFD platforms to stay in business (Deschamps, 2021). Since ODFD platforms charge commissions of up to 30 per cent, restaurateurs expressed concern about the impact of these commissions on restaurant revenues – an issue that arose before the pandemic but grew during lockdowns. Governments, keen on managing the economic impacts of public health restrictions, deployed temporary ODFD commission caps to help restaurants survive. Griswold (2022a) identified fifty instances of US jurisdictions enacting commission cap regulations, usually on a temporary basis, to address the state of emergency generated by pandemic-related restrictions on restaurant openings. In Canada, where mayors could only ask ODFD platforms to reduce their commissions, provincial governments in Ontario, British Columbia, Nova Scotia, Saskatchewan, and Quebec passed temporary commission caps, limiting commissions to between 10 and 20 per cent (Deschamps, 2021). The restaurant industry credits this measure with helping to manage economic distress and reduce business closures during this difficult time.

Labour challenges also spur calls for action, especially regarding employee classification, wages, and breaks. In the US, where some cities have wide-ranging power, both New York and Seattle stand out for their efforts to regulate wages in both ride-hailing and food delivery (CHS, 2022; NYC Consumer and Worker Protection, n.d.). In Canada, provincial governments have oversight for labour issues and minimum wage. In April 2022, Ontario's legislature passed the Digital Platform Workers' Rights Act, which is intended to provide gig economy workers with wage guarantees and additional worker protections. Labour advocates have challenged this legislation, however, suggesting that it does not go far enough to protect a fair wage for workers – in large part because it does not take into account the time spent waiting on the app for work, only the time spent in active work (McKenzie-Sutter, 2022).

Labour advocates argue that a regulated minimum wage could solve two problems. It would not only support labour but also help reduce the congestion and pollution associated with ODFD. Specifically, a regulated minimum wage could discourage platform firms from signing up large numbers of delivery workers when there might not be enough work to support sufficient delivery driver earnings. In principle, this single policy lever could address issues related to quality of work and wages, alongside congestion and excess emissions.

At present, municipal regulators in Canada's largest cities indicate that they do not plan to regulate local food delivery. A municipal regulation expert interviewed for this research suggested that cities see regulating food delivery as outside municipal licensing and regulation because this service does not directly affect consumer protection, health, and safety. Furthermore, municipal regulatory staff note that provincial regulations are already in place, such as the Highway Traffic Act, which governs transportation-related issues and the licensing and operation of vehicles.

However, a broader question remains unanswered. To what extent are cities or other jurisdictions in Canada aware of metrics including the number of restaurants participating in food delivery, the number of delivery workers hired, the impacts of delivery on restaurant revenues and employment, and the impacts of ODFD on congestion? Notably, the congestion impacts of ride-hailing have raised concerns (Jin et al., 2018) and instigated several reviews and updates of ride-hailing regulation. In New York City, for instance, data on congestion and emissions resulting from deadheading (driving a vehicle while waiting to match with a passenger) led the city to adjust ride-hailing governance (Hawkins, 2019). Yet studies on the urban impact of ODFD are lacking. Aside from efforts to cap commissions and limited efforts to improve working conditions, governments have done little to manage externalities such as the number of delivery workers, congestion, or curbside use.

Platform Pivots

Lockhart et al. (2021) use the term "platform inversion" to identify directional changes in the flow of people and goods during COVID-19. In this inversion, they explain, "the platform enabled reduction in urban travel via a massive reversal in flows of work, shopping and entertainment to the home rather than vice versa facilitated by the rapid growth of remote working, home entertainment, e-commerce and delivery platforms" (pg. 2). Although these changing activities helped reduce the circulation of people and therefore disease, platform inversions raised new challenges with respect to ODFD, including responses by governments, digital platforms, restaurants, and delivery workers.

As Uber's financial reports show (see Figure 10.1), digital platforms remained viable during the pandemic by effectively pivoting their businesses. Similarly, other leading global ODFD platforms such as DoorDash and SkipTheDishes saw their revenues and orders increase substantially. DoorDash revenues were up by 69 per cent in 2021 whereas orders for SkipTheDishes doubled in 2021 compared to 2019. However, increased revenues and orders do not necessarily mean that these food delivery firms were profitable.

In response to pandemic circumstances, ODFD platforms undertook two particular pivots. The first relates to shifting business focus. For Uber, shifting focus meant shifting the network from rides to food delivery. For some established ODFD platforms, the shift meant launching new types of delivery services, such as grocery delivery, while growing their existing delivery networks. Generally, it was assumed by many that multinational digital platforms faced fewer challenges because they had access to information spanning numerous places and because their nimble business models minimized direct employment. However, COVID-19 forced significant disruption and uncertainty on these firms, too. When governments initially enacted emergency measures, it was unclear whether food delivery could take place under lockdowns. When authorities permitted restaurants to operate on a delivery-only model, the decision clearly supported the viability of restaurants, but it also bolstered ODFD platforms. At the onset of COVID-19 lockdowns, ODFD platforms faced the challenge of rapidly scaling up activities, including managing a growing volume of orders. Firms also had to adjust to changing conditions within and among the three groups of stakeholders they served: restaurants, delivery workers, and consumers.

Second, and in response to government-mandated commission caps discussed earlier, ODFD platforms adjusted both the services they provided to restaurants and the associated fees. For instance, DoorDash introduced several pricing options to set up restaurants on its app. At a cost of 29 per cent commission for delivery orders and 15 per cent commission for pickup orders, DoorDash offered local marketing and a menu photo shoot. For restaurants that managed delivery on their own but relied on DoorDash for menu digitization, ordering, and payments, DoorDash charged a 12 per cent commission. Finally, for restaurants that had their own ordering platforms but required delivery drivers only, DoorDash charged an average flat fee of $8.50 per order (DoorDash, 2022). Similarly, Uber Eats offered three levels of service through plans named Lite, Plus, and Premium. Commissions ranged from 10 per cent for pickup orders and 20 per cent for delivery in the Lite Plan to 30 per cent for delivery orders in the Premium Plan (Uber Eats, 2022). Higher commission charges came with additional marketing supports and services, including enhanced placement on the Uber Eats app. By introducing different levels of support at different price points, ODFD platforms responded to commission cap requirements and innovated further on the services they offered.

Platform Hacks

Researchers have studied the ways that delivery workers experimented with delivery platforms to circumvent algorithms, surveillance efforts, and data sharing. While scholars commonly ascribe these hacking efforts to delivery workers (e.g., Gregory & Maldonado, 2020), our research also reveals that some restaurants and upstart delivery platforms used a form of platform hacking to assert agency. As these hacking efforts suggest, we may need to rethink our understanding of digital platforms as different actors are capable of capturing, and perhaps recapturing, power. Three examples help illustrate these platform hacks: when workers switched between driving passengers and delivering food during lockdowns to earn more from increasing delivery work, when local and grassroots platforms circumvented the perception of unfair market advantage ascribed to ODFD platforms, and when restaurants began leveraging ODFD platforms to their advantage.

For Uber Eats, the pivot from rides to food relied on the advantages of the existence of an extensive platform and network that operated across both mobility and logistics. When Uber switched from mobility to logistics activities, Griswold (2022b) suggested that Uber was able to pivot their activities from people to food in a relatively frictionless manner, especially because of a perception that delivering goods was safer than driving passengers. Furthermore, Griswold (2022b) points out that the process of both switching work, from driving people to delivering goods, and onboarding new delivery drivers was made easier because there is less regulatory oversight for the latter.

Globally, the number of Uber drivers decreased from 5 million at the end of 2019 to 3.5 million one year later (Uber, 2020, 2021). In Canada and elsewhere, COVID-19 restrictions also led to layoffs in many sectors, with notable employment declines in retail, accommodation, and food services. Not surprisingly, an uptick in delivery activities, including food delivery, drew new workers to delivery platforms (Sopher, 2022). During the pandemic, switching between rides and food was a form of hack that enabled and even encouraged delivery workers to keep working. In many instances, the growing use of ODFD platforms supported income opportunities for delivery workers. However, a lack of data prevents governments at all levels from understanding the extent of these pivots.

The pandemic also inspired grassroots actors to build their own delivery platforms to serve local restaurants, delivery workers, and customers. In an effort to address the negative externalities (e.g., high commissions, precarious gig work) associated with multinational ODFD platforms, entrepreneurs in Canadian cities developed and expanded restaurant ordering and delivery platforms (Krashinsky Robertson & Hannay, 2022; Trapunski, 2020). Examples include local food ordering and delivery platforms such as Vancouver's FromTo and Toronto-based Ambassador. While we can see these efforts as attempts to hack the multinational firms' domination, the new Canadian platforms remain small-scale outliers rather than large-scale disruptors.

Finally, as restaurants joined digital platforms at the height of COVID-19 restrictions, they learned to leverage the opportunities of ODFD. In particular, restaurants have learned to adapt to and benefit from the flexibility that ODFD platforms provided. For instance, restaurants can now choose to turn off online orders entirely during busy in-person dining times, they can choose to enable ODFD ordering and pickup only, or they can enable a full suite of services on particular days or times. In addition, once established on an ODFD platform, restaurants can choose to offer only ODFD services, with no in-person dining, during difficult periods such as pandemic closures or staff shortages. Although consumers now expect restaurants to have a digital presence and to offer ODFD convenience, restaurants can still decide when, whether, and how to engage with digital platforms. These choices provide the restaurant industry with some agency.

Clearly, the pandemic changed the way digital platforms operate and interact with cities, restaurants, workers, and consumers. Although we cannot predict whether these pandemic-inspired changes in mobility behaviour, ODFD offerings, and restaurant operation will endure, there is evidence to suggest an ongoing evolution of digital opportunities and challenges, and the need for continued guidance and support.

Conclusion

As this chapter has demonstrated, the pandemic-induced pivot from ride-hailing to food-hailing has important implications for mobility and the study of mobility in the twenty-first-century city. First, through the theme of governance and regulation, we found that while ride-hailing is governed as a ground transportation service, digital platform firms continue to resist simple categorization. Although debates on classification were tense in early stages of regulation, governments may want to revisit these early decisions given the relatively frictionless manner in which these companies were able to pivot from people to food. These quick shifts raise questions about how digital mobility platforms ought to be governed and whether existing regulations sufficed, especially as we witnessed the regulatory challenges around commission caps and labour rights and protections for gig economy workers. Second, through the theme of firm innovation (platform pivots), we found that directional changes in the flow of people and goods during COVID-19 led to significant process innovations at both large multinational platform firms and smaller businesses. For large multinational firms with deep access to information across the globe, they were able to quickly adjust to new regulations around commission caps, for example, and new labour regulations. Smaller firms were able to innovate in terms of service offerings and location. Finally, through the theme of platform hacks, we discovered how small businesses and gig workers hacked the platforms of larger digital firms as a way of asserting agency and recapturing some lost revenue. We saw how delivery workers, for example, experimented with delivery platforms to circumvent algorithms, surveillance efforts, and data sharing.

Our findings have implications for research more generally on mobility and cities. While mobility researchers typically separate personal mobility from goods delivery, the pandemic experience makes a case for studying these activities together. The digital turn, followed by the rise of digital platforms, has created an urgent need to understand the connections between the two.

Related to this, our findings show the importance of government action. Whereas municipal and state governments responded capably to digital platform disruptions, including those associated with ride-hailing, Canadian governments seem less prepared to address ongoing digital shifts. In food-hailing, platform firms continually push limits using technology, algorithms, and cloud-based networks. In response, municipalities need to maintain oversight and move faster than the traditional speed of government decision-making.

On the other hand, we learned that when governments did act decisively during the pandemic to protect the local business ecosystem (e.g., through capping commissions and/or new labour laws to protect gig workers), these decisions pushed multinational platform firms to adapt and innovate. By protecting the local business and civic and neighbourhood infrastructure (e.g., restaurants, food shops), governments may have helped cities transition back better once pandemic restrictions lifted. Only time will tell how effective some of these policies have been, but early days point to the critical role that governments play in supporting cities, including their role in ensuring a sustainable mobility future and vibrant city streetscape.

Acknowledgments

The authors acknowledge and are grateful for the research assistance provided by Sadman Hossein, Youjing Li, and Angelina Zahajko in the development of this chapter. Research for this paper was supported by the Social Sciences and Humanities Research Council Insight Grant 435-2019-0022.

REFERENCES

Ahuja, K., Chandra, V., Lord, V., & Peens, C. (2021, September). *Ordering in: The rapid evolution of food delivery*. McKinsey. https://www.mckinsey.com/industries/technology-media-and-telecommunications/our-insights/ordering-in-the-rapid-evolution-of-food-delivery?cid=other-eml-nsl-mip-mck&hlkid=ffcdf9915e494e0d9a6a3e3c12698a93&hctky=12944315&hdpid=03f1af8b-72b5-40a3-8afa-93a53ff4884b.

Allen, J., Piecyk, M., Cherrett, T., Juhari, M.N., McLeod, F., Piotrowska, M., Bates, O., Bektas, T., Cheliotis, K., Friday, A., & Wise, S. (2021, January). Understanding the transport and CO2 impacts of on-demand meal deliveries: A London case study. *Cities, 108*, Article 102973. https://doi.org/10.1016/j.cities.2020.102973.

Ash, J., Kitchin, R., & Leszczynski, A. (2018). Digital turn, digital geographies? *Progress in Human Geography*, *42*(1), 25–43. https://doi.org/10.1177/0309132516664800.

Attoh, K., Wells, K., & Cullen, D. (2019). "We're building their data": Labor, alienation, and idiocy in the smart city. *Environment and Planning D: Society and Space*, *37*(6), 1007–24. https://doi.org/10.1177/0263775819856626.

Beer, R., Brakewood, C., Rahman, S., & Viscardi, J. (2017). Qualitative analysis of ride-hailing regulations in major American cities. *Transportation Research Record*, *26150*(1), 84–91. https://doi.org/10.3141/2650-10.

Bissell, D. (2020, October). Affective platform urbanism: Changing habits of digital on-demand consumption. *Geoforum*, *115*, 102–10. https://doi.org/10.1016/j.geoforum.2020.06.026.

Brail, S. (2017, December). Promoting innovation locally: Municipal regulation as barrier or boost? *Geography Compass*, *11*(12), Article e12349. https://doi.org/10.1111/gec3.12349.

Brail, S. (2018). From renegade to regulated: The digital platform economy, ride-hailing and the case of Toronto. *Canadian Journal of Urban Research, 27*(2), 51–64. https://cjur.uwinnipeg.ca/index.php/cjur/article/view/132.

CBRE Research. (2019a). *US food in demand*. CBRE Research. http://cbre.vo.llnwd.net/grgservices/secure/US%20Food%20in%20Demand_April%202019.pdf?e=1615151390&h=fe7db68bb026687f568dfe5365b21fe4.

CBRE Research. (2019b). *US food in demand restaurants*. CBRE Research. http://cbre.vo.llnwd.net/grgservices/secure/US%20Food%20in%20Demand%20Restaurants%20November%202019.pdf?e=1614883543&h=e9ebf30334cbc81b23aa61b6827e6024.

Chang, A., & Miranda-Moreno, L. (2022, 9 March). Ride-hailing through the COVID-19 pandemic in New York City. *Findings*. https://doi.org/10.32866/001c.33160.

CHS. (2022, 1 June). Seattle sets minimum wage requirements for app-based delivery workers. *Capitol Hill Seattle News*. https://www.capitolhillseattle.com/2022/06/seattle-sets-minimum-wage-requirements-for-app-based-delivery-workers/.

City of Toronto. (2021). *The transportation impacts of vehicle-for-hire in the city of Toronto: October 2018 to July 2021*. https://www.toronto.ca/wp-content/uploads/2021/11/98cd-VFHTransportationImpacts2021-11-23.pdf.

Clewlow, R.R., & Mishra, G.S. (2017). *Disruptive transportation: The adoption, utilization, and impacts of ride-hailing in the United States. Institute of Transportation Studies, University of California*. Davis, Research Report UCD-ITS-RR-17–07.

Deschamps, T. (2021, 13 June). Restaurants press to keep delivery app caps as they rebound from COVID-19 pandemic. *The Globe and Mail*. https://www.theglobeandmail.com/business/article-restaurants-press-to-keep-delivery-app-caps-as-they-rebound-from-covid/.

DoorDash. (2022, 10 August). *Products and partnership plans | DoorDash for merchants*. https://get.doordash.com/en-ca/products.

Dudley, G., Banister, D., & Schwanen, T. (2017). The rise of Uber and regulating the disruptive innovator. *The Political Quarterly, 88*(3), 492–9. https://doi.org/10.1111/1467-923X.12373.

Ecker, Y., & Strüver, A. (2022). Towards alternative platform futures in post-pandemic cities? A case study on platformization and changing socio-spatial relations in on-demand food delivery. *Digital Geography and Society, 3*, Article 100032. https://doi.org/10.1016/j.diggeo.2022.100032.

Ferreira, D., Carmo, R.M., & Vale, M. (2022, September). Is the COVID-19 pandemic accelerating the platformisation of the urban economy? *AREA, 54*(3), 443–50. https://doi.org/10.1111/area.12785.

Flores, O., & Rayle, L. (2017). How cities use regulation for innovation: The case of Uber, Lyft and Sidecar in San Francisco. *Transportation Research Procedia, 25*, 3756–68. https://doi.org/10.1016/j.trpro.2017.05.232.

Goldstein, M. (2020, July 27). *What is the future for Uber and Lyft after the pandemic?* Forbes. https://www.forbes.com/sites/michaelgoldstein/2020/07/27/what-is-the-future-for-uber-and-lyft--after-the-pandemic/.

Graham, M. (2020). Regulate, replicate, and resist – The conjunctural geographies of platform urbanism. *Urban Geography*, *41*(3), 453–7. https://doi.org/10.1080/02723638.2020.1717028.

Gregory, K., & Maldonado, M.P. (2020). Delivering Edinburgh: Uncovering the digital geography of platform labour in the city. *Information, Communication & Society*, *23*(8), 1187–1202. https://doi.org/10.1080/1369118X.2020.1748087.

Griswold, A. (2022a, 28 July). *The food delivery fee compromise* [Substack newsletter]. Oversharing. https://oversharing.substack.com/p/the-food-delivery-fee-compromise.

Griswold, A. (2022b, 5 August). *Uber enters its Amazon era* [Substack newsletter]. Oversharing. https://oversharing.substack.com/p/uber-enters-its-amazon-era.

Hall, J.D., Palsson, C., & Price, J. (2018, November). Is Uber a substitute or complement for public transit? *Journal of Urban Economics*, *108*, 36–50. https://doi.org/10.1016/j.jue.2018.09.003.

Hawkins, A.J. (2019, 23 December). *Judge blocks NYC's law limiting Uber drivers 'cruising' for new passengers*. The Verge. https://www.theverge.com/2019/12/23/21035554/nyc-uber-lyft-law-limit-cruising-tlc-judge-blocked.

Hirschberg, C., Rajko, A., Schumacher, T., & Wrulich, M. (2016, 9 November). *The changing market for food delivery*. McKinsey & Company. https://www.mckinsey.com/~/media/McKinsey/Industries/Technology%20Media%20and%20Telecommunications/High%20Tech/Our%20Insights/The%20changing%20market%20for%20food%20delivery/The-changing-market-for-food-delivery-final.pdf.

Jin, S.T., Kong, H., Wu, R., & Sui, D.Z. (2018, June). Ridesourcing, the sharing economy, and the future of cities. *Cities*, *76*, 96–104. https://doi.org/10.1016/j.cities.2018.01.012.

Kenney, M., & Zysman, J. (2016, 29 March). *The Rise of the Platform Economy | Issues in Science and Technology*, *32*(3). https://issues.org/the-rise-of-the-platform-economy/.

Kim, A.J., Brown, A., Nelson, M., Ehrenfeucht, R., Holman, N., Gurran, N., Sadowski, J., Ferreri, M., Sanyal, R., Bastos, M., & Kresse, K. (2019, Spring). Planning and the so-called 'sharing' economy ... *Planning Theory & Practice*, *20*(2), 261–87. https://doi.org/10.1080/14649357.2019.1599612.

Krashinsky Robertson, S., & Hannay, C. (2022, May 6). How restaurants are hacking the food delivery system and trailblazing new takeout tools. *The Globe and Mail*. https://www.theglobeandmail.com/business/article-restaurants-finding-sustainable-alternatives-to-delivery-apps/.

Liang, X., & Andris, C. (2021). Measuring McCities: Landscapes of chain and independent restaurants in the United States. *Environment and Planning B: Urban Analytics and City Science*. https://doi.org/10.1177/23998083211014896.

Lockhart, A., Hodson, M., & McMeekin, A. (2021). *How digital platforms are reshaping urban mobility in a time of COVID-19 and after*. Sustainable Consumption Institute, University of Manchester, 41. https://documents.manchester.ac.uk/display.aspx?DocID=57580.

McGuirk, P., Dowling, R., Maalsen, S., & Baker, T. (2021). Urban governance innovation and COVID-19. *Geographical Research, 59*(2), 188–95. https://doi.org/10.1111/1745-5871.12456.

McKenzie-Sutter, H. (2022, 28 February). *Ontario to establish basic gig worker rights including $15 minimum wage*. Global News. https://globalnews.ca/news/8650029/ontario-gig-worker-rights-minimum-wage/.

NYC Consumer and Worker Protection. (n.d.). *Online third-party food delivery services*. Retrieved 17 August 2022 from https://www1.nyc.gov/site/dca/about/Third-Party-Food-Delivery-Services.page.

Ong, T. (2017, 20 December). *EU's top court rules that Uber is a transportation service*. The Verge. https://www.theverge.com/2017/12/20/16799970/eu-court-uber-regulation-taxi.

Pelzer, P., Frenken, K., & Boon, W. (2019, November). Institutional entrepreneurship in the platform economy: How Uber tried (and failed) to change the Dutch taxi law. *Environmental Innovation and Societal Transitions, 33*, 1–12. https://doi.org/10.1016/j.eist.2019.02.003.

Popan, C. (2021). Embodied precariat and digital control in the "gig economy": The mobile labor of food delivery workers. *Journal of Urban Technology*, 1–20. https://doi.org/10.1080/10630732.2021.2001714.

Ranchordas, S. (2015). Does sharing mean caring: Regulating innovation in the sharing economy. *Minnesota Journal of Law, Science and Technology, 16(1)*, 413–76. https://scholarship.law.umn.edu/mjlst/vol16/iss1/9.

Richardson, L. (2020a, May). Platforms, markets, and contingent calculation: The flexible arrangement of the delivered meal. *Antipode, 52*(3), 619–36. https://doi.org/10.1111/anti.12546.

Richardson, L. (2020b). Coordinating the city: Platforms as flexible spatial arrangements. *Urban Geography, 41*(3), 458–61. https://doi.org/10.1080/02723638.2020.1717027.

Scholl, C., & Kraker, J. de. (2021). Urban planning by experiment: Practices, outcomes, and impacts. *Urban Planning, 6*(1), 156–60. https://doi.org/10.17645/up.v6i1.4248.

Shapiro, A. (2020). *Design, control, predict: Logistical governance in the smart city*. University of Minnesota Press. http://ebookcentral.proquest.com/lib/utoronto/detail.action?docID=6404312.

Shapiro, A. (2022). Platform urbanism in a pandemic: Dark stores, ghost kitchens, and the logistical-urban frontier. *Journal of Consumer Culture, 23*(1), 168–87. http://journals.sagepub.com/eprint/2AYIRJ3SIYTGBV3XKBKY/full.

Sopher, B. (2022, 4 March). My life as a bike courier delivering food in the pandemic. *Toronto Star*. https://www.thestar.com/life/food_wine/opinion/2022/03/04/i-deliver-food-for-a-living-from-catching-covid-19-to-struggling-to-make-ends-meet-heres-what-the-pandemics-been-like.html.

Spicer, Z., Eidelman, G., & Zwick, A. (2019, March). Patterns of local policy disruption: Regulatory responses to Uber in ten North American cities. *Review of Policy Research, 36*(2), 146–67. https://doi.org/10.1111/ropr.12325.

Statista Market Forecast. (2022). *Online food delivery – Canada*. Statista. https://www.statista.com/outlook/dmo/eservices/online-food-delivery/canada.

Tabascio, A., & Brail, S. (2021, Summer). Governance matters: Regulating ride hailing platforms in Canada's largest city-regions. *The Canadian Geographer/Le Géographe canadien, 66*(2), 278–92. https://doi.org/10.1111/cag.12705.

Talamini, G., Li, W., & Li, X. (2022, September). From brick-and-mortar to location-less restaurant: The spatial fixing of on-demand food delivery platformization. *Cities, 128*, Article 103820. https://doi.org/10.1016/j.cities.2022.103820.

Temenos, C., & McCann, E. (2012). The local politics of policy mobility: Learning, persuasion, and the production of a municipal sustainability fix. *Environment and Planning A: Economy and Space, 44*(6), 1389–1406. https://doi.org/10.1068/a44314

Temenos, C., & McCann, E. (2013). Geographies of policy mobilities. *Geography Compass, 7*(5), 344–57. https://doi.org/10.1111/gec3.12063.

Tirachini, A., & del Río, M. (2019, October). Ride-hailing in Santiago de Chile: Users' characterisation and effects on travel behaviour. *Transport Policy, 82*, 46–57. https://doi.org/10.1016/j.tranpol.2019.07.008.

Trapunski, R. (2020, 13 May). Delivery wars: Toronto restaurants are taking the power back from Uber Eats. *Now Magazine*. https://nowtoronto.com/food-and-drink/delivery-wars-uber-eats-backlash-toronto.

Uber. (2020, 6 February). *2020 Investor presentation*. https://s23.q4cdn.com/407969754/files/doc_financials/2019/sr/InvestorPresentation_2020_Feb13.pdf.

Uber. (2021, 10 February). *2021 Investor presentation*. https://s23.q4cdn.com/407969754/files/doc_financials/2020/q4/InvestorPresentation2021.pdf.

Uber Eats. (2022, 11 August). *Pricing that works for your business*. https://merchants.ubereats.com/ca/en/pricing/.

van Doorn, N. (2020). At what price? Labour politics and calculative power struggles in on-demand food delivery. *Work Organisation, Labour & Globalisation*, *14*(1), 136–49. https://doi.org/10.13169/workorgalaboglob.14.1.0136.

van Doorn, N., & Badger, A. (2020). Platform capitalism's hidden abode: Producing data assets in the gig economy. *Antipode*, *52*(5), 1475–95. https://doi.org/10.1111/anti.12641.

Young, M., Allen, J., & Farber, S. (2020, January). Measuring when Uber behaves as a substitute or supplement to transit: An examination of travel-time differences in Toronto. *Journal of Transport Geography*, *82*, Article 102629. https://doi.org/10.1016/j.jtrangeo.2019.102629.

SECTION IV: GOVERNING MOBILITY

11 Integrated Mobility and the Governance of Urban Transit

ELENA GORACHINOVA, LISA HUH, AND DAVID A. WOLFE

Introduction

Smart mobility encompasses the development of connected and autonomous vehicles (C/AVs) that rely on platform-centred on-demand mobility solutions. The transition to smart mobility, driven by industry, promises new technologies to increase road safety and reduce emissions (Council of Canadian Academies 2021; Mukhtar-Landgren & Paulsson, 2020). At the same time, as others have pointed out, shared mobility solutions (e.g., mobility as a service or mobility-on-demand services, such as Uber) can also increase traffic congestion and undermine public transport providers (Ferreira et al., 2020; Groth, 2019; Pangbourne et al., 2020; see also Aoyama and Alvarez León, chapter 12 and Lorinc, chapter 14). Despite these threats, government strategies to manage the deployment of these technologies in public transit remain fragmented, and there is a lack of investment in building digital infrastructures. For example, most Canadian cities have begun implementing new mobility technologies in their public transit systems through one-off pilot projects. Yet, there is often a disconnect between the goals of the pilot projects and a city's transportation broader goals, given that initiatives are often driven by the private sector under the auspice of economic development objectives.

In this respect, there is an urgent need for governments to adopt smart mobility technologies in a way that expands access to, and the affordability of, public transit to create integrated mobility solutions. The Canadian Urban Transit Association defines integrated mobility as "the ability for people to move easily from place to place according to their own needs. This means taking a transit-centric approach which connects all modes of travel including active transportation and automobile travel, enabling door-to-door and seamless mobility throughout the city, that is accessible to everyone" (CUTA, 2017, p. 1). From this perspective, smart mobility solutions represent a subset of possible solutions available to achieve the broader objective of integrated mobility. However, relatively few scholars have explored these relationships in terms of how governments should steer the networks of actors, resources, and power involved. Failure to address these governance issues risks lock-in to a suboptimal public mobility system that will exacerbate the social and environmental problems currently challenging urban planners. In this chapter, we address the impacts of new forms of mobility by examining government's role in managing the transition to an integrated mobility future.

The chapter draws on a range of governance concepts to study the social organization of new mobility solutions across three case studies – Canada, the US, and Finland. By analysing smart mobility policy instruments and associated policy approaches, it offers local governments insight into ways to adopt new smart technologies to provide a higher-quality integrated transit service. Our broad objective is to understand how authorities in the three cases are managing the transition to integrated mobility. To achieve this goal, we examine how authorities are approaching three broad issues – digital infrastructure, C/AVs, and shared mobility – to identify gaps in these processes. The chapter analyses the extent to which initiatives in each case expand governance capacity with respect to new mobility technologies as well as the outcomes associated with these initiatives, notably decreased use of cars and increased access to transit. The following section outlines some governance issues that define the roles authorities play with respect to new urban mobility technologies. It demonstrates that the mode of governance selected to manage new forms of urban mobility has significant implications for the relative roles played by public and private transit providers in urban transit systems.

Governance Issues in Urban Mobility

Many of the challenges facing cities in adopting new mobility technologies exceed the scope and capabilities of their current organizations, institutional arrangements, and governance structures (Bolivar, 2016; Caragliu and Del Bo, 2012; Gil-Garcia et al., 2015,). While municipalities govern mobility systems, they operate in a broader political context that requires regional and national government support, which is not always forthcoming (Cardullo & Kitchin, 2019; Smith et al., 2019). In effect, the smart mobility transition involves multi-level governance, which is affected by a range of factors (see Aoyama and Alvarez León, chapter 12). Institutional factors include the degree of autonomy and/or the resources municipalities have available to support their decisions; other factors (such as geographical context, population, and institutional conditions) also affect the introduction of new technologies (Bache & Flinders, 2004; Hooghe & Marks, 2003; Ruhlandt, 2018).

According to the literature, the best way to evaluate emerging mobility partnerships is to determine how effectively they balance the interests of public and private actors (Davis, 2018). Wallsten et al. (2020) identify three distinct roles that governments play in the mobility space. In a leadership role, public authorities establish the framework for introducing new technologies and set the objectives that these technologies need to meet. In an enabling role, government facilitates private-sector innovation through its support of network and partnership initiatives. Finally, in a laissez-faire role, government largely remains on the sidelines. The enabling and laissez-faire approaches rely on the market, in the assumption it can implement new technologies in a way that meets public goals of affordable and sustainable transit. Each approach is linked to a different division of responsibility for transit provision. In the leadership approach, the public sector helps orchestrate new mobility systems, while in the enabling and laissez-faire approaches, public actors play a lesser role, with possibly detrimental effects on the sustainability and affordability of transit provision. Some observers express concern that market-driven or contractual agreements may diminish the public role because commercial actors will drive the development towards their primary goal of finding paying customers (Lyons, 2018).

The aim of this chapter is twofold: 1) to understand the degree to which public stakeholders assume a leadership, enabling, or laissez-faire role as they respond to these emerging modes of transit provision; and 2) to analyse how existing multi-level governance arrangements affect these outcomes. Decentralized governance structures in Anglo-American countries, with limited intergovernmental coordination and collaboration, may be less likely to implement new technologies on an effective scale given more adversarial relationships between public and private actors and siloed transportation organizations. More centrally coordinated jurisdictions, such as Finland, may be better positioned to facilitate partnerships needed to implement new mobility technologies in a way that benefits public transport goals. Finally, Finland has more favourable land-use and automobile-restrictive policies that make cars less attractive and encourage public transportation ridership – making it less likely that new transit technologies will favour automobiles (Buehler, 2009). To reiterate the point made above, the chapter analyses the extent to which initiatives being taken in each of the cases examined serve to expand governance capacity with respect to new mobility technologies as well as to realize improved outcomes associated with these initiatives, notably decreased use of cars and increased access to integrated forms of transit. The following sections examine existing governance structures in Canada, the US, and Finland, to understand how they condition the public response to technological developments in each of the three areas under consideration – digital infrastructure, C/AVs, and shared mobility.

Governance of Digital Infrastructure

Developing a robust data infrastructure is key to regulating new mobility services, such as mobility on demand (MOD) or mobility as a service (MaaS), so that they complement public goals, such as integrated transport. In the words of Seleta Reynolds, general manager of the Los Angeles Department of Transportation, governments must "put rules in place to prevent walled gardens or monopolies – where providers create their own closed ecosystem and don't share data and information with others … and ensure that there is some public accountability once people grow dependent on those services" (Goldsmith & Leger, 2020, p. 6).

Governmental authorities can adapt different regulatory levers to manage connected mobility marketplaces, while at the same time leveraging the data to inform public service delivery and improve public outcomes. For example, they can use trip data from dockless scooters to inform bus route planning and improve transit access (see Palm and Farber, chapter 6). However, this task is challenging because many governments are not accustomed to data-oriented planning; they neither synchronize nor regularly share data across departments or municipalities to coordinate region-wide transportation. Further, the introduction of neural networks in transportation planning accentuates the need to include civic organizations and citizens early in the process. Indeed, failure to do so may weaken accountability structures around data platforms in cities.

The following sections review recent initiatives in the three countries. We examine the extent to which governments at all levels have enhanced their governance capacity, such as by introducing laws and regulations.

Canada

In Canada, federal, provincial, and municipal levels of government share responsibility for transportation. The federal government lacks operational responsibility for the road network or for driver and vehicle licensing. Provinces are responsible for highway networks and may also transfer provincial funds to municipalities, notably for urban transit operations. Many jurisdictions have poorly integrated regional transportation networks (see Hutton, chapter 3, and Lorinc, chapter 14). With the provision of networks and services marked by limited intergovernmental collaboration, transit agencies must grapple with complex technological developments on their own.

Research on digital government in Canada has focused on how the public sector should be reimagined for a digital age – by becoming more "horizontal, entrepreneurial, data-driven, and user focused" (Clarke, 2020, p. 100). The federal government has adopted several programs to facilitate the introduction of C/AVs and to improve digital capabilities at the municipal level. Table 11.1 provides details on these programs.

REGIONAL/LOCAL ACTORS

Individual provinces undertake digital governance in unique ways. In some cases, the strategy is implemented by newly established government organizations (e.g., Ontario Digital Service), while in other provinces, existing ministries (e.g., BC's Ministry of Citizens' Services) assume responsibility. As with the federal government, the focus remains on open data, with less attention paid to data governance challenges. Provincial ministries of transportation are also developing new approaches for data governance. However, because of a lack of data literacy across departments, open data policies are often supply-driven and not based on the wishes and needs of users – at least, not citizen-users.

Municipalities and regions have also published reports outlining their role in the evolving transportation ecosystem (Digital Infrastructure Strategic Framework, 2022). Like the provinces, cities face challenges aligning open data policies with the needs of citizens or even establishing a clear transport-related goal for data sharing. In Toronto, a particular project – Sidewalk Toronto – intensified the pressure on governments to formulate data governance policies. By introducing a proprietary approach to personal data management, the project ignited a controversy over the appropriate governance of personal data (Artyushina, 2020; Morgan and Webb, 2020). While the City of Toronto continues to develop its digital infrastructure plan after the project's cancellation, its role in the emerging digital mobility ecosystem remains unclear. In other Ontario municipalities, open data is driven at the regional or municipal level. In York Region, for example, efforts are underway to establish a data co-op, comprised of nine municipalities, to share data and benefit from the collective investment in GIS technology, data, and people. However, due to the absence of provincial and federal guidelines on data sharing, York Region has been unable to get other municipalities or regional transportation agencies, such as Metrolinx, on board.

In Montreal, its Transportation Plan from 2008 encouraged the use of real-time public transit information to increase mobility (Bista et al., 2020). However, the city did not match these strategies with long-term transit investments until economic, social, and environmental problems became critical (Bakvis et al., 2009, 226). Only in the instance of Vancouver has TransLink, the regional transit authority, sought to open up data with the express goal of providing cities with information for setting policy, instead of simply enabling innovation among businesses (Wolff et al., 2019). This difference could result from variations in transportation governance structures between provinces. While in the case of TransLink, the board responds to a council of mayors for the region, Ontario's regional transit body, Metrolinx, reports to the provincial government. Overall, despite the attention to this issue in the CCA's Expert Panel Report (2021), multi-level governance approaches to a comprehensive data strategy are limited. To develop such a strategy, governments must work together, but provinces and municipalities are often reduced to playing an enabling role for private innovation.

US

In the US, actors at different levels of government are working to understand the opportunities and challenges around the sharing, analysis, and use of data collected as part of on-demand shared mobility and autonomous vehicle services. At the same time, governments must deal with governance obstacles, which stem from historical challenges determining the optimal mix of public and private provision as well as long-standing fragmentation in public-sector decision making (Winston, 2013).

The US has yet to establish a comprehensive strategy tackling issues of privacy or data use concerns. To inform its role, the Federal Highway Administration (FHWA) is working with the State Departments of Transportation (DOTs) and the National Academies of Sciences, Engineering, and Medicine (NASEM) to conduct research and implement new technologies and innovations. A voluntary data specification has also emerged out of NASEM's Transportation Research Board, allowing multiple organizations to manage a passenger's entire trip cycle, from trip request to trip delivery, with all necessary data. At the same time, there is no clear information regarding how a state-level department of transportation or MPO should be funding or implementing these guidelines (Green & Lucivero, 2018). Table 11.1 below offers more details on the national role.

REGIONAL/LOCAL ACTORS

The recent literature on the governance of urban transit systems focuses on the role of Metropolitan Planning Organizations (MPOs). As one of the most prevalent forums for regional planning in the United States, MPOs are responsible for planning and implementing federal transportation policy at the regional level (Gerber & Gibson, 2009). More than four hundred MPOs operate across the US in urbanized areas with populations greater than fifty thousand (Sciara, 2017). Contemporary MPOs, however, face a range of problems in practice (Fischer et al., 2020; Sciara, 2017). Although few, if any, cities have taken direct control of transit operations from MPOs, a growing number are making standalone transit plans, adding transit sections to larger transportation plans, and entering joint funding relationships with regional agencies to increase service.

Despite a lack of guidance from federal level authorities, some US states are implementing data governance policies that mandate transportation agencies to create a chief data officer position outside their IT departments to reduce data silos across the organization (Albee, Hamilton, & Chestnutt, 2020). However, a lack of coordination between transportation agencies still makes it difficult to use data to optimize transportation system performance. While MPOs are supposed to help local political and administrative authorities develop a unified regional vision and plan, they lack the funds or power to do so (Bond & Kramer, 2010; Crabbe et al., 2005; Gerber & Gibson, 2009).

However, unlike the case of Canada, there are more multi-level governance efforts to standardize and share transit data among public actors and between public and private actors (see Aoyama and Alvarez León, chapter 12). Some regional transit agencies are trying to establish data fusion systems across counties, as in the case of the Central Ohio Transit Authority.[1] Another example is the Los Angeles Department of Transportation (LADOT), which has developed the Mobility Data Specification (MDS). This open-source data specification allows digital communication between the public entities that manage streets

1 The data would run on a platform set up by the startup Waycare, which made a name for itself in the Las Vegas area helping transportation officials predict when and where traffic collisions were likely to occur.

and sidewalks and the organizations that use these thoroughfares to provide transportation services, including private operators. In turn, the standard, which is used by more than 120 cities around the world to plan transportation infrastructure, and support and regulate shared mobility services, is managed by the Open Mobility Foundation (Aoyama & Alvarez León, 2021).

Finland

In Finland, national policymaking that influences transport falls under five ministries: Transport and Communications; Environment; Finance; Employment and the Economy; and Education and Culture (Kivimaa, 2014). According to Kivimaa (2014), the Ministry of Transport and Communications (LVM) is largely responsible for transport systems and networks, the transport of people and goods, traffic safety, and issues relating to the environment, giving it significant power over the direction of new technologies. Municipalities are also important policymakers, with an independent role in land-use and regional transport planning. For example, Helsinki owns its metro and tram services (Hirschhorn et al., 2019).

REGIONAL/LOCAL ACTORS

In Finland, there is more regional coordination than in Canada or the US. In the Helsinki metropolitan area, ticketing for public transport has been fully integrated since the 1980s, and the Helsinki Regional Transport Authority (HSL) has been responsible for planning and overseeing public transport in the metropolitan area since 2010. HSL is also responsible for determining public transport fares, developing service plans, and setting routes and timetables (Hirschhorn et al., 2019). Helsinki's integrated public transport system has helped create a smartcard and mobile app ticketing system across the metropolitan area (Hirschhorn et al., 2019). Although HSL manages an expansive transit network, it plays a lesser role in steering the future of Helsinki's transport sector. Instead, public actors at the national level have taken a top-down approach to transport and have facilitated private-sector innovation. Table 11.1 below provides more details on the national role.

Following passage of the Act on Transport Services, HSL developed an operational contract that allows mobility service operators like MaaS Global (creator of the mobile application Whim) to freely use HSL's timetable and real-time data, route and disturbance information, and an open interface journey planner (Mukhtar-Landgren & Smith, 2019). The contract also lets mobility service operators resell HSL's single adult

Table 11.1. National-level activities: Canada, United States, Finland

Technology	National-level activities	Canada	United States	Finland
Digital infrastructure	National programs	**Federal Government** – Program to Advance Connectivity and Automation in the Transportation System (ACATS)	**Federal Highway Administration (FHWA)** – directed by Congress to improve data collection, management, and transparency; developed informational guide for State, Tribal, and Local Safety Data Integration **Department of Transportation (DOTs)** and **National Academies of Sciences, Engineering, and Medicine (NASEM)** works with FHWA– conducts research and directs implementation of technologies and innovations	**Finnish Ministry for Transport and Communication (LVM)** – introduction of 1st Intelligent Transport Strategy (ITS) in 2009, broadening scope of transport governance to cover all modes of transport; introduction of 2nd ITS 2013, advancing transport projects in real-time data collection, processing and distribution, open data, and integrated public transport; enactment of the Act on Transport Services in 2018, requiring all transport providers share operational data and enable third-party resale for single-tickets
	Governance	Enabling approach	Enabling approach	Leadership and enabling approach

(Continued)

Table 11.1. *Continued*

Technology	National-level activities	Canada	United States	Finland
C/AVs	National Programs	**Federal government–** released Guidelines for testing automated driving systems, and responsibilities at different levels of governance; Designed ACATS program to support innovation in C/AVs	**US Department of Transportation (USDOT)** – published the "Automated Vehicles 4.0" and "Automated Vehicles: Comprehensive Plan," call for a "consistent federal approach **US Department of Transportation's National Highway Traffic Safety Administration (NHTSA)** – has taken preliminary steps to centralize governance by seeking industry and stakeholder comments for how HAV testing, and safety should be regulated; released voluntary safety guidelines for the safe development and deployment of AV/CVs (SAE Automation Levels 3 through 5)	**National government** – developed Finland's Road Traffic Act, which called for the change in the colour of road markings (from yellow lines to white) to provide better for machine vision detection, and suggests that road data (e.g., signs, traffic lights, control devices) should be available for use by C/AV operators **Business Finland** – jointly funded pilot projects with municipalities and private actors like the Living Lab Bus, which was operated by VTT Technical Research Centre of Finland and tested a fleet of ten electric buses owned by HSL
	Governance	Enabling Approach	Enabling Approach	Leadership and Enabling Approach

MaaS	National programs	**National government**– controls funds that provinces and municipalities can apply for to invest in MoD or MaaS (currently has a minor role)	**Federal Transportation Agency (FTA)** – has been researching innovative MOD and public transit partnerships; developed the MOD Sandbox Demonstration program in 2016, which tests various concepts (e.g., smartphone applications and trip planners, integrated fare payment, first-and-last mile connections to public transportation and paratransit). All vehicles (used as part of shared mobility services) must comply with safety standards established by the Consumer Product Safety Commission and all other federal, state and city safety standards.	**Finnish Funding Agency for Innovation (Tekes)** – invested in MaaS Global in its early stages **LVM** – created an enabling environment for MaaS by proposing reforms to Finland's transport legislation; successfully pushed for the enactment of the Act on Transport Services in 2018; promoted MaaS in Finland by participating in new mobility workshops led by ITS Finland; established a "new mobility" think tank; co-founded MaaS-Alliance (a public-private partnership aimed at facilitating the diffusion of MaaS) **Tekes and LVM** – jointly funded 8 pre-studies for MaaS from 2015-2016
	Governance	Laissez-faire approach	Enabling approach	Leadership and enabling approach

tickets (Mukhtar-Landgren & Smith, 2019). However, individual operators must determine their own pricing scheme and control the data processed through their services (Mukhtar-Landgren & Smith, 2019). Despite these efforts to facilitate data sharing, the long-term effect of the Act on Transport Services is undetermined. While HSL's operational contract may increase transparency among local public transport operators and private mobility service operators, it remains unclear whether municipal authorities are benefiting from the partnership.

Cross-Country Comparison

With respect to data generated from shared mobility solutions, gaps in data governance seem evident in all three cases, with some exceptions. The focus is on public entities opening their datasets to private operators, while the same responsibility is not placed on the private entities. Because of this focus, data subjects neither understand where their individual data is located, processed, and used nor have any control over it. Azarmi and Resnick (2020) argue that lack of transparency can erode community trust and lead to unwillingness to use smart mobility technologies and services (p. 20). This mistrust is greater in Europe because US-based institutions provide many building blocks of smart mobility, and their usage requires personal data to be transferred to the US – a country with less-protective regulation on data privacy.

C/AVs

Authorities have generally taken an enabling approach to governing C/AVs, aiming to create opportunities for the private sector to take the lead and innovate, and often framing these opportunities as an aspect of economic development policy. Nevertheless, governments are taking more of a leadership role in response to the backlash cities have faced for playing this enabling role. These unanticipated consequences have sparked some reflection. However, officials have directed less discussion towards the political obstacles to policies needed to regulate the impact of C/AVs, despite the fact these obstacles could reinforce car dependence and undermine transit in the long run (Shao, 2020). In addition, the absence of rules regarding the collection of data from AV companies prevents policymakers from understanding how smart mobility technology works and the performance capabilities of products from different firms perform. This approach leaves the automotive industry in the driver seat, simultaneously presenting automation as an innovation with the potential to revitalize the market for private vehicles while creating effective ways to undermine alternatives to car dominance.

Canada

In 2018, Canada was ranked seventh on KPMG's autonomous vehicle readiness index (*Autonomous Vehicle Readiness Index*, 2018). By 2019, it had slipped five spots to twelfth place and remained there in 2020 (*2019 Autonomous Vehicles Readiness Index*, 2019). Canada's drop is partly due to the low score on infrastructure, particularly the lack of 4G coverage and electric vehicle charging stations across the country. At the same time, the index gives Canada one of the highest ratings for government-funded C/AV pilots and industry partnerships, with much of its work focused on collaboration. Furthermore, Canada shares the Great Lakes vehicle manufacturing cluster with the US, with the industry employing more than 125,000 people nationally and assembling more than two million vehicles a year (Council of Canadian Academies, 2021). Table 11.1 above presents details on the federal government's role in promoting the adoption and diffusion of C/AV technology.

The provinces and municipalities both play significant roles. The provinces are responsible for the legislative framework for C/AV testing and deployment in their respective jurisdictions, and enacting legislation that incorporates federal vehicle safety requirements. As well, they manage driver licensing, vehicle registration and insurance, rules of the road, and changes to highway infrastructure to support C/AV deployment. Finally, municipalities execute the legislative and regulatory framework created by provinces and territories, including C/AV safety enforcement. They also make land-use planning decisions and operate transit systems.

Despite these responsibilities, policymakers and industry actors have been slow to establish specific regulations or governance structures for autonomous systems. As automotive firms achieve higher levels of vehicle autonomy – where control shifts to the vehicle and the infrastructure supporting it – those decisions, and the responsibility, will transfer to a new set of actors, and there is a lack of guidance for how to make them (Mordue et al., 2020).

REGIONAL/LOCAL POLICIES

Despite the absence of broad standards governing C/AVs, several measures have gradually been introduced. On 1 January 2019, O. Reg. 517/18: *Pilot Project – Automated Vehicles* came into force in Ontario. Under this regulation, with authorization, level 3 automated vehicles (with the presence of a human driver) can be driven on Ontario public roads. Quebec has a similar legal regime. Transport Canada's ACATS program has funded several C/AV pilot projects on the subnational level (see Table 11.1). Various provinces have also invested in developing and testing C/AVs. Québec has been more focused on funding research related to C/AVs, including the technical monitoring of demonstration projects, with some funding also allocated to support pilot projects that help municipalities understand the potential of AVs (Kelly, 2021).

Most initiatives seek to strengthen the digital sector or renew the automotive industry. These goals are especially true in Ontario, where much of the investment in C/AVs has been led by the provincially funded Ontario Vehicle Innovation Network (OVIN) (Gorachinova, 2021). OVIN has "uploaded" the C/AV policy field to a network of regional technology development sites – including cities, regional economic development agencies, regional innovation centres, and universities – with incentives to form alliances that would allow regions to support companies advancing C/AV technologies. However, the funded projects do not always accord with municipal transportation goals as they intend to make cities appealing locations for economic investment and experimentation (Gorachinova, 2021). Cities must rely on OVIN for funding and cannot independently support companies that are aligned with long-term municipal transportation goals. For example, the City of Toronto has established transportation innovation zones (that is, specific sites for testing alternative transportation approaches and emerging technologies) (Toronto, 2023). Actors who want to experiment in the space must obtain funding through OVIN or federal level programs instead of directly from the city.

US

A comprehensive C/AV specific regulatory structure has yet to emerge at either the federal or state level in the United States (see Aoyama and Alvarez León, chapter 12). High automated vehicle (HAV) manufacturers and suppliers lack definitive guidance because there are no comprehensive federal requirements for roadway testing protocols, minimum safety criteria, or vehicle design. Developers and investors are free to back their preferred technologies and seek permission to test and prove those technologies on the nation's roads. Table 11.1 provides more details on the federal government role in the US.

Regional/Local Policies

Given that the federal government has been slow to enact a regulatory or policy framework for C/AVs, even around basic safety, US states and cities are taking a more active role (Brown et al., 2018). Studies in the US investigate the extent to which C/AV initiatives have helped US states, cities, and municipalities to meet transportation goals and priorities (McAslan et al., 2021). Although C/AV pilot projects are the most common initiative, regulations and rules related to testing are growing in number.

State and local governments have been at the forefront of regulatory efforts, taking an accommodating approach to driverless vehicle safety and testing. Numerous state-specific regulations have emerged, creating a patchwork regulatory scheme that differs state to state and changes nearly every month (*Autonomous Vehicles: Legal and Regulatory Developments in the United States*, 2021). With respect to pilot projects, cities adopt significantly different approaches, which often lack coherent policy goals (Chatman & Moran, 2019; McAslan et al., 2021). The pilot project's goals rarely match the city's transportation goals, and cities fail to anticipate how the pilot project's findings could support a long-term vision for C/AVs within their overall mobility system. Instead, C/AV pilot goals, focused on introducing C/AV technology to the public and promoting economic development, may benefit C/AV companies more than they benefit cities. McAslan et al. (2021) postulate this outcome is due, in part, to the lack of available funding, which is a major constraint on long-term transportation-related planning for C/AVs.

Other experiments involve private flexible-route C/AV testing and passenger service pilots, led by companies such as Waymo (owned by Alphabet), Cruise (owned by General Motors), Uber, and Lyft. Waymo has begun its "Waymo One" service in Arizona, which provides rides in C/AVs to members of the public who are part of the company's "early rider program" (Korosec, 2018; Krafcik, 2018). In Las Vegas, Lyft has partnered with Aptiv to pilot a small fleet of C/AV sedans (Ackers, 2019). An interesting experiment is under way in Boston, where the city wants to establish itself as an innovation hub for C/AV development (Aoyama and Alvarez León, 2021). In partnership with private firms and other local organizations, Boston is testing vehicles as well as experimenting with connected transportation infrastructure. For example, the city is working with Aptiv, an MIT spin-off that is generating high-definition maps of several districts and creating software that simulates complex traffic scenarios. It is also partnering with another company that develops sensors for C/AVs. Researchers view Boston as "building a multi-scalar urban governance framework to guide the adoption of autonomous vehicles" (Aoyama and Alvarez León, 2021, 7).

Despite these promising initiatives, the research surveyed above suggests public-sector employees in US cities believe there is a lack of coordination between cities and C/AV companies regarding testing and pilot C/AV ride-hailing services. Even when the public and private sectors have formed explicit partnerships, government workers view the information shared by C/AV companies on their operations as inadequate for planning purposes.

Finland

In Finland, national authorities have been enabling C/AV governance, using legislative changes to create more favourable conditions for C/AV testing on public roads. Rather than steering the development of new mobility options, national efforts are geared towards the provision of an enabling environment for private actors. C/AV pilots in Finland focus on shared-use opportunities (e.g., connected bus pilots or autonomous shuttle pilots). Main actors include the Finnish Transport and Communications Agency (Traficom), Business Finland, Forum Virium Helsinki, and the VTT Technical Research Centre of Finland (see Table 11.1). Business Finland has jointly funded pilot projects with municipalities and private actors like the Living Lab Bus (Living Lab Bus 2016–2019, n.d.). The Living Lab Bus pilot, operated by VTT Technical Research Centre of Finland, tested a fleet of ten electric buses owned by HSL (Heino et al., 2018).

REGIONAL/LOCAL POLICIES

Municipal authorities have also taken an enabling approach to C/AV governance, initiating smaller-scale pilots, and participating in larger-scale EU-led pilot projects. The main actors include cities responsible for granting special permissions for pilot routes. Between July 2015 and August 2015, Vantaa participated in an EU-funded C/AV pilot organized by CityMobil2 (Hunter, 2018). The objective of CityMobil2 was to assess how municipalities could integrate C/AVs into public transportation (Ainsalu et al., 2018). The Vantaa pilot project had the highest ridership among nine EU cities participating in the program, a success attributed to Vantaa's high ridership to the Housing Fair and the system's connection to a metro transit station. Local transport authorities at the Helsinki Regional Transport Authority have led small-scale pilot programs to test shared mobility services. Between 2013 and 2015, HSL conducted the Kutsuplus pilot program, a transit service of autonomous Wi-Fi–equipped minibuses serving users between city bus stops, but the city ultimately cancelled it due to funding and scalability issues (Heikkila, 2014; Hensher, 2017).

Between 2016 and 2018, Finland also participated in an EU-funded interregional project called Sohjoa Baltic (Ainsalu et al., 2018). Trials, led by Metropolia University of Applied Sciences and Forum Virium Helsinki, took place on three routes in Espoo, Helsinki, and Tampere (Ainsalu et al., 2018). The project had two objectives: to test how autonomous shuttles would adapt in winter conditions and to study how the public perceived C/AVs. Findings concluded that C/AV technology was not yet advanced enough to adapt to winter conditions and that the slow speed of buses (12 km/h) made shuttles more suitable to travel on pedestrian and bicycle lanes (Ainsalu et al., 2018).

Cross-Country Comparison

The mobility transition and the emergence of the C/AV industry are prompting cities to operate in a more multi-scalar governance context. They are attempting to mediate across multiple stakeholders, ranging from non-profit organizations to universities (Aoyama & Alvarez León, 2021). The roles that they adopt depend on political relations between the public, private corporations, urban institutions, and other governance scales. Within our case studies, it remains unclear whether C/AVs can reduce car use, with some studies concluding they will mostly substitute for walking and cycling.

Shared Mobility

A growing number of public transportation agencies are partnering with shared mobility (MOD or MaaS) companies, raising the question of what role these companies should play in providing public transport. MOD refers to an innovative transportation concept where consumers can access mobility, goods, and services on demand by dispatching or using shared mobility. MOD passenger mobility can include bike-sharing, car-sharing, microtransit, ride-sharing (i.e., carpooling and vanpooling), scooter sharing, shuttle services, urban air mobility, and public transportation. MOD courier services can include app-based delivery services (known as courier network services), robotic delivery, and aerial delivery (e.g., drones).

In Europe, another form of multimodal transportation known as MaaS is emerging (see Lorinc, chapter 14). Although MOD and MaaS share certain similarities, such as an emphasis on multimodal integration (physical co-location of services, fare payment, and digital integration), the concepts are different. MaaS focuses on mobility aggregation and subscription services, often facilitated through a smartphone application or website (Shaheen & Cohen, 2020a, 2020b). Both services share an underlying assumption: as consumers gain access to more shared forms of mobility, they will be more inclined to transition to a car-free lifestyle.

Canada

In Canada, the federal government has not played a major role in regulating different forms of shared mobility, such as car-sharing, microtransit, or bike-sharing. It does control funds that the provinces and municipalities can apply for to invest in MOD or MaaS transportation (see Table 11.1). Municipal government staff have noted, however, that cities can find it difficult to obtain funding to support shared mobility initiatives, such as car-sharing pilots or infrastructure to encourage shared parking (Ditta et al., 2016).

REGIONAL/LOCAL POLICIES

Demand for more responsive forms of public transit is growing in Canada. Municipalities are searching for alternative ways to provide transit to customers in metropolitan areas that are traditionally under serviced. Demand-responsive forms of transit that utilize shared mobility technologies are thus coming to be viewed as a possible alternative to overcome this policy dilemma (Klumpenhouwer, 2020; see Palm and Farber, chapter 6).

When it comes to regulating shared mobility services, there is significant variation across the country, given pre-existing regulatory and governance structures. For instance, while Toronto treats ground transportation regulation as a local responsibility, in Montreal and Vancouver, ride-hailing is primarily regulated at the provincial level (Tabascio & Brail, 2021). Across the board, studies suggest that "decisions around ride hailing are often political and decoupled from ridership and other transportation concerns" (Tabascio & Brail, 2021, p. 1). Instead, ride-hailing regulation is driven by a desire to attract and retain innovative firms (Brail, 2018). Reports also suggest that regional transportation agencies (such as Metrolinx in Ontario) have contributed to the growth of some of these mobility services, including Bike Share Toronto (for a detailed study of Bike Share Toronto, see McNee and Miller, chapter 8) – which received $4.9 million to double its network by adding 1,000 bikes and 120 stations, extending its reach into new parts of the city (Ditta et al., 2016).

Municipalities have also experimented with shared mobility technologies and platforms to improve their responsiveness to citizen needs, compared to fixed-transit routes. In the Greater Toronto and Hamilton Area (GTHA), the *2041 Regional Transportation Plan* (RTP) identified MOD as key for optimizing the transportation system (*2041 RTP*, 2017, p. 37). Dozens of Canadian municipalities have turned to on-demand transit technology companies, many of which have headquarters in Canada and offer a new, more personalized form of transit (i.e., on-demand transit). Klumpenhouwer (2020) points out that transit systems could enhance their efficiency in low-demand areas or times by directly responding to passengers' smartphone requests. Montreal has also used on-demand transit to expand its reach into low-density areas.

Although cities have attempted to integrate MOD services into their transit system, users must often pay separately for each service, making the combination less convenient to use. Another challenge is the lack of regionally integrated fare payment, which makes it more difficult for customers to use separate transit services across the region. For example, the Greater Toronto and Hamilton Area has ten autonomous transit agencies, one provincial agency, and one regional transit service. Within this complex arrangement, the agencies have difficulty coordinating their services in a way that allows customers the easy use of an integrated payment system.

US

Recognizing the importance of multimodal transportation, the Federal Transportation Agency (FTA) has been researching innovative MOD and public transit partnerships. In 2016, the FTA developed the MOD Sandbox Demonstration program to explore opportunities and challenges for public transportation related to technology-enabled mobility services. In particular, it studies ways that public transit can learn from, build on, and interface with innovative transportation modes from user, business model, technology, and policy perspectives (see Table 11.1).

REGIONAL/LOCAL POLICIES

As on-demand mobility providers, such as Uber and other companies, launched across the US, they initially flouted existing regulations before public officials could implement measures to regulate them. As of August 2016, thirty-nine states and the District of Columbia had implemented such measures, although municipal regulation of Uber varies substantially (Ditta et al., 2016). The introduction of these technologies has largely been driven by private actors, with public transportation goals often secondary to their interests. Although some shared mobility companies have agreed to recognize consumer protection regulations, other measures have proven to be more contentious. For instance, Uber has "vigorously opposed those it fears will restrict the easy entry of drivers and the supply of cars on the road, like fingerprint-based background checks, vehicle caps, and, in the most extreme case, full bans" (Collier et al., 2018, p. 921). And even those regulations that do exist, such as ones in San Francisco, were agreed to based on the involvement of a small number of regulatory agencies (Davis, 2018).

At the same time, local authorities are experimenting with partnerships with mobility-on-demand companies to make transit more accessible. Researchers identify six types of partnerships between public transportation and MOD service providers, among them 1) first- and last-mile partnerships, where a public-sector partner subsidizes an MOD service operator to provide services to or from a public transit stop or low-density area; 2) service and public transit replacement partnerships that subsidize an MOD provider to offer service in a lower-density area; and 3) paratransit partnerships that leverage MOD services to supplement or replace an existing paratransit service. In their study of the Los Angeles MOD program, especially on the impact of subsidized ride-hailing to and from transit, Brown et al. (2021) find that the MOD program neither advanced nor undermined transit access among disadvantaged transportation populations and neighbourhoods.

In some instances, municipal regulations have been more proactive with respect to shared mobility. For example, the Seattle Department of Transportation has prescribed specific permitting systems to regulate bike-sharing and ride-sharing services. According to Moscholidou and Pangbourne (2020), Seattle's clear regulatory position can help cities and their public transportation systems deal with and manage the potential pressure from new market forces.

The fragmented nature of transportation organizations in the US presents a challenge for introducing shared mobility schemes (Schweiger, 2017). Many transit agencies operate independently and may not coordinate their services with private or public providers. Despite these institutional challenges, cities are experimenting with the provision of these services. For instance, Pittsburgh just became the first US community where every resident can feasibly and affordably trade their private cars for an app with the launch of its long-awaited Move PGH pilot. The transit app allows residents to pay bus fares; rent micromobility vehicles such as electric bikes, mopeds, and scooters; or find someone to carpool with. Despite the involvement of private service providers, the city will manage the program and deal with any problems that may emerge between different transit providers.

However, studies of smaller-scale on-demand demonstration projects funded by the FTA suggest some of the challenges that may emerge because of public-private partnerships. For example, the Chicago MOD Sandbox demonstration encountered a problem with the acquisition of a primary project partner/vendor by another company, which then delayed the pilot project (Cohen et al., 2021).

Finland

Finland has adopted a top-down approach to mobility governance, with policies at the national level guiding actions at the municipal level. National authorities have taken an enabling approach, using hard policy instruments to lower or remove barriers to entering the mobility market (e.g., Act on Transport Services) and soft policy instruments to promote market-driven MaaS (e.g., funding MaaS-related studies) (Smith et al., 2019). The policy instruments used to promote MaaS include funding pilots and pre-studies. These instruments tend to direct local and regional actors to comply with national-level objectives for promoting MaaS, rather than steering the development of MaaS themselves (Smith et al., 2019).

REGIONAL/LOCAL POLICIES

At the regional level, authorities have adopted an enabling approach to MaaS governance, complying with national-level objectives (Smith et al., 2019). HSL has created operational contracts at the regional level to enable platforms such as Whim to purchase tickets and resell them. Despite state efforts to promote MaaS and carve out a niche for the state in the international market, neither Finland nor Helsinki may benefit economically from an enabling environment for private actors in the MaaS ecosystem, as the majority of MaaS Global's shareholders are foreign (Veeneman et al., 2018). Further, there has been ongoing conflict between Whim and the city-run HSL. With its own mobile ticketing app, HSL had little incentive to open the market to MaaS providers, and MaaS Global accused it of refusing to share easy access to its popular monthly transit passes (Carey, 2021).

Private actors like MaaS Global have clearly benefited from the enabling environment created by Finland's national government and the regional transport authorities. Helsinki may have introduced HSL's generic contract as a response to top-down initiatives to enable innovations in MaaS. Yet it is currently unclear whether the transport authority's operational contract increased collaboration between public transport operators and MaaS providers or improved transparency between public and private stakeholders. Current data indicate that

Whim's monthly subscription packages support 1.8 million out of 374 million trips made in Helsinki each month (Pangbourne et al., 2020). However, given the cost of subscriptions, transport inequity could become a pressing concern once MaaS is widely adopted.

Cross-Country Comparison

Across the three case studies, there is growing concern that shared mobility services offered by private profit-seeking companies may weaken public transit and further privatize the urban mobility sector. At the same time, the ability to regulate private providers depends on pre-existing governance capacity at the local level. When the national-level government intervenes, as in the case of Finland, and encourages public actors to collaborate with private ones in the delivery of MaaS services, local stakeholders may feel threatened and fear public transit may be in peril. Overall, local authorities seem best positioned to plan for and integrate shared mobility services, and reforms may be necessary to give them the power to do so.

Conclusion

This chapter describes the current state of policy experimentation at the municipal and local level across the three cases examined. First, the case studies highlight the need to consider issues of data governance, C/AVs, and shared mobility as deeply interconnected and interdependent. Without close attention to data governance, government can fully leverage neither C/AVs nor shared mobility solutions to create an integrated public transport system that decreases citizens' dependence on cars. Adopting smart mobility solutions in a comprehensive manner requires that both national and local governments collaborate to establish clearer data governance guidelines, instead of leaving the task to municipalities alone.

Second, the chapter demonstrates that economic development motives do not always align with public transit goals. The mismatch between the two means that public stakeholders need to dedicate their efforts towards integrating smart mobility solutions with urban transit goals, rather than assume that economic development initiatives will support green and affordable transit. This is a challenging objective, given the financial constraints faced by many public stakeholders, especially at the local level. The case of Finland shows that even when governance is more comprehensive and centralized (i.e., a leadership approach), conflicts and uncertainties around mobility technologies (i.e., MaaS) may emerge when governments design regulation with economic development rather than transportation goals in mind.

Furthermore, despite the importance of multi-level governance, certain tasks, and decisions, such as the implementation and oversight of MaaS systems, may best be guided by local-level rather than by national-level actors, as in the case of Finland. When cities or municipalities lack the governance capacity or funding to tackle these issues, it may be necessary to introduce institutional changes to empower them to adopt more than an enabling role. Finally, the case studies illuminate some of the threats inherent in smart mobility technologies, especially when private actors are relied upon to provide solutions.

As the chapter emphasizes, there are no definitive answers regarding the right combination of stakeholders necessary for the successful implementation of smart mobility technologies. Instead, the need for greater attention to governance arrangements in the provision of smart mobility services could guide policy stakeholders as they consider next steps. The deployment of smart mobility solutions thus far has not demonstrated a positive impact on public transport, even when governments take a leadership role and governance structures are robust. As an increasing number of private actors enter the public realm of transit, more unforeseen obstacles may emerge and destabilize transit (e.g., a smart mobility provider going bankrupt). Overall, we find limited consensus in terms of what national or local governments should do regarding new transportation technologies, but we see a greater role for multi-level governance that empowers municipalities. There is a pressing need for more coordinated, collaborative, and multi-scalar processes to guide the implementation of these technologies.

Acknowledgment

The research for this paper was funded by the Social Sciences and Humanities Research Council of Canada and Infrastructure Canada under Research Grant No. 872-2020-1029 as well as SSHRC Grants No. 895-2018-1006 and 435-2019-0022. The authors are indebted to the funding agencies for their research support.

REFERENCES

ACATS Funded Projects 2018–2021. (2021). Transport Canada.

Ackers, M. (2019, 22 January). Aptiv, Lyft use latest tech on self-driving fleet in Las Vegas. *Las Vegas Review-Journal*. https://www.reviewjournal.com/traffic/aptiv-lyftuse-latest-tech-on-self-driving-fleet-in-las-vegas-1578938/.

The Administration's Priorities for Transportation Infrastructure (testimony of Pete Buttigieg). https://www.transportation.gov/administrations-priorities-transportation-infrastructure.

Ainsalu, J., Arffman, V., Bellone, M., Ellner, M., Haapamäki, T., Haavisto, N., Josefson, E., Ismailogullari, I., Lee, B., Madland, O., Madžulis, R., Müür, J., Mäkinen, S., Nousiainen, V., Pilli-Sihvola, E., Rutanen, E., Sahala, S., Schønfeldt, B., Smolnicki, P.M., Soe, R.-M. ... & Åman, M. (2018). State of the art of automated buses. *Sustainability, 10*(9), Article 3118. https://doi.org/10.3390/su10093118.

Aoyama, Y., & Alvarez León, L F. (2021, December). Urban governance and autonomous vehicles. *Cities, 119*, Article 103410. https://doi.org/10.1016/j.cities.2021.103410.

Artyushina, A. (2020, December). Is civic data governance the key to democratic smart cities? The role of the urban data trust in Sidewalk Toronto. *Telematics and Informatics, 55*, Article 101456. https://doi.org/10.1016/j.tele.2020.101456.

Autonomous Vehicles: Legal and Regulatory Developments in the United States. (2021, May). Jones Day. https://www.jonesday.com/en/insights/2021/05/autonomous-vehicles-legal-and-regulatory-developments-in-the-us.

Autonomous Vehicles Readiness Index. (2018). KPMG. Retrieved 5 March 2024 from https://assets.kpmg.com/content/dam/kpmg/tw/pdf/2018/03/KPMG-Autonomous-Vehicle-Readiness-Index.pdf.

AVIN Ecosystem analysis and roadmap 2020. (2020). Deloitte LLP & Ontario Centers of Excellence (OCE). https://www2.deloitte.com/ca/en/pages/consumer-industrial-products/articles/avin-ecosystem-analysis-and-roadmap.html.

Azarmi, M., & Resnick, N. (2020, 25 June). *Smart-enough cities*. Center for Democracy and Technology. Retrieved 5 March 2024 from https://cdt.org/wp-content/uploads/2020/06/2020-06-25-CDT-Mobility-Data-Whitepaper-full-FINAL.pdf.

Bache, I., & Flinders, M. (Eds.). (2004). *Multi-level governance*. Oxford University Press.

Beauregard, R.A. (1995). Theorizing the global–local connection. In P.L. Knox & P.J. Taylor (Eds.), *World cities in a world-system* (pp. 232–48). Cambridge University Press.

Bakvis, H., Brown, D.M., & Baier, G. (2009). *Contested federalism: Certainty and ambiguity in the Canadian federation*. Oxford University Press.

Barns, S. (2018, March). Smart cities and urban data platforms: Designing interfaces for smart govern. *City, Culture and Society, 12*, 5–12. https://doi.org/10.1016/j.ccs.2017.09.006.

Bista, S., Hollander, J. B., & Situ, M. (2021, March). A content analysis of transportation planning documents in Toronto and Montreal. *Case Studies on Transport Policy, 9*(1), 1–11. https://doi.org/10.1016/j.cstp.2020.06.007.

Bolívar, J. (2016). Mapping dimensions of governance in smart cities. In *Proceedings of the 17th international digital government research conference on digital government research* (pp. 312–24). https://doi.org/10.1145/2912160.2912176.
Bond, A., & Kramer, J. (2010). Governance of metropolitan planning organizations: Board size, composition, and voting rights. *Transportation Research Record*, *2174*(1), 19–24. https://doi.org/10.3141/2174-03.
Brail, S. (2018). From renegade to regulated: The digital platform economy, Ride-hailing and the case of Toronto. *Canadian Journal of Urban Research*, *27*(2), 51–64. https://cjur.uwinnipeg.ca/index.php/cjur/article/view/132/67.
Brown, A., Manville, M., & Weber, A. (2021, June). Can mobility on demand bridge the first-last mile transit gap? Equity implications of Los Angeles' pilot program. *Transportation Research Interdisciplinary Perspectives*, *10*, Article 100396. https://doi.org/10.1016/j.trip.2021.100396.
Brown, J., Rodriguez, G., & Hoang, T. (2018, December). *Federal, state, and local governance of automated vehicles*. https://epm.ucdavis.edu/sites/g/files/dgvnsk296/files/inline-files/AV%20Gov_IssuePaper_FINAL_TH_2018_14_12%20%281%29.pdf.
Buehler, R. (2009). Promoting public transportation: Comparison of passengers and policies in Germany and the United States. *Transportation Research Record: Journal of the Transportation Research Board*, *2110*(1), 60–8. https://doi.org/10.3141/2110-08.
Canadian Urban Transit Association (CUTA). (2017). *Integrated mobility: Implementation toolbox* [online]. Available from https://cutaactu.ca/wp-content/uploads/2021/01/Integrated-mobility-toolkit.pdf.
Caragliu, C. Del Bo, & Nijkamp, P. (2011). Smart cities in Europe. *Journal of Urban Technology*, *18*(2), 65–82. https://doi.org/10.1080/10630732.2011.601117.
Cardullo, P., & Kitchin, R. (2019). Smart urbanism and smart citizenship: The neoliberal logic of 'citizen-focused' smart cities in Europe. *Environment and Planning C: Politics and Space*, *37*(5), 813–30. https://doi.org/10.1177/0263774X18806508.
Carey, C. (2021, 8 September). *MaaS faces its make-or-break moment*. Cities Today. https://cities-today.com/maas-faces-its-make-or-break-moment/.
Chatman, D.G., & Moran, M. (2019, August). *Autonomous vehicles in the United States: Understanding why and how cities and regions are responding*. (UC-ITS-2019-13). University of California Institute of Transportation Studies. https://escholarship.org/uc/item/29n5w2jk.
Clarke, A. (2020). Data governance: The next frontier of digital government research and practice. In E. Dubois & F. Martin-Bariteau (Eds.), *Citizenship in a connected Canada: A research and policy agenda* (pp. 97–117). University of Ottawa Press.
Cohen, A., Shaheen, S., Broader, J., Martin, E., & Brown, L. (2021, June). *Mobility on demand (MOD) sandbox demonstration: Chicago Transit Authority (CTA) Ventra-Divvy integration case study* (dot:56482). *FTA Report No. 0196*. https://doi.org/10.21949/1520682.

Cohen, T., Stilgoe, J., & Cavoli, C. (2018). Reframing the governance of automotive automation: Insights from UK stakeholder workshops. *Journal of Responsible Innovation, 5*(3), 257–79. https://doi.org/10.1080/23299460.2018.1495030.

Cohen, T., Stilgoe, J., Stares, S., Akyelken, N., Cavoli, C., Day, J., Dickinson, J., Fors, V., Hopkins, D., Lyons, G., Marres, N., Newman, J., Reardon, L., Sipe, N., Tennant, C., Wadud, Z., & Wigley, E. (2020, July). A constructive role for social science in the development of automated vehicles. *Transportation Research Interdisciplinary Perspectives, 6*, Article 100133. https://doi.org/10.1016/j.trip.2020.100133.

Collier, R.B., Dubal, V.B., & Carter, C.L. (2018). Disrupting regulation, regulating disruption: The politics of Uber in the United States. *Perspectives on Politics, 16*(4), 919–37. https://doi.org/10.1017/S1537592718001093.

Council of Canadian Academies (CCA). (2021, 22 March). *Choosing Canada's automotive future*. The Expert Panel on Connected and Autonomous Vehicles and SharedMobility, Council of Canadian Academies. https://cca-reports.ca/reports/connected-and-autonomous-vehicles-and-shared-mobility/.

Crabbe, A.E., Hiatt, R., Poliwka, S.D., & Wachs, M. (2005, 22 August). Local transportation sales taxes: California's experiment in transportation finance. *Public Budgeting & Finance, 25*(3), 91–121. https://doi.org/10.1111/j.1540-5850.2005.00369.x.

Albee, M., Hamilton, I., & Chestnutt, C. (2020). *Data governance: Ohio's people, processes, and technology*. (FHWA-SA-20-059). https://rosap.ntl.bts.gov/view/dot/58089.

Davis, D.E. (2018). Governmental capacity and the smart mobility transition. In G. Marsden & L. Reardon (Eds.), *Governance of the smart mobility transition* (pp. 105–22). Emerald Publishing Limited. https://doi.org/10.1108/978-1-78754-317-120181007.

Digital Infrastructure Strategic Framework. (2022, March). City of Toronto. Retrieved 27 February 2024 from https://www.toronto.ca/wp-content/uploads/2022/03/9728-DISFAcc2.pdf.

Ditta, S., Urban, M.C., & Sunil, J. (2016, 19 August). *Sharing the road: The promise and perils of shared mobility in the GTHA*. (No. 124). Mowat Centre. https://mowatcentre.munkschool.utoronto.ca/sharing-the-road/.

Ferreira, A., von Schönfeld, K.C., Tan, W., & Papa, E. (2020). Maladaptive planning and the pro-innovation bias: Considering the case of automated vehicles. *Urban Science, 4*(3), 41. https://doi.org/10.3390/urbansci4030041.

Fischer, L.A., Ray, R.S., & King, D.A. (2020). Who decides? Toward a typology of transit governance. *Urban Science, 5*(1), 6. https://doi.org/10.3390/urbansci5010006.

Gerber, E.R., & Gibson, C.C. (2009). Balancing regionalism and localism: How institutions and incentives shape American transportation policy.

American Journal of Political Science, *53*(3), 633–48. https://doi.org/10.1111/j.1540-5907.2009.00391.x.

Gil-Garcia, G.R., Pardo, T.A., & Nam, T. (2015), What makes a city smart? Identifying core components and proposing an integrative and comprehensive conceptualization. *Information Polity*, *20*(1), 61–87. https://doi.org/10.3233/IP-150354.

Goldsmith, S., & Leger, M. (2020, February). *Effectively managing connected mobility marketplaces*. Roy and Lila Ash Center for Democratic Governance and Innovation. https://ash.harvard.edu/publications/effectively-managing-connected-mobility-marketplaces.

Gorachinova, E. (2021). *Varieties of embeddedness: Essays on technological transition in automotive regions*. University of Toronto.

Green, M., & Lucivero, A. (2018). *Data governance & data management case study: Case studies of select transportation agencies*. US Department of Transportation.

Groth, S. (2019). Multimodal divide: Reproduction of transport poverty in smart mobility trends. *Transportation Research Part A: Policy and Practice*, *125*, 56–71. https://doi.org/10.1016/j.tra.2019.04.018.

Guidelines for regulating shared micromobility. (2019). NACTO. https://nacto.org/wp-content/uploads/2019/09/NACTO_Shared_Micromobility_Guidelines_Web.pdf.

Heikkilä, S. (2014). *Mobility as a service – a proposal for action for the public administration* [MA thesis, Aalto University]. https://aaltodoc.aalto.fi/server/api/core/bitstreams/a56fb4a5-a299-485d-8457-009c4d47f421/content.

Heino, I., Kostiainen, J., Lahti, J., Linna, J., & Pihlajamaa, O. (2018, September). Living Lab Bus platform for IoT service development in public transport context. In *25th ITS World Congress*. https://livinglabbus.fi/ITSWC18-papers/EU-TP1590-Heino.pdf.

Hensher, D.A. (2017, April). Future bus transport contracts under a mobility as a service (MaaS) regime in the digital age: Are they likely to change? *Transportation Research Part A: Policy and Practice*, *98*, 86–96. https://doi.org/10.1016/j.tra.2017.02.006.

Hirschhorn, F., Paulsson, A., Sørensen, C.H., & Veeneman, W. (2019, December). Public transport regimes and mobility as a service: Governance approaches in Amsterdam, Birmingham, and Helsinki. *Transportation Research Part A: Policy and Practice*, *130*, 178–91. https://doi.org/10.1016/j.tra.2019.09.016.

Hooghe, L., & Marks, G. (2003). Unraveling the state, but how? Types of multi-level governance. *American Political Science Review*, *97*(2), 233–43. https://doi.org/10.1017/S0003055403000649.

Hunter, A. (2018). *Approaching autonomous shuttle pilot programs in public transportation* (Rep.). Retrieved from https://krex.k-state.edu/bitstream/handle/2097/38906/AliciaHunter2018.pdf?sequence=5.

Kelly, B. (2021). Montreal's driverless bus pilot project offers glimpse into the future. *Montreal Gazette*. https://montrealgazette.com/news/local-news/montreals-driverless-bus-pilot-project-offers-glimpse-into-the-future#:~:text=Though%20there%20will%20be%20no,will%20be%20allowed%20on%20board.

Kivimaa, P. (2014). Government-affiliated intermediary organisations as actors in system-level transitions. *Research Policy*, *43*(8), 1370–80. https://doi.org/10.1016/j.respol.2014.02.007.

Klumpenhouwer, W. (2020). *The state of demand-responsive transit in Canada*. University of Toronto.

Korosec, K. (2018, 5 December). Waymo launches self-driving car service Waymo One. *TechCrunch*. http://social.techcrunch.com/2018/12/05/waymo-launches-selfdriving-car-service-waymo-one/.

Krafcik, J. (2018, 5 December). Waymo One: The next step on 0ur self-driving journey. Medium - Waymo Team (blog). https://medium.com/waymo/waymo-one-the-nextstep-on-our-self-driving-journey-6d0c075b0e9b.

Living Lab Bus 2016–2019). (n.d.). *Tampere universities*. https://www.tuni.fi/en/research/living-lab-bus-2016-2019.

Lyons, G. (2018). Getting smart about urban mobility – Aligning the paradigms of smart and sustainable. *Transportation Research Part A: Policy and Practice*, *115*, 4–14. https://doi.org/10.1016/j.tra.2016.12.001.

McAslan, D., Najar Arevalo, F., King, D.A., & Miller, T.R. (2021). Pilot project purgatory? Assessing automated vehicle pilot projects in US cities. *Humanities and Social Sciences Communications*, *8*(1), 325. https://doi.org/10.1057/s41599-021-01006-2.

Mordue, G., Yeung, A., & Wu, F. (2020). The looming challenges of regulating high level autonomous vehicles. *Transportation Research Part A: Policy and Practice*, *132*, 174–87. https://doi.org/10.1016/j.tra.2019.11.007.

Morgan, K., & Webb, B. (2020). Googling the city: In search of the public interest on Toronto's 'smart' waterfront. *Urban Planning*, *5*(1), 84–95. https://doi.org/10.17645/up.v5i1.2520.

Moscholidou, I., & Pangbourne, K. (2020). A preliminary assessment of regulatory efforts to steer smart mobility in London and Seattle. *Transport Policy*, *98*, 170–7. https://doi.org/10.1016/j.tranpol.2019.10.015.

Mukhtar-Landgren, D., & Paulsson, A. (2020). Governing smart mobility: Policy instrumentation, technological utopianism, and the administrative quest for knowledge. *Administrative Theory & Praxis*, 1–19. https://doi.org/10.1080/10841806.2020.1782111.

Mukhtar-Landgren, D., & Smith, G. (2019, June 25). Perceived action spaces for public actors in the development of Mobility as a Service. *European Transport Research Review*, *11*(1). https://doi.org/10.1186/s12544-019-0363-7.

Pangbourne, K., Mladenović, M.N., Stead, D., & Milakis, D. (2020). Questioning mobility as a service: Unanticipated implications for society and governance. *Transportation Research Part A: Policy and Practice, 43*(2), 135–53. https://doi.org/10.1016/j.tra.2019.09.033.

Pierre, J. (2019). Multilevel governance as a strategy to build capacity in cities: Evidence from Sweden. *Journal of Urban Affairs, 41*(1), 103–16. https://doi.org/10.1080/07352166.2017.1310532.

Ruhlandt, R.W.S. (2018). The governance of smart cities: A systematic literature review. *Cities, 81*, 1–23. https://doi.org/10.1016/j.cities.2018.02.014.

Schweiger, C. (2017). *Bringing mobility as a service to the United States: Accessibility opportunities and challenges*. ITS World Congress 2017, Montreal, Quebec Canada.

Sciara, G.-C. (2017). Metropolitan transportation planning: Lessons from the past, institutions for the future. *Journal of the American Planning Association, 83*(3), 262–76. https://doi.org/10.1080/01944363.2017.1322526.

Shaheen, S., & Cohen, A. (2020a). Mobility on demand in the United States: From operational concepts and definitions to early pilot projects and future automation. In E. Crisostomi, B. Ghaddar, F. Häusler, J. Naoum-Sawaya, G. Russo, & R. Shorten (Eds.), *Analytics for the sharing economy: Mathematics, engineering and business perspectives* (pp. 227–54). Springer International Publishing. https://doi.org/10.1007/978-3-030-35032-1_14.

Shaheen, S., & Cohen, A. (2020b). Mobility on demand (MOD) and mobility as a service (MaaS): Early understanding of shared mobility impacts and public transit partnerships. In *Demand for Emerging Transportation Systems* (pp. 37–59). Elsevier. https://doi.org/10.1016/B978-0-12-815018-4.00003-6.

Shao, S. (2020). Iterative autonomous vehicle regulation and governance: How distributed regulatory experiments and inter-regional coopetition within federal boundaries can nurture the future of mobility. *University of Illinois Journal of Law, Technology and Policy*. https://heinonline.org/HOL/LandingPage?handle=hein.journals/jltp2020&div=15&id=&page=.

Smith, G., Sarasini, S., Karlsson, I.C.M., Mukhtar-Landgren, D., & Sochor, J. (2019). Governing mobility-as-a-service: Insights from Sweden and Finland. In M. Finger & M. Audouin (Eds.), *The governance of smart transportation systems* (pp. 169–88). Springer International Publishing. https://doi.org/10.1007/978-3-319-96526-0_9.

Tabascio, A., & Brail, S. (2021, Summer). Governance matters: Regulating ride hailing platforms in Canada's largest city-regions. *The Canadian Geographer / Le Géographe Canadien, 66*(2), 278–92. https://doi.org/10.1111/cag.12705.

Toronto, C.O. (2023, 14 December). *Transportation innovation zones*. City of Toronto. https://www.toronto.ca/services-payments/streets-parking-transportation/transportation-projects/transportation-innovation-zones/.

Traficom. (2019, 2 January). *The Finnish Transport and Communications Agency Traficom started operations*. Retrieved 11 December 2021 from https://www.traficom.fi/en/news/finnish-transport-and-communications-agency-traficom-started-operations.

The 2041 Regional Transportation Plan. (2017). https://www.transportation.gov/administrations-priorities-transportation-infrastructure.

2019 Autonomous Vehicle Readiness Index. (2019). KPMG International. Retrieved 5 March 2024 from https://assets.kpmg.com/content/dam/kpmg/xx/pdf/2019/02/2019-autonomous-vehicles-readiness-index.pdf.

Veeneman, W., Van der Voort, H., Hirschhorn, F., Steenhuisen, B., & Klievink, B. (2018). PETRA: Governance as a key success factor for big data solutions in mobility. *Research in Transportation Economics*, *69*, 420–9. https://doi.org/10.1016/j.retrec.2018.07.003.

Wagner, F., Alarcon-Rubio, D., Grigaliūnas, S., & Syrusaitė, D. (2021). *On Data Privacy, Governance and Portability: Turning Obstacles into Opportunities*. Trafi.

Wallsten, A., Sørensen, C.H., Paulsson, A., & Hultén, J. (2020). Is governing capacity undermined? Policy instruments in smart mobility futures. In A. Paulsson & C.H. Sørensen (Eds.), *Shaping smart mobility futures: Governance and policy instruments in times of sustainability transitions* (pp. 153–68). Emerald Publishing Limited. https://doi.org/10.1108/978-1-83982-650-420201009.

Winston, C. (2013). On the performance of the US transportation system: Caution ahead. *Journal of Economic Literature*, *51*(3), 773–824. https://doi.org/10.1257/jel.51.3.773.

Wolff, H., Possnig, C., & Petersen, G. (2019). *An open data framework for the new mobility industry in Metro Vancouver*. https://www.translink.ca/-/media/translink/documents/plans-and-projects/programs-and-studies/translink-tomorrow/new-mobility-lab/finalized-reports/wolff-maas-report-2019.pdf?sc_lang=en&hash=9664C6431858038F827C421DC75ED23C.

12 Regulating Autonomous Vehicles: Lessons from US Cities[1]

YUKO AOYAMA AND LUIS F. ALVAREZ LEÓN

Introduction: Autonomous Vehicles and the Future of Cities

Recent literature on the future of smart cities points to the importance of autonomous vehicles (AVs) in urban transformation (Cohen & Hopkins, 2019; Cooper et al., 2019; Duarte & Ratti, 2018; Inkinen et al., 2019). AVs are part of a broader set of trends in urban mobility, which emphasize digitally mediated and increasingly automated on-demand transport options (Kuhnert et al., 2018) and the data infrastructure that supports them. Centred on ride-hailing provided by transportation network companies such as Uber, the current system of on-demand transport relies on shared vehicles driven by human drivers. However, developments throughout the past decade have opened up the possibility of a shift from individually owned and operated vehicles towards shared use of AVs for a variety of purposes across various countries (Alvarez León & Aoyama, 2022). In this context, future trends of urban mobility lean on two interrelated concepts. Mobility-as-a-service (MaaS) represents a shift from personally owned vehicles to services that provide mobility solutions. CASE (Connected, Autonomous, Shared and Electric) integrates new mobility service models, changing from combustion engines to electric motors, and from passenger-operated to connected and autonomous vehicles.

While MaaS and CASE depict two different perspectives on the future of urban mobility, they both hinge on the development of a physical- and cyber-infrastructural platform that can coordinate and connect customers with service providers. More broadly, MaaS and CASE are

---

1 A version of this article was published as Aoyama, Y. and Alvarez León, L.F. (2021). Urban governance and autonomous vehicles. *Cities*, *119*, article 103410.

just two of the most prominent possibilities of the new urban mobility sector, much of which depends on the deployment of AVs. Although AVs are still in testing phase, with their development plagued by continued delays, this mobility innovation has far-reaching implications, particularly for urban areas. For example, fully functioning autonomous vehicles could offer many uses, such as driving an elderly person to a hospital, connecting isolated or mobility-impaired populations, and adapting to passengers' behaviour, preferences, and routines. To optimize routes, an AV could combine services such as freight delivery and passenger taxi. Finally, an AV adoption would also impact land-use and traffic-flow patterns by reducing the need for parking spaces and traffic lights. However, these scenarios – which vary widely in their feasibility, practicality, and implications – depend on future adoption and reliable performance. Because AVs will have widespread effects on data collection, cybersecurity, privacy, public safety, and urban infrastructure, the prospect of adoption is already influencing how cities approach regulating urban innovation in the mobility sector.

In light of the high stakes, technological complexity, and unprecedented nature of the mobility revolution, developing regulatory frameworks to govern autonomous vehicles is a significant challenge. As a recent Rand Corporation study demonstrated, AVs need to be tested for billions of miles – in the authors' estimate, equivalent of hundreds of years – to generate statistically significant results for lower accident rates (Kalra & Paddock, 2016). In addition to this, the debates on liability and attribution when it comes to errors in complex systems that involve autonomous machines and a vast network of human actors (users, firms, programmers, executives, regulators, pedestrians, etc.) are far from settled, and have indeed prompted a slowdown in the roll-out of truly autonomous vehicles – as opposed to vehicles with automated features and partially autonomous functions. In this evolving context, cities, states, and federal authorities must design and plan forward-looking and adaptive regulations. These rules must both guide and evolve with technology while generating new knowledge, ensuring safety, and updating vehicular requirements, including driving, parking, and licensing.

With the growing relevance of AVs in urban mobility, the roles of cities need to be reassessed in the context of evolving multi-scalar regulatory frameworks. Urban mobility involves complex webs of alliances, investments, and partnerships of public and private-sector actors (Alvarez León & Aoyama, 2022; Aoyama & Alvarez León, 2021). As traditional modes of owning and operating cars give way to new arrangements, urban traffic patterns and other key urban dynamics will see significant transformations. Cities must anticipate these transformations and re-examine how they plan public infrastructure, provide public transit, and make budgetary decisions.

In this chapter, using evidence from American cities, we analyse urban governance within a multi-scalar regulatory framework to better understand the conditions shaping the emerging AV deployment. While cities are key in this mobility revolution, they perform their roles in a broader context where federal and state governments negotiate new rules, regulations, and standards. In the next section, we describe multi-scalar regulatory frameworks for the future of urban mobility. We then develop a typology of roles that cities play as part of multi-scalar governance as they plan for AVs and other mobility innovations.

A Multi-Scalar Regulatory Framework for Autonomous Vehicles

The literature on multi-scalar governance has varying intellectual roots, ranging from the socio-technical tradition represented by multi-level perspectives (Geels, 2002, 2010) to actor-network theory (see, for example, Caprotti et al., 2020). Attention to multi-scalar governance is vital in developing solutions that address issues across administrative boundaries. Transportation is one such case, where regulatory frameworks developed within and across administrative units have shaped the development of automotive technologies, specifications, and markets (see Gorachinova, Huh, and Wolfe, chapter 11).

Due to the radical nature of a shift from human to autonomous driving, AV development and deployment will inevitably affect all levels of government. As central nodes of population, infrastructure, and vehicular concentration, cities face unique challenges in this technological transition. Federal and state regulations are in rapid transition to accommodate the emerging urban mobility sector. At the federal level, the Department of Transportation issues voluntary guidelines for autonomous vehicle development and deployment (Brown et al., 2018). States, by contrast, take a greater role in issuing mandatory regulations for autonomous vehicle operations, using existing transportation regulations (including traffic laws, driver licensing, vehicle registration/inspection, and insurance). To coordinate efforts across states and keep track of numerous developments, the National Conference of State Legislators (NCSL) developed an online database that tracks state bills on autonomous vehicles. As this database reveals, legislative actions at the state level depict a dramatically changing landscape. According to the NCSL (2023), 22 states enacted 49 bills related to autonomous vehicles in 2023, up from 15 states and 18 bills in 2018. Since 2012, 47 states and the District of Columbia have introduced 634 pieces of legislation, and, to date, 42 states have enacted 136 bills.

In addition to the patchwork of autonomous vehicle legislative actions at the state level, policy at the federal scale has been discontinuous across administrations, and often internally inconsistent. In January 2017 (US Department of Transportation, 2020), the Obama administration, through the United States Department of Transportation (DoT), designated 10 proving ground locations across 8 states (2 in California, 1 each in Florida, Iowa, Maryland, Michigan, North Carolina, Pennsylvania, Texas, and Wisconsin). However, the Trump administration reportedly withdrew federal support on these sites (Bean & Courtney, 2018). Furthermore, the DoT and the Trump administration jointly expressed support for voluntary consensus standards by the private sector, against criticisms by the National Transportation Safety Board (NTSB). With the start of the Biden administration, however, a different government agency, the National Highway Traffic Safety Administration (NHTSA), has taken a more active stance in regulating the development of autonomous vehicles, paying particular attention to Tesla's "autopilot" function and the company's inadequate accident reporting (Barlett, 2022; Siddiqui, 2022).

Despite the federal government's enormous influence, state actions have been instrumental in triggering nationwide regulatory frameworks – with important precedents in the private sector. State-level regulations can prompt a process known as "technology forcing," which directly influences the product specifications, industrial processes, and strategic choices of automakers (Tillemann, 2016). The State of California, with its regulations for vehicle emissions, offers one salient example. Since the late 1960s, California has used the research and recommendations of the California Air Resource Board (CARB) to set stringent environmental standards that require automakers to develop new technologies to reduce carbon emissions. After protracted resistance and appeals to the federal government, the automakers eventually met many of the state's standards. These regulations led the auto industry to make changes not only in California but in the entire United States market as it is more economical to develop a "50-state car" that automakers can sell nationwide. The CARB's most recent vote to phase out the sale of new gasoline-powered vehicles by 2035 is another groundbreaking plan.

Thus, state-level regulatory gatekeepers can pressure the largest single territorial market in the United States. California's laws have not only shaped automobile environmental standards but also reconfigured the national market itself. The regulatory battles between CARB and the automakers resulted in substantial R&D towards fuel efficiency, hybrid engines, and improved batteries. Moreover, state-level regulations around vehicular safety, liability/insurance, and emissions have structured the governance frameworks that shape the US car market and auto industry. Therefore, although national standards are key to the growth of the AV industry (Anderson et al., 2016), certain progressive state regulations may serve as benchmarks. In the emerging mobility transition, policymakers can apply these lessons to address new issues raised by AV technologies, such as privacy and cybersecurity.

Similarly, a separate legislation is in place for ride-hailing. As mobility services are increasingly offered through a platform, ride-hailing may develop in tandem with autonomous driving, particularly within the MaaS model (see Lorinc, chapter 14). Accordingly, a regulatory analysis for mobility transition will require not only multi-scalar but also multidimensional contextualizing (see Gorachinova, Huh, and Wolfe, chapter 11). Cities and counties have traditionally led the way in regulating the taxi business, but with the onset of ride-hailing (e.g., Uber, Lyft), state regulations have been implemented along with the federal legislations to regulate this emerging sector. These regulations specify the nature of background checks and the type of licences required for the drivers as well as such matters as data-sharing and wheelchair accessibility.

Cities are part of this multi-scalar regulatory structure, and mobility technologies like AV take place under conditions shaped by state and federal authorities. With their concentrations of population, capital, and vehicular traffic, cities remain at the centre of debates on AVs. For one, regulations for testing and licensing are developing at the state level, potentially leading to the lack of national coordination (Fagnant & Kockelman, 2015). For another, cities are best positioned to adapt regulations given local contexts such as land use, zoning, and road infrastructure (e.g., crosswalks, lanes, and curb space) as well as traffic flows and public right-of-way rules. Furthermore, the introduction of AVs can have direct impacts on cities: from projections of reduced energy consumption and traffic congestion to reduced demand for parking (Litman, 2017). Efficiencies generated by AVs, along with the new modes of use and ownership, may also affect taxation, public transit fees, tolls, vehicle sale taxes, municipal parking, and registration fees (Corwin & Pankratz, 2017) As part of multi-scalar governance, cities must balance their own rules of the road with those of higher-level administrative hierarchies. For this reason, we must examine the strategies of engagement that cities can develop as they manage the deployment of new mobility technologies, such as AVs.

A Typology of Cities' Engagements with Autonomous Vehicles

Cities are taking the leadership in experimenting with and regulating AVs (Rainwater & Dupuis, 2018). They do so in a number of ways. Urban policymakers not only oversee the development of infrastructure but also craft regulations on certifying automated driving systems (ADS) and determine rules for liabilities and insurance against damages (Brown et al., 2018; Lewis et al., 2017). Furthermore, in the United States, cities have spearheaded the public-sector effort by developing testing sites for AVs. The testing sites typically involve the city, the land for the site, and private-sector firms involved in the AV market.

The federal level does not offer a straight path, and this absence has delayed a national regulatory framework and limited support for local projects. Yet cities are forging ahead with AV testing. According to a Bloomberg/Aspen Institute Atlas of cities around the world preparing for AVs, 136 cities were identified as "setting goals, mobilizing resources, and providing oversight and evaluation for AV efforts" in 2019. US cities make up the majority of existing AV testing sites on this list (Bloomberg Philanthropies, 2019). Many of these testing sites represent partnerships between the public, private, and non-profit sectors. Figure 12.1, below, shows a breakdown of planned and operational deployments of connected vehicles in the United States, as reported by the Department of Transportation. Of the 171 total projects shown in the map, 70 were operational, and 101 were in planning stages as of June 2022.

This widespread implementation of connected vehicle projects across the United States dovetails with the active role of cities, which are incorporating AVs in urban regulations. This, in turn, highlights the need to understand how cities engage with AVs and under what conditions. To address this need, we present the following typology, with four modes of engagement, which we consider complementary rather than mutually exclusive: regulator, promoter, mediator, and data catalyst.

City as Regulator

In this classic public-sector role, cities regulate AV testing. For instance, cities often decide where testing can take place, when, and under what conditions (for example, during traffic jams, at night, in certain weather conditions, etc.) typically through its transportation division and its unit in charge of smart city initiatives. As was the case of Uber at its inception, ride-hailing was initially viewed as an illegal taxi service, leading

Figure 12.1. Planned and operational connected vehicle deployments in the United States

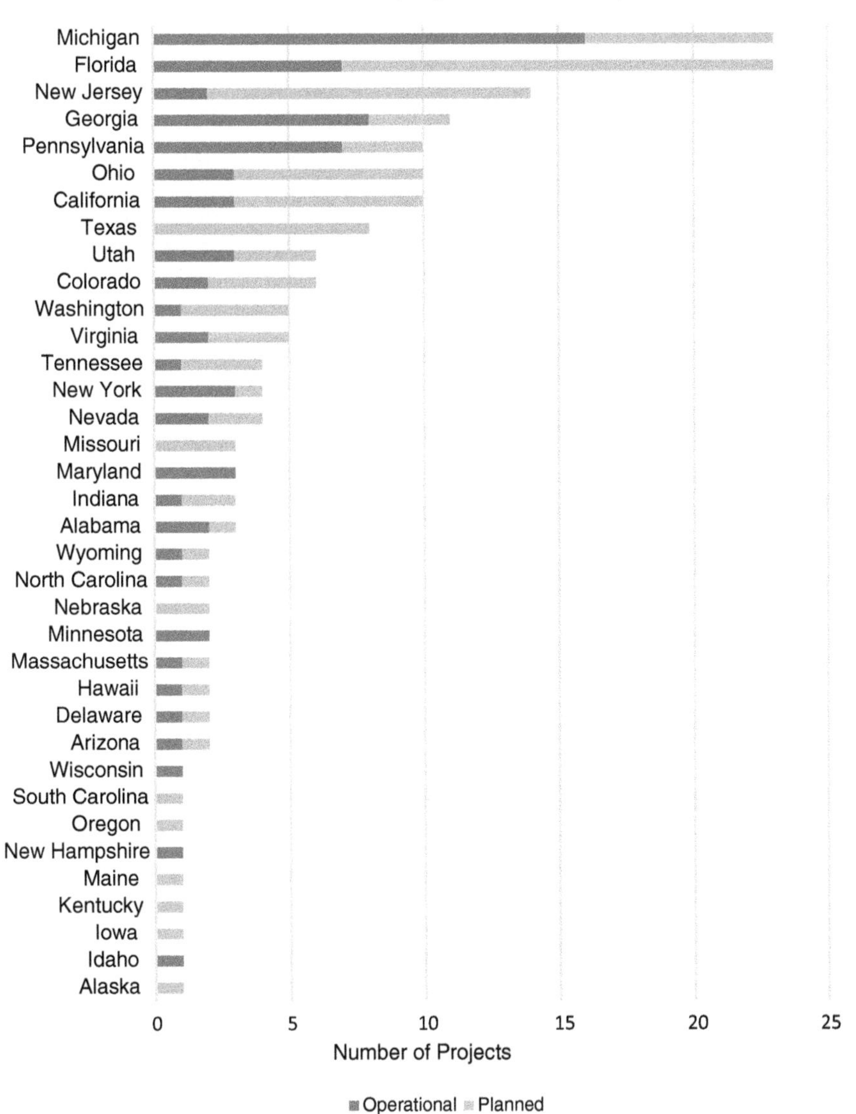

Source: USDOT, 2022 https://www.transportation.gov/sites/dot.gov/files/2022-06/CVlocations_USA_map_06.29.2022.jpg

San Francisco Metro Transit Authority (SFMTA), which regulates the taxi sector, to issue a cease and desist order (James, 2018). Now known as transportation network companies (TNCs), ride-hailing companies are increasingly regulated at the level of the city, which often imposes stricter rules than the federal and state regulations. The experience of cities and the regulation of TNCs highlights the urgency of addressing the unprecedented safety, security, and governance challenges brought about by AVs. And, while these challenges are only beginning to come into view, early indications suggest that careful regulation will be necessary to avoid widespread harms.

One such indication appeared in a tragic incident that took place in Tempe, Arizona, on 18 March 2018, when an Uber vehicle on autonomous mode killed pedestrian Elaine Herzberg. This first documented fatality from an AV crash led to widespread public condemnation and investigation into the company as well as debate about policies regarding autonomous driving. The incident tragically highlighted the need for regulatory frameworks to safeguard the public from accidents caused by technological or human error, and to ensure road safety as well as cybersecurity. In this particular case, a federal investigation by the National Transportation Safety Board split the responsibility for Herzberg's death between the company (Uber) and the human operator in charge of monitoring the vehicle at the time of the accident (Domonoske, 2020).

As companies begin to combine ride-hailing and AVs, cities will take on a larger role in regulating urban mobility. Cities will not only facilitate public transit but also shape the deployment of new mobility options and negotiate with other levels of government to craft rules and mitigate harms. Furthermore, these new mobility options will rely on public infrastructure (e.g., roads), affect city life, and collect large volumes of data – implications that will require the public sector's active involvement.

City as Promoter

Cities have long been viewed as "growth machines" that derive economic value from, among other key urban activities, the intensification of land use (Molotch, 1976). The literature on entrepreneurial cities (Hall & Hubbard, 1996; Jessop, 1998, 2019; Jessop & Sum, 2000; Roberts & Schein, 1993; While et al., 2004) and entrepreneurial urbanism (Peck, 2014; Ward, 2003a, 2003b) demonstrates that urban redevelopment has become both a political act involving financial risks and a performative act. As such, cities are already active in promoting the new urban mobility services by providing testing sites for AVs.

The City of Pittsburgh provides a fitting example of a promoter. Over the past decade, Pittsburgh has embraced self-driving cars as a competitive advantage, making early efforts to attract AV development and deployment. These efforts have been inspired by Carnegie Mellon's pioneering Robotics Institute and recent investment from technology firms, along with offices opened by Google, Apple, and Facebook. Capitalizing on its competitive advantage, Pittsburgh became the location for a critical mass of firms in the AV industry, including Delphi, Argo AI, and Aurora. Of these firms, Uber has proven the most salient, since it made a significant investment by building its Advanced Technologies Group Center in 2015, with a forty-two-acre "simulated town" testing facility for self-driving cars located along the Monongahela River (Goldberg, 2017).

The greater Ann Arbor area, Michigan, also exemplifies the city as promoter. As a testbed for AV systems, the city played a crucial role in bringing together a coalition of partners (including Michigan's Department of Transportation, the University of Michigan, the US Department of Transportation, and industry participants) to foster the development of AVs. The University of Michigan Transportation Research Institute (UMTRI) operates the Mobility Transportation Center (MTC), a joint industry-university R&D collaboration, and runs MCity, an experimental ground involving thirteen hectares of test courses. In addition to MCity, the Ann Arbor area became the host of American Center for Mobility (ACM) in 2016, a much larger testing ground on the site of a closed GM plant in Willow Run. While MCity focuses on research, ACM serves as a site for testing applied technologies, acquiring certification from the National Highway Traffic Safety Administration (NHTSA), and developing a technical labour force. Also, in 2012, Ann Arbor established a public road test area called Connected Vehicle Safety Pilot Model Deployment (SPMD) in the northeastern part of the city in collaboration with UMTRI and the US Department of Transportation. MTC plans to expand SPMD with vehicle-to-vehicle (V2V) and vehicle-to-infrastructure (V2I) communication networks.

City as Mediator

With respect to new mobility technologies, cities may also act as mediators, coordinating across a broad range of stakeholders. In this role, cities function as part of a broad consortium that coordinates interests across public, private, and non-profit sectors for a range of technological, logistical, and infrastructural requirements for adopting new mobility technologies, such as AV testing. Public-private partnerships have long become a vehicle for urban regeneration (Kort & Klijn, 2011), community development (Stephenson Jr., 1991), and transport (Estache et al., 2008), with some attributing to such partnerships the devolution of the state and subsequent neo-liberalism (Purcell, 2008), while others observing the limits of the public- and the private-sector functioning independently, with the growing role of the "hybrid domain" (Aoyama, 2016).

The mediator role is also multi-scalar. For instance, regional transportation agencies that operate regional commuter rails and highway maintenance may also develop a plan and regulations for AV testing and operation. In some ways, AV testing sites are grounds for experimentation not only with novel technology but also with collaborative governance among levels of government.

Two organizations embody this mode of engagement. The first is Open Mobility Foundation, a global coalition led by cities that coordinates an effort to develop new digital tools to organize mobility options and data among various cities (Open Mobility Foundation, 2019). It partnered with OASIS, an open-source and software standard industry, to do this work. The second is the non-profit American Automobile Association (AAA), which purchased the largest autonomous vehicle testing site in the US from the local (county) authority, Contra Costa Transportation Authority (Descant, 2019).

These public-private partnerships may involve universities, some with not just a local presence but also international stakeholders. For example, DriveOhio is a partnership between the public sector (four cities), private industries, and research organizations for smart mobility. It has attracted a combined investment of $500 million to develop and test AV technologies, and it is currently constructing 164 miles of roadway (Ohio Department of Transportation, 2020). MCity, part of Ann Arbor's successful promotion of autonomous vehicles, provides another example. Located within the University of Michigan, it includes fifty-nine industry partners such as APTIV, Denso, Econolite, Ford, GM, Honda, Intel, LG, StateFarm, Toyota, and Verizon, and boasts forty-four hundred hours of testing in its facility (University of Michigan, 2020).

Several other organizations use partnerships, mediated by cities, to advance their AV goals. For example, GoMentum in the San Francisco Bay area, California, in addition to automobile companies and ride-hailing companies, brings together a variety of groups. International partners include Baidu, the University of Waterloo, Tongji University, Singapore's Ministry of Transport, ITS Australia Intelligent Transport Systems, and ITS New Zealand. Local partners include transport authorities such as Intelligent Transportation Society of California,

Bay Area Rapid Transit (BART), and Bay Area Air Quality Management District (Gomentum Station, 2020). In other examples, Toyota has partnered with Stanford, MIT, and the University of Michigan to develop AVs. Uber has partnered with the University of Arizona, specifically for mapping and optical safety, and hired forty engineers from Carnegie Mellon Robotics Lab. An Irish AV firm, Valeo, tests its system at the National University of Ireland at Galway for rainy weather testing.

City as Data Catalyst

Smart cities, which initially emerged as "a high-tech variation of the 'entrepreneurial city'" (Hollands, 2008, p. 303), have since evolved from wired to wireless with ubiquitous cyber access, along with the rise of big data and cloud computing (Batty, 2012, 2013; Kitchin, 2014; Townsend, 2013). In this evolution, smart cities are functioning as public-facing data catalysts involved in complex partnerships among public agencies, private technology companies, and research institutions. As a data catalyst, a city must collect data on its residents' transport behaviour as well as manage data governance, including ensuring cybersecurity, privacy, and safety.

Significantly, cities must also combine data generated by the public and private sectors. In the public sector, cities provide important information for AVs, from physical/geographical data on roads, traffic lights, crosswalks, curbs, and street lighting to flow data on vehicle traffic patterns and transit demand. In the private sector, automobiles are already a rich source of consumer data (Fowler, 2019; Hanvey, 2019). Indeed, without regulations for data security, drivers and passengers are subject to systematic privacy infringement. And since the manufacturer – not the operator – owns the data generated by vehicles, automakers have amassed considerable data, including how people use infotainment, when and where they travel, and how they drive their cars (patterns of speed, braking, etc.). The data are highly valuable not only to the automakers and their suppliers but also to insurance brokers and other third parties.

Since cities are embedded in multi-scalar regulatory frameworks, their particular roles and capabilities are shaped through interactions with other governance scales. For example, the European Union has been aggressive in developing regulations such as the General Data Protection Regulation, or GDPR, that increase personal data privacy and cybersecurity protections (Lim & Taeihagh, 2018). For cities to function as data catalysts, such regulations are essential. Japan, China, and Singapore are also following suit with a number of regulations on data privacy and cybersecurity. In the United States, the National Highway Traffic Safety Administration (NHTSA) provides recommendations for responding to cyber incidents and conducting Voluntary Safety Self-Assessments. The SPY Car Act, first introduced to the Senate by Massachusetts Senator Ed Markey in 2015, requires a Cyber Dashboard with up-to-date cybersecurity information. At the level of the state, California was the first to legislate a data collection disclosure agreement for AV manufacturers in 2012, and other states (such as Georgia, Massachusetts, Michigan, and Tennessee) soon followed. However, AV manufacturers are generally reluctant to share data with each other and with local authorities (see Vinodrai, chapter 4).

Another example is the City of Boston, which was selected by the World Economic Forum in 2016 as a focus city for policy and pilot development for autonomous vehicles. Boston devised a new transportation strategy vision, with active use of AVs, called Go Boston 2030. Adopting the Vision Zero commitment of "Zero deaths. Zero injuries. Zero disparities. Zero emissions. Zero stress," Go Boston 2030 seeks to rehabilitate the city's vehicular reputation. Among US cities, Boston has some of the worst commuter stress, due partly to weather and roads but also to notorious driving practices, an affinity for double parking, and poorly maintained lane markings. Boston has begun AV testing in the Seaport district by partnering with an MIT spinoff, nuTonomy (now called Aptiv after being purchased by Delphi Automotive in 2017), starting with a high-definition mapping of the test area in South Boston and the Seaport area in July 2018. Along with developing software that simulates complex traffic scenarios, Aptiv is conducting manoeuvre tests designed to stress-test the software. The city has also partnered with Optimus Ride, another MIT spinoff that focuses on obstacle detection. Optimus Ride has developed sensors with overlapping field-of-view to ensure multiple layers of redundancy for detecting obstacles, and it has created software to detect, track, and predict obstacles. These innovations can catalyse the collection of data crucial to AVs. Using Boston's unique research and design team, called the "Mayor's Office of New Urban Mechanics" (established in 2010), the city actively explores partnerships. Its shared research agenda goes beyond testing vehicles to business model development, experimenting with connected transportation infrastructure, boosting civic engagement, conducting research on autonomous mobility, and studying AV's implications on Boston's workforce.

Conclusion: Urban Regulations for the Future of Urban Mobility

The rise of AVs is injecting innovation, entrepreneurship, and new challenges into urban mobility. In some contexts, these innovations are disrupting what used to be the monopoly of the public sector. Our conceptual typology is intended to address crucial questions about the roles of cities as they engage with urban innovation within multi-scalar regulatory frameworks. During a transformative mobility transition, governments must build adaptive regulatory frameworks that can deliver public goods and apply lessons learned from the past century of automobile-dependent urban development. While this typology can be useful for cities and other actors to understand the possible forms of engagement, it can also shed light on the balance of priorities and trade-offs required to pursue a specific path for future mobility options, such as AVs.

While introduction of smart infrastructure is vital for cities to become more sustainable, liveable, and inclusive, socio-technical transitions fail without effective collaborative governance among stakeholders. In the era of Internet of things (IoT), artificial intelligence (AI), and mobility platformization, cities play a key role in ensuring civic priorities and coordinating multi-scalar (state, federal) regulatory regimes with multi-sectoral stakeholders. Beyond conventional analysis in public administration, the overwhelming majority of public-private partnerships are understood as the contracting out of services to the private sector. In the new era of virtual constructions of cities, what is needed is a robust theorization that involves strong civic engagements in collaborative governance as well as up-to-date empirical case studies of implementation challenges. Our evidence suggests that cities hold key positions in shaping the deployment of AVs. Through varied roles, they actively insert themselves into multi-scalar frameworks where new rules, regulations, and standards are negotiated and enacted.

Building on the experience of TNCs and ride-hailing services, cities must be prepared to address the transformative potential of new mobility technologies. In particular, they must decide the terms on which these technologies are incorporated into the lives of urban residents and commuters. Our typology of cities reveals the permeability and flexibility of urban engagements with AVs, connecting in various ways with other governments at multiple scales (e.g., state, federal) as well as universities, private actors (e.g., automakers, start-ups), and NGOs to establish new alliances and partnerships. Although these partnerships will vary widely across urban experiences, they nevertheless must be responsive to the needs and voices of people in cities. Therefore, rather than being passive recipients of innovation and technological change, cities must be active participants and engage with the new mobility in roles that may include, combine, or transcend the four we describe.

Where might this mobility revolution lead? Will it erode public services, expand a market for urban mobility services, and/or accelerate the shift towards platform-centric business models? The answers will depend on the dynamics between key stakeholders, mediated by highly contingent, often geographically specific factors in cities around the world.

These questions become even more urgent in the face of acute disruptions to urban life, such as the multidimensional impacts of the COVID-19 pandemic and climate change. At its height, the pandemic dramatically reduced the use of public transit and ride-hailing services, and the rates of recovery have been variable at best. It also interrupted the global supply chain of semiconductor chips, which in turn led to rising prices and an acute shortage of new vehicles, including electric and autonomous. As industries are currently developing electric vehicles and AVs in tandem, the greater awareness of climate change may expedite the introduction of autonomous functions. While it is too early to know how either the pandemic or climate change will shape the future of AVs, we must analyse the role of cities as they prepare for this transition. Ultimately, we must understand how cities are engaging with new mobility technologies, precisely because such technologies (and the resources directed to them) may – or may not – help cities adapt to ongoing and future disruptions.

REFERENCES

Alvarez León, L.F., & Aoyama, Y. (2022). Industry emergence and market capture: The rise of autonomous vehicles. *Technological Forecasting and Social Change, 180*, Article 121661. https://doi.org/10.1016/j.techfore.2022.121661.

Anderson, J.M., Kalra, N., Stanley, K.D., Sorensen, P., Samaras, C., & Oluwatola, T.A. (2016). *Autonomous vehicle technology: A guide for policymakers*. Retrieved from https://www.rand.org/pubs/research_reports/RR443-2.html.

Aoyama, Y. (2016). Business in the public domain: The rise of social enterprises and implications for economic development planning. In *Insurgencies and revolutions* (pp. 125–36). Routledge. https://doi.org/10.4324/9781315545011-21.

Aoyama, Y., & Alvarez León, L.F. (2021). Urban governance and autonomous vehicles. *Cities, 119*, 103410. https://doi.org/10.1016/j.cities.2021.103410.

Barlett, J.S. (2022, 9 June). NHTSA expands Tesla autopilot investigation. *Consumer Reports*. https://www.consumerreports.org/car-safety/nhtsa-expands-tesla-autopilot-investigation-a7977631326/.

Batty, M. (2012). Smart cities, big data. *Environment and Planning B: Planning and Design, 39*(2), 191–3. https://doi.org/10.1068/b3902ed.

Batty, M. (2013). Big data, smart cities and city planning. *Dialogues in Human Geography, 3*(3), 274–9. https://doi.org/10.1177/2043820613513390.

Bean, R., & Courtney, S. (2018, 4 October). Trump plan is said to drop 10 US self-driving car test sites. *Bloomberg*. https://www.bnnbloomberg.ca/trump-plan-is-said-to-drop-10-u-s-self-driving-car-test-sites-1.1147287.

Bloomberg Philanthropies. (2019). *Bloomberg Aspen initiative on cities and autonomous vehicles / Global atlas of AVs in cities*. Bloomberg. Retrieved from https://avsincities.bloomberg.org/global-atlas/; accessible at https://www.bloomberg.org/press/bloomberg-philanthropies-launches-first-ever-autonomous-vehicles-map-living-inventory-cities-planning-driverless-future/.

Brown, A., Rodriguez, G., & Hoang, T. (2018). *Federal, state, and local governance of automated vehicles*. Issue Paper. UC Davis Institute of Transportation Studies & Policy Institute for Energy, Environment and the Economy. Retrieved from https://epm.ucdavis.edu/sites/g/files/dgvnsk296/files/inline-files/AV%20Gov_IssuePaper_FINAL_TH_2018_14_12%20%281%29.pdf.

Caprotti, F., Essex, S., Phillips, J., de Groot, J., & Baker, L. (2020, December). Scales of governance: Translating multiscalar transitional pathways in South Africa's energy landscape. *Energy Research & Social Science, 70*, Article 101700. https://doi.org/10.1016/j.erss.2020.101700.

Cohen, S.A., & Hopkins, D. (2019). Autonomous vehicles and the future of urban tourism. *Annals of Tourism Research, 74*, 33–42. https://doi.org/10.1016/j.annals.2018.10.009.

Cooper, P., Tryfonas, T., Crick, T., & Marsh, A. (2019). Electric vehicle mobility-as-a-service: Exploring the "tri-opt" of novel private transport business models. *Journal of Urban Technology, 26*(1), 35–56. https://doi.org/10.1080/10630732.2018.1553096.

Corwin, S., & Pankratz, D. (2017). *Forces of change: The future of mobility*. Retrieved from tps://www2.deloitte.com/content/dam/insights/us/articles/4328_Forces-of-change_FoM/DI_Forces-of-change_FoM.pdf.

Descant, S. (2019, 4 January). AAA acquires largest Autonomous vehicle test site in the country. *Government Technology*. Retrieved from https://www.govtech.com/fs/automation/AAA-Acquires-Largest-Autonomous-Vehicle-Test-Site-in-the-Country.html.

Domonoske, C. (2020, 25 February). *Tesla river was playing game before deadly crash. But Tesla software failed, too*. NPR. Retrieved from https://www.npr.org/2020/02/25/809207519/tesla-driver-was-playing-game-before-deadly-crash-but-tesla-software-failed-too.

Duarte, F., & Ratti, C. (2018). The impact of autonomous vehicles on cities: A review. *Journal of Urban Technology, 25*(4), 3–18. https://doi.org/10.1080/10630732.2018.1493883.

Estache, A., Juan, E., & Trujillo, L. (2008). *Public-private partnerships in transport*: The World Bank. https://doi.org/10.1596/1813-9450-4436.

Fagnant, D.J., & Kockelman, K. (2015). Preparing a nation for autonomous vehicles: Opportunities, barriers and policy recommendations. *Transportation Research Part A: Policy and Practice*, *77*, 167–81. https://doi.org/10.1016/j.tra.2015.04.003.

Fowler, G.A. (2019, 17 December). What does your car know about you? We hacked a Chevy to find out. *Washington Post*. Retrieved from https://www.washingtonpost.com/technology/2019/12/17/what-does-your-car-know-about-you-we-hacked-chevy-find-out/.

Geels, F.W. (2002, December). Technological transitions as evolutionary reconfiguration processes: A multi-level perspective and a case-study. *Research Policy*, *31*(8–9), 1257–74. https://doi.org/10.1016/S0048-7333(02)00062-8.

Geels, F.W. (2010, May). Ontologies, socio-technical transitions (to sustainability), and the multi-level perspective. *Research Policy*, *39*(4), 495–510. https://doi.org/10.1016/j.respol.2010.01.022.

Goldberg, M. (2017, 18 October). *Uber built a miniature fake city in Pittsburgh to test self-driving cars*. The Drive. Retrieved from https://www.thedrive.com/tech/15241/uber-built-a-miniature-fake-city-in-pittsburgh-to-test-self-driving-cars.

Gomentum Station. (2020). *Gomentum station: Test drive the future*. Retrieved from http://gomentumstation.net/.

Hall, T., & Hubbard, P. (1996). The entrepreneurial city: New urban politics, new urban geographies? *Progress in Human Geography*, *20*(2), 153–74. https://doi.org/10.1177/030913259602000201.

Hanvey, B. (2019, 20 May). Your car knows when you gain weight. *New York Times*. https://www.nytimes.com/2019/05/20/opinion/car-repair-data-privacy.html.

Hollands, R.G. (2008). Will the real smart city please stand up? *City*, *12*(3), 303–20. https://doi.org/10.1080/13604810802479126.

Inkinen, T., Yigitcanlar, T., & Wilson, M. (2019). Smart cities and innovative urban technologies. *Journal of Urban Technology*, *26*(2), 1–2. https://doi.org/10.1080/10630732.2019.1594698.

James, O. (2018, 24 July). *Uber and Lyft are lobbying states to prohibit local regulation*. Mobility Lab. Retrieved from https://mobilitylab.org/2018/07/24/uber-and-lyft-are-lobbying-states-to-prohibit-local-regulation/.

Jessop, B. (1998). The narrative of enterprise and the enterprise of narrative: Place marketing and the entrepreneurial city. In A.M Orum (Ed.), *The entrepreneurial city: Geographies of politics, regime and representation* (pp. 77–99). Wiley. https://doi.org/10.1002/9781118568446.eurs0090.

Jessop, B. (2019). Entrepreneurial city. *The Wiley Blackwell Encyclopedia of Urban and Regional Studies*, 1–10. https://doi.org/10.1002/9781118568446.eurs0090.

Jessop, B., & Sum, N.-L. (2000). An entrepreneurial city in action: Hong Kong's emerging strategies in and for (inter) urban competition. *Urban Studies, 37*(12), 2287–313. https://doi.org/10.1080/00420980020002814.

Kalra, N., & Paddock, S.M. (2016). Driving to safety: How many miles of driving would it take to demonstrate autonomous vehicle reliability? *Transportation Research Part A: Policy and Practice, 94*, 182–93. https://doi.org/10.1016/j.tra.2016.09.010.

Kitchin, R. (2014). The real-time city? Big data and smart urbanism. *GeoJournal, 79*(1), 1–14. https://doi.org/10.1007/s10708-013-9516-8.

Kort, M., & Klijn, E.H. (2011). Public–private partnerships in urban regeneration projects: Organizational form or managerial capacity? *Public Administration Review, 71*(4), 618–26. https://doi.org/10.1111/j.1540-6210.2011.02393.x.

Kuhnert, F., Sturmer, C., & Koster, A. (2018). *Five trends transforming the automotive industry*. Retrieved from https://www.pwc.com/gx/en/industries/automotive/assets/pwc-five-trends-transforming-the-automotive-industry.pdf.

Lewis, P., Rogers, G., & Turner, S. (2017). *Beyond speculation: Automated vehicles and public policy*. Eno Center for Transportation. Retrieved from https://www.enotrans.org/etl-material/beyond-speculation-automated-vehicles-public-policy/.

Lim, H., & Taeihagh, A. (2018). Autonomous vehicles for smart and sustainable cities: An in-depth exploration of privacy and cybersecurity implications. *Energies, 11*(5), Article 1062. https://doi.org/10.3390/en11051062.

Litman, T. (2017). *Autonomous vehicle implementation predictions*. Victoria Transport Policy Institute Victoria, Canada.

Molotch, H. (1976). The city as a growth machine: Toward a political economy of place. *American Journal of Sociology, 82*(2), 309–32. https://doi.org/10.1086/226311.

National Conference of State Legislatures (NCSL). (2023). *Autonomous vehicles: Self-driving vehicles enacted legislation*. Retrieved from https://www.ncsl.org/transportation/autonomous-vehicles.

Ohio Department of Transportation. (2020). *DriveOhio: The future of smart mobility*. Retrieved from https://drive.ohio.gov/wps/portal/gov/driveohio.

Open Mobility Foundation. (2019). *Global coalition of cities launches the 'Open Mobility Foundation'* [Press release]. Retrieved from https://www.openmobilityfoundation.org/wp-content/uploads/2019/10/OpenMobilityFoundationLaunch_NewsRelease_25June2019_final2-1.pdf.

Peck, J. (2014). Entrepreneurial urbanism: Between uncommon sense and dull compulsion. *Geografiska Annaler: Series B, Human Geography, 96*(4), 396–401. https://doi.org/10.1111/geob.12061.

Purcell, M. (2008). *Recapturing democracy: Neoliberalization and the struggle for alternative urban futures*. Routledge.

Rainwater, B., & Dupuis, N. (2018, 23 October). *While federal and state governments take a back seat, cities are driving the regulation of autonomous vehicles*. Bloomberg. Retrieved from https://www.citylab.com/perspective/2018/10/cities-lead-regulation-driverless-vehicles/573325/.

Roberts, S.M., & Schein, R.H. (1993). The entrepreneurial city: Fabricating urban development in Syracuse, New York. *The Professional Geographer*, *45*(1), 21–33. https://doi.org/10.1111/j.0033-0124.1993.00021.x.

Siddiqui, F. (2022, 9 June). Federal investigators step up probe into Tesla Autopilot crashes. *Washington Post*. https://www.washingtonpost.com/technology/2022/06/09/tesla-autopilot-probe/.

Stephenson Jr, M.O. (1991). Whither the public-private partnership: A critical overview. *Urban Affairs Quarterly*, *27*(1), 109–27. https://doi.org/10.1177/004208169102700106.

Tillemann, L. (2016). *The great race: The global quest for the car of the future*: Simon and Schuster.

Townsend, A.M. (2013). *Smart cities: Big data, civic hackers, and the quest for a new utopia*: W.W. Norton & Company.

University of Michigan. (2020). *MCity*. Retrieved from https://mcity.umich.edu/.

US Department of Transportation. (2020). *USDOT AV Proving Grounds*. Retrieved from http://nationalavpg.com.

Ward, K. (2003a). Entrepreneurial urbanism, state restructuring and civilizing ' New' East Manchester. *Area*, *35*(2), 116–27. https://doi.org/10.1111/1475-4762.00246.

Ward, K. (2003b). The limits to contemporary urban redevelopment. 'Doing'entrepreneurial urbanism in Birmingham, Leeds and Manchester. *City*, *7*(2), 199–211. https://doi.org/10.1080/1360481032000136778.

While, A., Jonas, A.E., & Gibbs, D. (2004). The environment and the entrepreneurial city: Searching for the urban 'sustainability fix' in Manchester and Leeds. *International Journal of Urban and Regional Research*, *28*(3), 549–69. https://doi.org/10.1111/j.0309-1317.2004.00535.x.

13 The Evolution of Ride-Hailing Regulation in Canadian Cities: COVID-19 and Policy Convergence

AUSTIN ZWICK, MISCHA YOUNG, AND ZACHARY SPICER

Introduction

Regulating transportation-for-hire services has been a long-standing responsibility of most municipal governments. Spicer et al. (2019) found that Uber's rapid ascent represented both a market disruption and a policy disruption, with municipalities undertaking sweeping regulatory changes to accommodate the new firm. When Uber entered most North American markets, the company initially branded itself as a technology platform that merely connected passengers and operators, arguing that it should not conform to existing local transportation-for-hire regulations or be considered a taxi operator. Most municipalities did not agree with this interpretation and began enforcement against the upstart company (Spicer et al., 2019). The popularity of ride-hailing among the public, however, forced many municipal leaders to reconsider. Soon new regulatory classes for ride-hailing emerged, with most North American cities legalizing this form of transport (Spicer et al., 2019). As this regulation can be uneven across cities and subject to change in some jurisdictions, researchers have further examined how cities have enacted and evaluated regulation over time.

This chapter builds on the growing body of academic literature on the Canadian experience with municipal regulation of the ride-hailing industry. For instance, Zwick and Spicer (2018) outlined the benefits and social costs of ride-hailing and covered the early political and policy steps being taken in Canadian cities. Young (2019) warned that pre-emptively legalizing ride-hailing may lead to embedded players making future regulatory action difficult. Brail (2021) wrote about Canada's emerging ride-hailing regulatory environment nationwide, while Donald & Sage (2021) focused on the political and policy processes at play in a case study on Kingston, Ontario. Woodside et al. (2021) found that municipalities were scaling back on direct oversight due to a lack of enforcement mechanisms and the acceptance of ride-hailing into the private market of transportation-for-hire services. These works have contributed to a growing body of research on ride-hailing regulation in both Canada (Brail, 2018; Woodside, 2021) and abroad (Reilly and Lozano-Paredes, 2019; Yu et al., 2020). Projects like the Digital Mobilities Lab (www.digitalmobilitieslab.ca) have further added to this knowledge by producing digital regulatory maps that elaborate on regulatory approaches to ride-hailing across Canada. The COVID-19 pandemic has prompted a series of studies examining the impact on ridership (Meredith-Karam et al., 2021; Monahan and Lamb, 2022; Morshed et al., 2021;), the health of Uber as a firm and its shift in corporate strategy (Chi et al., 2022; Katta et al., 2020), and drivers' perception of Uber's pandemic response measures (Saksida et al., 2022). However, studies to date have not yet analysed the regulatory changes made by municipalities in response to the pandemic. In this chapter, we add to the literature by examining the evolving regulatory environment through various stages over time and during a crisis-induced reconsideration ignited by the COVID-19 pandemic.

Our analysis builds on Tabascio and Brail's (2022) framework, which examined the thirty largest Canadian cities according to eight policy criteria (listed in Table 13.2). We review the same cities with the same criteria, but we split municipal regulatory actions into three time periods that exhibit different patterns in their policymaking processes: initial responses (January 2016–December 2017), which exhibited policy experimentation; subsequent changes (January 2018–February 2020), which demonstrated policy learning; and pandemic-induced reactions (March 2020–May 2022), which revealed policy permissiveness. Our goals are twofold: (1) to understand the reasons municipalities review and update previously enacted ride-hailing regulations, and (2) to explain intermunicipal dynamics of policy learning with regards to ride-hailing. Our findings indicate that cities initially took individualized approaches, then modified their regulatory actions to align overall with those of other municipalities and to satisfy external stakeholders. We also find that policymaking mainly ceased during the pandemic, halting the regulatory evolution and creating a policy stasis – meaning a condition when policy innovation slows down within a given policy space and practices tend to reproduce themselves across jurisdictions (see Dunn, 2007; Hoppe, 2011). Zwick et al. (2022) discuss media attention and agenda-setting in relation to ride-hailing, noting that major events can stir municipalities to take concurrent regulatory action during time-limited legislative windows. When other issues become more pressing, policy items can fall off policymakers' agenda. By the time COVID-19 struck, most municipalities were reaching a regulatory consensus around the boundaries of acceptable operating standards for ride-hailing firms. The pandemic shifted the focus of regulators to other areas, leaving ride-hailing firms to institute new rules through the pandemic on their own (for a response to the COVID-19 pandemic by ride-hailing firms, see Brail and Donald, chapter 10). We conclude that the era of debates over the major components of ride-hailing regulation ended due to policy convergence, not COVID-19, despite the prima facie appearance of an exogenous shock.

This chapter proceeds in several sections. We first review literature on policy learning and regulatory change. We then construct a temporal framework for understanding the regulatory progression in Canadian cities with respect to policy learning processes, and we chart findings from thirty Canadian cities over three time periods. The following section dives into how and why COVID-19 caused a regulatory stasis. The final section ends the chapter with insights on policy learning and the future of platform economy regulation in Canada.

Policy Learning and Regulatory Change

Municipal policy and regulatory systems are not static. In fact, they often experience change and, like many policy systems, are sensitive to external events that demand response from local decision-makers (Howlett and Ramesh, 2002; Williams, 2009). Determining how and why this change occurs can be challenging, but a variety of theories can help researchers to make sense of the causes. In this section, we review some leading theories and explore how scholars have applied them to ride-hailing regulation in the past, providing a view into the nature of policy change, disruption, and continuity across jurisdictions.

Governments often make decisions informed by others' experience. This process, referred to as policy learning, takes place when policymakers experience common problems and examine the response of comparable jurisdictions (May, 1992). Decision-makers often engage in a learning process to better understand why a policy was implemented, what impact the policy had, what contextual factors led to the policy's failure or success, and what advantages their own jurisdiction could gain by adopting the policy (Stritch, 2005). This learning can take several forms. For example, decision-makers can deliberately try to account for both policy success and failure, or they can take a more organic approach, which gradually allows for consensus on certain policy ideas (Bennett and Howlett, 1992).

This learning may lead to a policy convergence, where several jurisdictions come towards a similar policy or regulatory position (Heichel et al., 2005). If a consensus approach emerges, other actors are likely to emulate it, creating a convergence point and producing some homogeneity in policy response. Past research on ride-hailing has shown the likely presence of a learning environment, where regulators converged upon a common approach over time (see Spicer et al., 2019). At this early point, regulators were unsure how to approach ride-hailing firms because the firms defied classification as taxi services and refused to abide by existing transportation-for-hire policies. This hesitation created an environment of policy uncertainty, leading to a scattered policy response. As a regulatory path emerged, uncertain actors adopted common regulatory elements. This shared response probably resulted from Uber's continual lobbying efforts (Tzur, 2019), which highlighted the approach (favourable to the company) of comparable jurisdictions (Beer et al., 2017). Once decision-makers reached this convergence point, change became incremental.

Policy change, however, can occur suddenly. Unexpected events – often referred to as "exogenous shocks" in policy literature – may realign policy systems or policy thinking, thereby setting up the conditions for institutional or organizational change (Williams, 2009). These types of events may open the possibility of a "policy window," which is an opportunity to bring about policy change when an issue suddenly captures the attention of decision-makers (Lieberman, 2002). These windows increase the likelihood that governments will align the problem with policy action. According to many researchers, the COVID-19 pandemic represents one such policy window (see Auener et al., 2020; Beland and Marier, 2020; Dupont et al., 2020; Minkler et al., 2020). This theorization makes intuitive sense, as the pandemic fundamentally affected economic, social, and political systems globally (DeLeo et al., 2021). In such an environment, policy options that seemed impossible suddenly become possible, allowing governments to test policy ideas for a period of time (Howlett, 1998). When windows open, however, they often require a "policy entrepreneur" to lead the policy change (Mintrom, 1997).

The Iterative Policy Learning Process for Platform Mobility Services in Canada

As we will show, municipalities in Canada have undergone an iterative policy-learning process with ride-hailing. This process provides insight into how Canadian municipal policy outcomes, at least with regards to platform economy regulation, are symbiotic rather than independent. Our first research step, using the same thirty cities and eight policy categories created by Tabascio and Brail (2022), was to update the data to a document analysis on every municipal act about ride-hailing passed between January 2016 and May 2022. In total, we reviewed thirty-five regulatory acts from the thirty municipalities' websites, and the analysis is up to date as of May 2022.

We then placed those thirty-five regulatory acts into three distinct periods of ride hailing regulation. An *Initial* period was defined by policy experimentation (January 2016–December 2017), a *Subsequent* period was defined by policy learning (January 2018–February 2020), and a crisis-induced *Pandemic* period was brought on by COVID-19 (March 2020–May 2022). During this third period, municipal policymaking came to a stasis. We theorize that this was in response to this exogenous shock. Thereafter, we read and examined each regulatory act for its relation to the eight policy categories. As Tabascio and Brail (2022) focused on describing the variation within each policy category, we instead focused only on whether the city made a policy change within the category or not (marked as a Y for yes and an N for no) during each time period. Driver requirements, for example, include background checks, licensing obligations, and driver training. Vehicle requirements include safety inspections, display of corporate emblems, and physical attributes of the vehicle. For a full description of what each category entails, see Tabascio and Brail (2022). Appendix A provides the full list of cities included and the time periods when they took regulatory action.

Table 13.1 demonstrates how cities revisited their regulatory frameworks in an iterative process that led to policy convergence in nearly all regulatory categories. The numbers displayed indicate what percentage of cities, from our sample of thirty, had policies in place in each of the categories during that time period. In our sample, twenty-seven out of thirty cities (90%) passed a municipal regulation on ride-hailing at least once during the full study period, between January 2016 and May 2022. Between the *Initial* and *Subsequent* period, the data shows that cities began coalescing around almost all policy areas: driver requirements (57% to 83%), vehicle requirements (57% to 80%), charging a fee per firm (57% to 87%), a fee per ride (50% to 80%), and data sharing (53% to 83%). Insurance requirements (63% to 87%) represents the only policy category that all cities acted on. On the flip side, fares (30% to 60%) and licence fees per vehicle (17% to 27%) are the only two policy categories that did not engage a supermajority of cities. Though the exact details of the policy category (e.g., size of the fees) may vary between

Table 13.1. Cumulative percentage of cities with ride-hailing regulation by time period

	Initial (Jan. 2016–Dec. 2017)	Subsequent (Jan. 2018–Feb. 2020)	Pandemic (March 2020–May 2022)
Driver requirements	57%	83%	87%
Vehicle requirements	57%	80%	83%
Insurance requirements	63%	87%	87%
Fares	30%	60%	60%
Fee per firm	57%	87%	90%
Licence fee per vehicle	17%	27%	30%
Fee per ride	50%	80%	80%
Data sharing	53%	83%	87%

municipalities, the areas of policy regulation clearly converged over time. As this convergence would not have occurred if stakeholders and municipalities had not been aware of changes happening in other jurisdictions, it thus exhibits policy-learning behaviour.

From the *Initial* period to the *Subsequent* period, municipalities revisited their ride-hailing regulations for various reasons. For each city, city council minutes, news articles, and reports from municipal bureaucracies discussed why the updated regulations were under consideration. When we reviewed these documents, the following reasons for revisiting became recurring themes:

- To create a permanent program after a successful pilot program
- To update driver licensing requirements, particularly around insurance requirements
- To update vehicle requirements, particularly around vehicle inspections
- To update and change fares
- To increase equity between taxi and transportation network service regulations

Typically, during the *subsequent* regulation period, cities legally defined taxi and transportation network services as distinct transportation-for-hire services (if they had not done so already) and then updated them both to follow similar rules. Halifax (2019), for example, updated its regulations to apply certain provisions to both services. Background checks must occur every four months, ensuring that drivers have no felony convictions and have not appeared on child abuse registries in the last ten years. Vehicles must be less than ten years old. Finally, drivers must not talk on cell phones while on fare commission, and they must accept credit/debit payment. The legislation also called for a further staff report on taxi meter fares and limousine standards (Halifax Regional Council, 2019). Similarly, Edmonton's subsequent regulation increased monetary penalties and enforcement on transportation network services that flouted municipal rules, including rules for valid driver and vehicle licences, mechanical certification, and information displayed, such as the driver's name and a current photo (Riebe, 2018). The City of Winnipeg (2019) added to its background check requirements and modified its fare schedule for both taxis and transportation network services to bring the two industries into alignment. This is what policy learning in action looks like.

We can attribute some of this convergence to cities passing their first round of regulations in the *Subsequent* period. However – as Table 13.2 shows – nine out of fourteen (64%) cities that passed a regulation in the *Initial* period amended their regulation in the *Subsequent* period, with two cities (14%) passing a third regulatory action. Out of the eight municipalities that passed their first ride-hailing regulation between January 2018 and February 2020 (*Subsequent* period), only one city (Saskatoon) revisited that regulation and did so under a unique circumstance in which a newly elected city council modified the regulation passed by the previous council six months earlier. Out of the entire sample of thirty cities, only one city (Surrey) passed its first regulatory action during the *Pandemic* period (March 2020–May 2022), but it had planned this regulation before the pandemic in response to British Columbia's legalization of ride-hailing in January 2020. Three cities out of our sample never passed a municipal ride-hailing regulation (Burlington, Markham, and Richmond Hill), all in Ontario.

Only two cities (Toronto and Edmonton) arrived at their current regulatory frameworks through three regulatory acts. In both cases, councils had previously passed a ride-hailing policy that asked their bureaucracy to report regularly on regulations and impacts. We can view this approach as a form of political compromise that automatically reinserts the issue into the legislative agenda going forward. For example, the City of Toronto (2021) report begins by summarizing the findings in its 2019 report, then discusses how the pandemic affected ride-sharing's numbers, and concludes by recommending future studies on congestion, curbside impacts, and road safety to keep the process going. Whether city councils choose to act on these reports and

Table 13.2. Number of regulatory steps by Canadian municipalities over time

	Regulatory steps			
Time period of first regulatory act	One step	Two steps	Three steps	TOTAL
Initial (Jan. 2016–Dec. 2018)	5	7	2	14
Subsequent (Jan. 2019–Feb. 2020)	11	1	–	12
Pandemic (Mar. 2020–May 2022)	1	–	–	1
TOTAL	17	8	2	27

turn them into legislative action would then depend on city council's priorities at those times. In this scenario, the iterative policymaking process (as well as the consequential policy learning) might have continued if not for the pandemic.

As ride-hailing regulation is an interplay between municipal and provincial governments, we investigated timing and iteration patterns among provinces. Our data show that different patterns emerged throughout Canada (see Table 13.3). For more information on the municipal-provincial relationship, see Tabascio and Brail (2022). Policy learning does not require municipalities to copy each other's bylaws verbatim but rather to learn from others' experiences when shaping their own approach. We see this kind of learning play out in different ways across provinces.

In Alberta, the two largest cities, Edmonton and Calgary, both took a multi-step regulatory process that involved overhauling taxi and transportation network service regulations to create a level playing field in terms of fares and insurance requirements between the ride-hailing drivers and taxi drivers. Determining what each participant considered fair dominated the discourse, and it took multiple attempts before stakeholders were satisfied. City council minutes within Manitoba and Saskatchewan tell a similar story.

In Quebec, three municipalities (Laval, Longueil, and Montreal) began pilot programs as part of a provincial program in 2016, extended them in 2018, and eventually made them permanent in province-wide legislation by Bill 17 in 2019. The provincial research and reporting on this pilot program embody the policy-learning concept, with final legislation learning from the program's previous iterations. In response to this province-led approach, some Quebec cities in our sample took a two-step process. The rest, not part of the original pilot, took only one step, but they still benefited from their peers' shared experience and thus exhibited policy learning.

Table 13.3. Number of cities ride-hailing regulatory steps by province in study

	Regulatory steps Group B				
Province	None	One step	Two steps	Three steps	TOTAL
AB	0	0	1	1	2
BC	0	4	0	0	4
MB	0	0	1	0	1
NS	0	1	0	0	1
ON	3	7	3	1	14
QC	0	3	3	0	6
SK	0	1	1	0	2
TOTAL	3	16	9	2	30

In Ontario, the provincial government allowed municipalities to regulate ride-hailing on their own. This hands-off approach led to a wide array of responses in geographically proximate municipalities within our sample. In brief, Ontario had both the greatest level of policy experimentation, shown by the number of acts passed, and the greatest amount of policy inaction, shown by three cities of Greater Toronto's urban fringe (Burlington, Markham, and Richmond Hill), which never passed ride-hailing regulations at all. Meanwhile, the smaller cities outside the Greater Toronto Area carved out their own path. As media coverage and council minutes reveal, the cities in the Greater Toronto Area often cited policies in neighbouring jurisdictions, learned from each other, and generally aligned their policies in an iterative manner, leading to multi-step processes.

Even though ride-hailing firms made an early foray into the province, British Columbia was the last Canadian province to officially legalize ride-hailing in January 2020. This long-anticipated move gave municipalities in the Greater Vancouver area time to work through a regional government organization, Metro Vancouver, to coordinate a mutually agreed-upon framework that each city council then passed individually. Though all cities in British Columbia passed regulations in a single step, their documented processes leaned heavily on studying ride-hailing regulations in other cities – a clear indication of policy learning.

The COVID-19 Regulatory Turn

COVID-19 has led to the death of over 14 million people worldwide since being declared a global pandemic in March 2020 (WHO, 2022). In addition to its dire health implications and the strain it has placed on health-care systems, the pandemic has also affected the global economy and transformed individuals' daily routines and travel needs. The COVID-19 virus is primarily transmitted through contact with infected individuals (WHO, 2021), leading many governments to limit social interactions through gathering restrictions, school and work closures, lockdowns, and stay-at-home orders. As expected, these measures have curtailed the transmission of the virus but have also significantly reduced daily mobility. Indeed, an analysis of data from 128 countries during the earlier months of the pandemic found that one week after implementing lockdown measures, countries saw mobility decline, on average, by 26 per cent (IMF, 2020).

Despite these measures, governments still permitted essential travel, and those who continued to travel seem to have favoured private vehicles over shared forms of mobility, such as public transit and ride-hailing, as this choice allowed them to avoid contact with other passengers and reduced their risk of contracting the virus. This preference for private vehicles revealed itself in aggregate anonymized cell-phone data collected by Apple in the United States during the earlier months of the pandemic. The data recorded a 76 per cent drop in routing queries for transit but only a 45 per cent drop in driving queries during that same time period (Jewell et al., 2021). A similar trend appears to have occurred in Canada, which saw its largest city, Toronto, temporarily lay off twelve hundred of its transit agency employees amid an 85 per cent drop in transit ridership during the pandemic. Another major city's transit agency, Vancouver's TransLink, lost an estimated $75 million per month due to lower transit ridership levels (Smart, 2020).

The ride-hailing industry did not escape this decline and experienced a similar or an even more abrupt reduction in ridership (see Brail and Donald, chapter 10). As governments across North America issued restriction measures, ride-hailing companies such as Uber and Lyft saw their business decrease by approximately 75 per cent (Goldstein, 2020; Hawkins, 2020). A study in Toronto found that among those who had previously used ride-hailing, over half had avoided this form of travel since the start of the pandemic (Loa et al., 2020). In the city of Vancouver, which had just legalized ride-hailing in January 2020, companies also saw a steep decline in ridership, going from 420,245 monthly trips in February 2020 to 138,985 monthly trips in April 2020, a 67 per cent decrease (Hara Associates, 2021). Strict lockdown measures and the ensuing closure of bars and restaurants during the height of the pandemic clearly explain part of the decline in ridership. However, other responses to the pandemic, such as working remotely from home and buying more goods online (Young et al., 2022), also decreased ride-hailing demand.

In addition, many people during the pandemic felt reluctant to be confined in a vehicle with a person they did not know (i.e., the driver). Ride-hailing companies sought to address this concern by instructing their drivers to wear masks and by providing drivers with hand sanitizer and disinfectant wipes for their vehicles (Uber, 2022a). Companies also reduced the maximal capacity of trips to avoid having passengers sit in the front seat and to establish some distance between the driver and passengers. Finally, to regain consumer confidence, ride-hailing companies suspended all shared services (e.g., UberPool and Lyft Share) during the pandemic to minimize contact between strangers (Bellon, 2020a).

In most cities, operators made and monitored these efforts on their own because few municipalities passed regulations that explicitly required ride-hailing companies to take these steps. Most municipalities either passed COVID-19 general safety guidelines or participated in provincial action, which they tried to enforce even when they lacked a legal mechanism. Ride-hailing operators were quick to treat these municipal guidelines as corporate rules, even if cities did not legally or explicitly compel companies to do so. One example is the *Reopening Ontario Act*, which required all passengers to wear masks and provided drivers with training on how to communicate with and accommodate people unable to wear masks (Toronto, 2022). To support government efforts in limiting the spread of the virus, Uber and Lyft also displayed messages to remind customers to travel only when necessary (Uber, 2020). In addition, Uber introduced a feature on its apps to inform Canadians of the safety and effectiveness of the approved COVID-19 vaccines and to help customers find vaccination sites nearby (Uber, 2022b). Both Uber and Lyft also provided data for contact-tracing purposes to Canadian health organizations (Bellon, 2020b). All these measures introduced by ride-hailing companies in Canadian cities during the pandemic were primarily to reduce passengers' fear of contracting the virus and to support – or at least appear supportive of – municipal guidelines.

What is also striking during this *Pandemic* period is the extent to which ride-hailing fell off regulators' policy agenda. In the two years before the pandemic, as many as twenty-one of the thirty cities we surveyed (70%) had either introduced or reviewed ride-hailing policies. Once the pandemic started in March 2020, the number of regulations – notwithstanding temporary COVID-19-related measures – fell to a mere three cities (Edmonton, Toronto, and Surrey). Moreover, as Surrey implemented its ride-hailing policy on 9 March 2020, in direct response to the Province of British Columbia's legalization of ride-hailing in January 2020, we argue that this policy occurred before the *Pandemic*

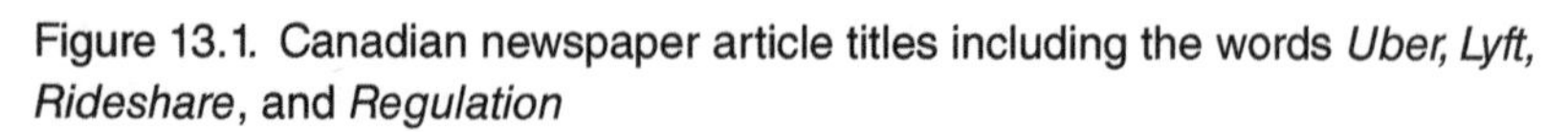

Figure 13.1. Canadian newspaper article titles including the words *Uber, Lyft, Rideshare*, and *Regulation*

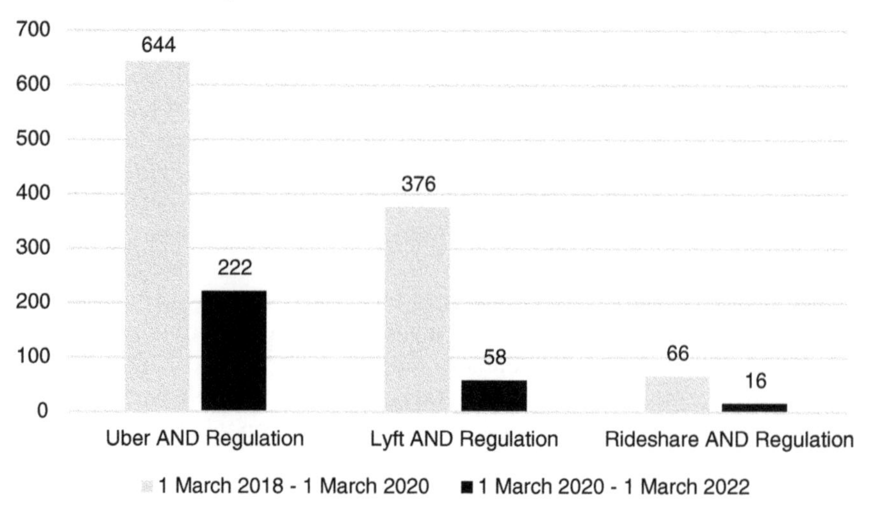

period. Regulating ride-hailing therefore seems no longer top of mind for policymakers.

This drop in political interest may have resulted from less discussion in the news media or public discourse. Indeed, using ProQuest Canadian Newsstream, we examined the frequency of Canadian newspaper article titles that included the words *Uber, Lyft, Rideshare,* and *Regulation* in the two years before and two years after the COVID-19 pandemic began, and we found a stark decline in news media attention (see Figure 13.1). Figures 13.2, 13.3, and 13.4 further display Google search interest for the terms *Rideshare, Lyft,* and *Uber* in Canada between 20 June 2018 and 20 June 2022. A notable decline in Google searches for these terms appears at the onset of the pandemic in March 2020 (denoted by a black vertical line), which we believe indicates a decrease in the public's interest.

By dominating the airwaves, the COVID-19 pandemic has shifted media attention away from rid- hailing, and this shift, we surmise, has redirected public interest away from the regulation debate. As Canadians are no longer using or hearing about ride-hailing as often since the pandemic started, they have turned their attention elsewhere, to seemingly more pressing issues, and governments have responded by leaving ride-hailing regulations in place, considering the status quo sufficient. In other words, municipal policymakers had more pressing

Figure 13.2. Google search interest for the term *Rideshare* in Canada

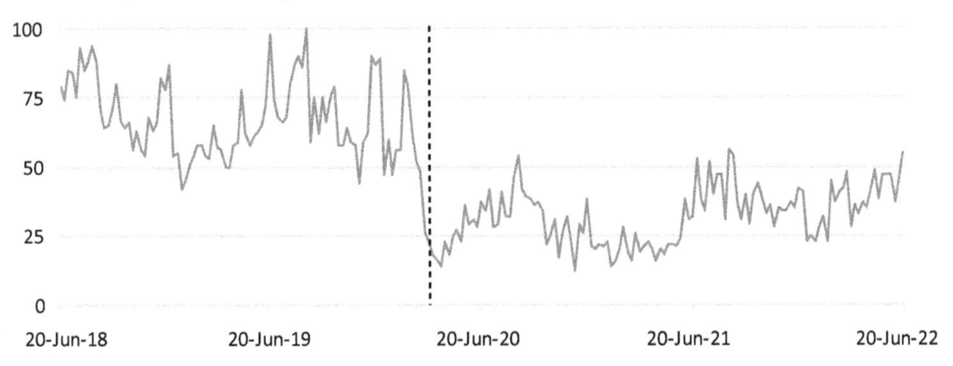

Figure 13.3. Google search interest for the term *Lyft* in Canada

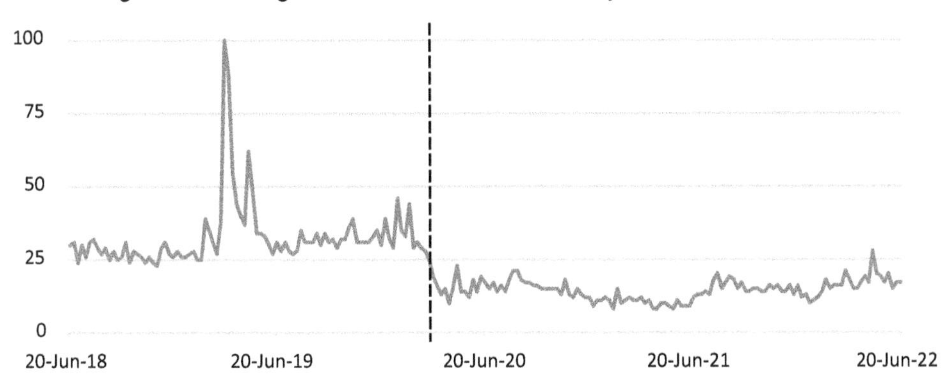

Figure 13.4. Google search interest for the term *Uber* in Canada

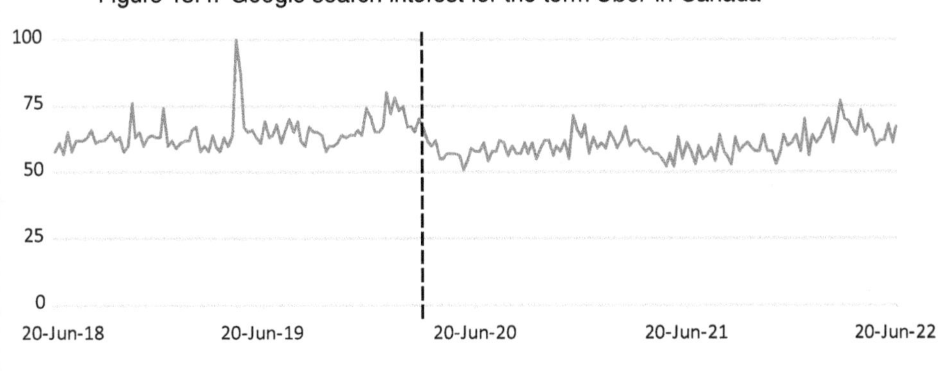

issues as well. What can be said for certain is that the exogenous shock of the pandemic created a period of policy stasis, a void that ride-hailing firms filled with self-regulatory actions. Only the future will tell whether the halt in policy learning and emulation was a temporary phenomenon or whether it brought about a premature and permanent end to municipal policymaking in the transportation-for-hire industry. It will be interesting to see whether regulators continue (or not) to take a more hand-off approach with regard to ride hailing regulations.

Conclusion and Discussion

Since the arrival of ride-hailing in Canada in 2014, Canadian municipalities took nearly a decade to legalize and regulate this form of transport. Although cities took different roads to get there, the destination was approximately the same, with remarkable policy convergence around driver, vehicle, and insurance requirements. This chapter shows that when we view the Canadian municipal ride-hailing regulatory experience from a top-down perspective, we can see a significant pattern. Most cities travelled from a period of policy experimentation and quick adoption based on local context, through a period of policy learning in which ride-hailing policies grew increasingly similar, to a *Pandemic* period where regulatory innovation slowed.

When Uber and Lyft first arrived in various markets, regulators (with some exceptions) mainly took a zero-tolerance approach to the upstart firms, enforcing existing transportation-for-hire regulations and arguing that ride-hailing operations looked enough like traditional taxi services that the two should share a common regulatory approach (see Spicer et al., 2019). This approach began to change as the popularity of ride-hailing grew. Specific regulations came into place based on local politics and context, mainly because municipal leaders still did not know how to classify firms like Uber and Lyft (Spicer et al., 2019). Ride-hailing was still new, and regulators were caught in a whirlwind of politics and protest. Technology played the role of disrupter to markets and regulators alike. As more cities regulated ride-hailing, a policy pathway emerged where municipal decision-makers began to adopt regulatory components that proved successful elsewhere (Spicer et al., 2019). Ride-hailing thus offered a prime example of municipal policy learning.

The pace of ride-hailing policy actions in Canada gained momentum between the *Initial* period (seventeen regulatory actions in total) and the *Subsequent* period (twenty-one regulatory actions in total), and this pace might have continued were it not for the pandemic. Instead, we found that only three cities passed ride-hailing regulations between March 2020 and May 2022 – a substantial decrease compared to the earlier time periods. A question thus arises: why did regulation cease? We have a couple of hypotheses.

On the one hand, it could be argued that an era of policy permissiveness began by an exogenous shock. Several municipalities passed pilot-period regulations with a fixed expiry to force their own councils to revisit the regulatory frameworks in short order, based on what did and did not work. When these councils passed the subsequent regulations, rules became permanent. However, that process did not mean that cities and provinces did not plan on reviewing their regulations again in the future. Without the pandemic, it is possible that the public and key stakeholders might have continued pushing city councils to take further action for the same reasons as the *Subsequent* period in our dataset (e.g., fairness between taxis and ride-hailing companies, enhanced vehicle and driver safety standards, etc.).

On the other hand, we argue that the regulatory window on rid-hailing was already closing, regardless of the pandemic, because of policy convergence. We find this interpretation more convincing as it aligns with the evidence laid out above. The early experimentation of ride-hailing regulation had turned into sustained policy learning, which eventually settled the major questions about necessary components. In order to design a more successful model for themselves, many cities started their administrative policymaking process by detailing what other similar municipalities were doing, and the resulting reports turned into bills for city council to review, amend, and pass, taking local political circumstances into consideration. Early movers had to take multiple regulatory steps, while last movers only needed to study what the early movers had done, and all ended up in the same place. In other words, when the exogenous shock of the pandemic hit, the enacted regulatory actions may have become "good enough" to satisfy most stakeholders (though rarely all). Thus, ride-hailing could be pushed off the municipal agenda by more pressing concerns – if not COVID-19, then something else. Once policy convergence happened, with one municipality's regulations aligning with those of most others, further work was not urgent. The remaining inequities and inadequacies of existing regulatory frameworks were no longer important enough to consume a city council's limited time and capacity.

We conclude that the era of debates over the major components of ride-hailing regulation ended due to policy convergence, not COVID-19, despite the prima facie appearance of an exogenous shock. We expect to see future tweaking of ride-hailing policy, for example,

Toronto introduced a cap on ride-hailing licences in October 2023, but not overhauls of major components. No municipality in the dataset tried to overturn any major component of ride-hailing regulation once it was in place.

One potential exception to this current period of regulatory stasis is the regulation of greenhouse gas (GHG) emissions of ride-hailing vehicles by municipalities as climate emergency legislation continues to move forward. There are currently several studies that measure ride-hailing's impact on GHG emissions, and they overwhelmingly conclude that the industry adds to traffic congestion (Erhardt et al., 2019) and consequentially increases emissions (Barnes et al., 2020; Rodier, 2018; Shen et al., 2020). Mayors in Metro Vancouver are calling for stricter vehicle standards for ride-hailing vehicles to reduce emissions beyond that of regular passenger vehicle regulations set by the province and federal government and in turn offset the increase already caused by this industry (Saltman, 2020). California's Clean Miles Standards requires ride-hailing firms to meet increasingly stringent GHG emission and electrification targets (California Air Resources Board, 2021), and it is only a matter of time before more Canadian municipalities follow suit.

Lately, regulatory efforts surrounding ride-hailing seem to have shifted from local issues (e.g., passenger safety, permits, etc.) to more national ones such as employment status and worker conditions. The enactment of Assembly Bill 5 in California may best exemplify this shift. The bill, which sought to establish worker protection in the gig economy (including ride-hailing), culminated in the subsequent passing of Proposition 22, a ballot initiative led by app-based companies to exclude ride-hailing and food delivery services from the bill. Spending over $200 million, Uber, Lyft, and other tech companies eventually convinced Californians to adopt Proposition 22, but the ruthlessness of the campaign and sheer sums of money spent demonstrate how far these companies are willing to go to avoid having their drivers classified as employees (Murphy, 2020). This legal conflict also illustrates the next likely regulatory battleground, with several other states looking into similar legislation (Campbell, 2021) and others already implementing minimum wage requirements for ride-hailing drivers (New York City, 2022).

Comparable efforts are happening in Canada as well. The Province of Ontario plans to legislate a fifteen-dollar per hour minimum wage for gig workers (Gray, 2022), and ride-hailing drivers themselves are asking the government to consider them as employees (Crawley, 2021). In response, ride-hailing companies are already lobbying governments to change the law so that it exempts gig workers from employee protections (Marx, 2021). This debate, still in its nascent stages, has already gained considerable media attention and is poised to garner more due to its implications for a growing precarious labour force and its ramifications for Canadian workers' rights, benefits, and pay. As these issues are within the national government's jurisdiction, we conclude that ride-hailing regulation has run its course at the municipal scale, whereas the national/international scale still has room for policy experimentation and learning, particularly surrounding the industry's labour practices and environmental impact.

Acknowledgments

The authors would like to thank Melissa Marchetti, Andrew Regalado, and Jessica Patti, who served as research assistants on this project.

REFERENCES

Auener, S., Kroon, D., Wackers, E., van Dulmen, S., & Jeurissen, P. (2020). COVID-19: A window of opportunity for positive health care reforms. *International Journal of Health Policy and Management*, *9*(10): 419–22. https://doi.org/10.34172/ijhpm.2020.66.

Barnes, S.J., Guo, Y., & Borgo, R. (2020). Sharing the air: Transient impacts of ridehailing introduction on pollution in China. *Transportation Research Part D: Transport and Environment*, *86*, Article 102434. https://doi.org/10.1016/j.trd.2020.102434.

Beer, R., Brakewood, C., Rahman, S., & Viscardi, J. (2017). Qualitative analysis of ride-hailing regulations in major American cities. *Transportation Research Record*, *26150*(1), 84–91. https://doi.org/10.3141/2650-10.

Béland, D., & Marier, D. (2020). COVID-19 and long-term care policy for older people in Canada." *Journal of Aging & Social Policy*, *32*(4–5), 358–64. https://doi.org/10.1080/08959420.2020.1764319.

Bellon, T. (2020a). *Uber suspends pooled rides in US, Canada to limit coronavirus spread*. Reuters. https://www.reuters.com/article/health-coronavirus-uber-idCAL8N2BA0GE.

Bellon, T. (2020b). *Uber offers COVID-19 contact tracing help amid chaotic US response*. Reuters. https://www.reuters.com/article/idUSKCN24L17X.

Bennett, C.J., and Howlett, M. (1992). The lessons of learning: Reconciling theories of policy learning and policy change. *Policy Sciences*, *25*(3): 275–94. https://doi.org/10.1007/BF00138786.

Brail, S. (2018). From renegade to regulated: The digital platform economy, ride-hailing and the case of Toronto." *Canadian Journal of Urban Research*, 27(2), 51–63. https://www.jstor.org/stable/26542036.

Brail, S. (2021). Ride hailing in Canadian cities. In A. Zwick & Z. Spicer (Eds.), *The platform economy and the smart city: Technology and the transformation of urban policy*. McGill-Queen's University Press.

California Air Resources Board (2021, 20 May). *California requires zero-emissions vehicle use for ridesharing services, another step toward achieving the state's climate goals*. https://ww2.arb.ca.gov/news/california-requires-zero-emissions-vehicle-use-ridesharing-services-another-step-toward.

Campbell, H. (2021). Everything you should know about AB5 & its impact on Uber. *Ride share guy*. https://therideshareguy.com/ab5-end-of-rideshare/

Chi, H.-R., Ho, H.-P., & Lin, P.-K. (2022). Double-loop strategies of Uber, Lyft, and Didi during COVID-19. *JABS*, *8*(1), 35–40. https://doi.org/10.20474/jabs-8.1.5.

City of Toronto. (2021, 21 November). The transportation impacts of vehicle for hire in the city of Toronto October 2018 to July 2021. Data Analytics Unit. https://www.toronto.ca/wp-content/uploads/2021/11/98cd-VFHTransportationImpacts2021-11-23.pdf.

City of Winnipeg. (2019). By-law No. 101/2019. https://clkapps.winnipeg.ca/dmis/docext/ViewDoc.asp?DocumentTypeId=1&DocId=7641.

Crawley, M. (2021, 21 October). *Uber drivers, gig workers pressure Ontario government for employee status*. CBC News. https://www.cbc.ca/news/canada/toronto/ontario-gig-workers-uber-lyft-skip-employees-1.6216443.

DeLeo, R.A., Taylor, K., Crow, D.A., & Birkland, T.A. (2021). During disaster: Refining the concept of focusing events to better explain long-duration crises. *International Review of Public Policy*, *3*(1): 5–28. https://doi.org/10.4000/irpp.1868.

Donald, B., & Sage, M. (2021). Taking Kingston for a regulatory ride? Uber's entrance into Kingston, ON. In A. Zwick & Z. Spicer, *The platform economy and the smart city: Technology and the transformation of urban policy* (pp. 147–64). McGill-Queen's University Press.

Dunn, W.N. (2007). *Public policy analysis: An introduction*. Prentice Hall.

Dupont, C., Oberthür, S., & von Homeyer, I. (2020). The Covid-19 crisis: A critical juncture for EU climate policy development? *Journal of European Integration*, *42*(8), 1095–1110. https://doi.org/10.1080/07036337.2020.1853117.

Erhardt, G.D., Roy, S., Cooper, D., Sana, B., Chen, M., & Castiglione, J. (2019, 8 May). Do transportation network companies decrease or increase congestion? *Science Advances*, *5*(5). https://doi.org/10.1126/sciadv.aau2670.

Goldstein, M. (2020, 27 July). What is the future for Uber and Lyft after the pandemic? *Forbes*. https://www.forbes.com/sites/michaelgoldstein/2020/07/27/what-is-the-future-for-uber-and-lyft--after-the-pandemic/?sh=4045bb8c3bc8.

Gray, J. (2022, 28 February). Ontario to legislate $15 minimum wage for gig workers. *The Globe and Mail*. https://www.theglobeandmail.com/canada/article-ontario-to-legislate-minimum-wage-for-gig-workers.

Halifax Regional Council. (2019, 17 September). *Vehicle for hire review*. Item No. 15.1.6. https://www.halifax.ca/sites/default/files/documents/city-hall/regional-council/190917rc1516pres.pdf.

Hara Associates. (2021). *Economic effects of Covid-19 on the BC passenger transportation industry*. https://www.ptboard.bc.ca/documents/20210922-Economic-Effectsof-Covid-19.pdf.

Hawkins, A. (2020). *Uber's response to COVID-19: Face masks, selfies, and fewer people in the car*. The Verge. https://www.theverge.com/2020/5/13/21257432/uber-face-mask-driver-rider-require-selfies-maximum-passengers.

Heichel, S., Pape, J., & Sommerer, T. (2005). Is there convergence in convergence research? An overview of empirical studies on policy convergence. *Journal of European public policy*, *12*(5), 817–40. https://doi.org/10.1080/13501760500161431.

Hoppe, R. (2011). *The governance of problems: Puzzling, powering and participation*. The Policy Press.

Howlett, M. (1998). Predictable and unpredictable policy windows: Institutional and exogenous correlates of Canadian federal agenda-setting. *Canadian Journal of Political Science*, *31*(3), 495–524. https://doi.org/10.1017/S0008423900009100.

Howlett, M., & Ramesh, M. (2002). The policy effects of internationalization: A subsystem adjustment analysis of policy change. *Journal of Comparative Policy Analysis, Research and Practice*, *4*, 31–50. https://doi.org/10.1023/A:1014971422239.

International Monetary Fund. (2020). *World economic outlook, October 2020: A long and difficult ascent*. https://www.imf.org/en/Publications/WEO/Issues/2020/09/30/world-economic-outlook-october-2020.

Jewell, S., Futoma, J., Hannah, L., Miller, A.C., Foti, N.J., & Fox, E.B. (2021). It's complicated: Characterizing the time-varying relationship between cell phone mobility and COVID-19 spread in the US. *NPJ digital medicine*, *4*(1), Article 152. https://doi.org/10.1038/s41746-021-00523-3.

Katta, S., Badger, A., Graham, M., Howson, K., Ustek-Spilda, F., and Bertolini, A. (2020). (Dis) embeddedness and (de) commodification: COVID-19, Uber, and the unravelling logics of the gig economy. *Dialogues in Human Geography*, *10*(2), 203–7. https://doi.org/10.1177/2043820620934942.

Lieberman, J.M. (2002). Three streams and four policy entrepreneurs converge: A policy window opens. *Education and Urban Society*, *34*(4), 438–50. https://doi.org/10.1177/0012450203400400 3.

Loa, P., Hossain, S., Liu, Y., Mashrur, S.M., & Habib, K.N. (2020, August). *How has COVID-19 impacted ride-sourcing use in the Greater Toronto Area?* University of Toronto Transportation Research Institute.

Marx, P. 2021. *Uber's proposed revamp of provincial labour laws would cement the notion that gig workers are not employees*. CBC News. https://www.cbc.ca/news/opinion/opinion-gig-workers-uber-1.5957452.

May, P.J. (1992). Policy learning and failure. *Journal of Public Policy, 12*(4), 331–54. https://doi.org/10.1017/S0143814X00005602.

Meredith-Karam, P., Kong, Hl, Wang, S., & Zhao, J. (2021, December), The relationship between ridehailing and public transit in Chicago: A comparison before and after COVID-19. *Journal of Transport Geography, 97*, Article 103219. https://doi.org/10.1016/j.jtrangeo.2021.103219.

Minkler, M., Griffen, J., & Wakimoto, P. (2020). Seizing the moment: Policy advocacy to end mass incarceration in the time of COVID-19. *Health Education & Behavior, 47*(4), 514–18. https://doi.org/10.1177/1090198120933281.

Mintrom, M. (1997). Policy entrepreneurs and the diffusion of innovation. *American Journal of Political Science, 41*(3): 738–70. https://doi.org/10.2307/2111674.

Monahan, T., & Lamb, C.G. (2022). Transit's downward spiral: Assessing the social-justice implications of ride-hailing platforms and COVID-19 for public transportation in the US. *Cities, 120*, Article 103438. https://doi.org/10.1016/j.cities.2021.103438.

Morshed, S.A., Khan, S.H., Tanvir, R.B., & Nur, S. (2021). Impact of COVID-19 pandemic on ride-hailing services based on large-scale Twitter data analysis. *Journal of Urban Management, 10*(2), 155–65. https://doi.org/10.1016/j.jum.2021.03.002.

Murphy, C. (2020, 12 November). Uber bought itself a law. Here's why that's dangerous for struggling drivers like me. *The Guardian*. https://www.theguardian.com/commentisfree/2020/nov/12/uber-prop-22-law-drivers-ab5-gig-workers.

New York City. (2022, 11 February). *Mayor Adams delivers raise for essential gig workers*. https://www1.nyc.gov/office-of-the-mayor/news/072-22/mayor-adams-delivers-raise-essential-gig-workers#/0.

Reilly, K., & Lozano-Paredes, L.H. (2019). Ride hailing regulations in Cali, Colombia: Towards autonomous and decent work. In P. Nielsen & H.C. Kimaro (Eds.), *International Conference on Social Implications of Computers in Developing Countries* (pp. 425–35). Springer.

Riebe, N. (2018, 2 April). *Uber and taxis in Edmonton followed the rules better in 2017, city says*. CBC News.https://www.cbc.ca/news/canada/edmonton/edmonton-city-council-uber-1.4602320.

Rodier, C. (2018). *The effects of ride hailing services on travel and associated greenhouse gas emissions*. National Center for Sustainable Transportation. https://trid.trb.org/view/1509164.

Saksida, T., Mihelic, K.T., Maffie, M., Culiberg, B., and Merkuz, A. (2022). Should I stay or should I go? Ride-hail driver reactions to Uber's Covid-19 response. *Academy of Management Proceedings, 1*, Article 11520. https://journals.aom.org/doi/abs/10.5465/AMBPP.2022.11520abstract.

Saltman, J. (2020, 29 June). Metro mayors request GHG requirements for ride-hailing operators. *Vancouver Sun*. https://vancouversun.com/news/metro-mayors-request-ghg-requirements-for-ride-hailing-operators.

Smart, A. (2020). *Coronavirus impact: Could Canadians end up using cars more, taking transit less?* CTV News. https://bc.ctvnews.ca/coronavirus-impact-could-canadiansend-up-using-cars-more-taking-transit-less-1.4929258.

Shen, H., Zou, B., Lin, J., & Liu, P. (2020). Modeling travel mode choice of young people with differentiated E-hailing ride services in Nanjing China. *Part D: Transport and Environment*, *78*, Article 102216. https://www.sciencedirect.com/science/article/abs/pii/S1361920919301865.

Spicer, Z., Eidelman, G., & Zwick, A. (2019). Patterns of local policy disruption: Regulatory responses to Uber in ten North American cities. *Review of Policy Research*, *36*(2), 146–67. https://doi.org/10.1111/ropr.12325.

Stritch, A. (2005). Power resources, institutions and policy learning: The origins of workers' compensation in Quebec. *Canadian Journal of Political Science/Revue canadienne de science politique*, *38*(3), 549–79. https://doi.org/10.1017/S0008423905040606.

Tabascio, A., & Brail, S. (2022). Governance matters: Regulating ride hailing platforms in Canada's largest city-regions. *The Canadian Geographer*, *66*(2), 278–92. https://doi.org/10.1111/cag.12705.

Toronto. (2022). *COVID-19 guidance for taxi and ride share vehicles*. Toronto Public Health. https://www.toronto.ca/wp-content/uploads/2020/03/8d19-COVID-19-Guidance-for-Taxi-Ride-Share-Employers-Drivers-Vehicle-Owners.pdf.

Tzur, A. (2019). Uber Über regulation? Regulatory change following the emergence of new technologies in the taxi market. *Regulation & Governance*, *13*, 340–61. https://doi.org/10.1111/rego.12170.

Uber. (2020). *Supporting cities and communities around the world*. https://www.uber.com/newsroom/supporting-our-communities/.

Uber. (2021). *The impact of Uber in Canada*. https://ubercanada.publicfirst.co/.

Uber. (2022a). *Health and safety supplies – Distribution locations*. https://www.uber.com/en-CA/blog/hssdistribution/.

Uber. (2022b). *Helping the world move safely again*. https://www.uber.com/us/en/coronavirus/.

WHO. (2021). *Coronavirus disease (COVID-19): How is it transmitted?* https://www.who.int/news-room/questions-and-answers/item/coronavirus-disease-covid-19-how-is-it-transmitted.

WHO. (2022). *14.9 million excess deaths associated with the COVID-19 pandemic in 2020 and 2021*. https://www.who.int/news/item/05-05-2022-14.9-million-excess-deaths-were-associated-with-the-covid-19-pandemic-in-2020-and-2021.

Williams, R. n. (2009). Exogenous shocks in subsystem adjustment and policy change: The credit crunch and Canadian banking regulation. *Journal of Public Policy*, *29*(1), 29–53. https://doi.org/10.1017/S0143814X09001007.

Woodside, J. (2021). Just rides: Ride-hailing, the capabilities approach and the just city. *Canadian Journal of Urban Research, 30*, 85–100. https://cjur.uwinnipeg.ca/index.php/cjur/article/view/356.

Woodside, J., Moos, M., & Vinodrai, T. (2021). Private car, public oversight: Municipal regulation of ride-hailing platforms in Toronto and the Greater Golden Horseshoe. *Canadian Planning and Policy / Aménagement et politique au Canada, 2021*(1), 146–65. https://doi.org/10.24908/cpp-apc.v2021i01.14362.

Young, Mischa. (2019). Ride-hailing's impact on Canadian cities: Now let's consider the long game. *The Canadian Geographer/Le Géographe canadien, 63*(1), 171–5. https://doi.org/10.1111/cag.12514.

Young, M., Soza-Parra, J., & Circella, G. (2022). The increase in online shopping during COVID-19: Who is responsible, will it last, and what does it mean for cities? *Regional Science Policy & Practice, 14*(S1), 162–78. https://doi.org/10.1111/rsp3.12514.

Yu, J.JJ., Tang, C.S., Shen, Z.-J.M., & Chen, X.M. (2020). A balancing act of regulating on-demand ride services. *Management Science, 66*(7), 2975–2992. https://doi.org/10.1287/mnsc.2019.3351.

Zwick, A.. and Spicer, Z. (2018). Good or bad? Ridesharing's impact on Canadian cities. *The Canadian Geographer/Le Géographe canadien, 62*(4), 430–6. https://doi.org/10.1111/cag.12481.

Zwick, A., Spicer, Z., and Young, M. (2022). Moving ideas? The news media's impact on ridehailing regulation in Canadian cities. *Journal of Urban Affairs, 46*(2), 356–72. https://doi.org/10.1080/07352166.2022.2053332.

Appendix A

City	Province	Population (2016)	INITIAL regulation: Y/N	SUBSEQUENT regulation: Y/N	PANDEMIC regulation: Y/N
Brampton	ON	593,638	N	Y	N
Burlington	ON	183,314	N	N	N
Burnaby	BC	232,755	N	Y	N
Calgary	AB	1,239,220	Y	Y	N
Edmonton	AB	932,546	Y	Y	Y
Gatineau	QC	276,245	Y	Y	N
Greater Sudbury	ON	161,531	Y	Y	N
Halifax	NS	403,131	N	Y	N
Hamilton	ON	536,917	Y	N	N
Kitchener	ON	233,222	Y	N	N
Laval	QC	422,993	Y	Y	N
London	ON	383,822	Y	Y	N
Longueuil	QC	239,700	Y	Y	N
Markham	ON	328,966	N	N	N
Mississauga	ON	721,599	Y	Y	N
Montreal	QC	1,704,694	Y	Y	N
Oakville	ON	193,832	Y	N	N
Ottawa	ON	934,243	Y	N	N
Quebec City	QC	531,902	Y	Y	N
Regina	SK	215,106	N	Y	N
Richmond	BC	198,309	N	Y	N
Richmond Hill	ON	195,022	N	N	N
Saskatoon	SK	246,376	N	Y	N
Sherbrooke	QC	161,323	N	Y	N
Surrey	BC	517,887	N	N	Y
Toronto	ON	2,731,571	Y	Y	Y
Vancouver	BC	631,486	N	Y	N
Vaughan	ON	306,233	Y	N	N
Windsor	ON	217,188	N	Y	N
Winnipeg	MB	705,244	Y	Y	N

14 Mobility as a Service: A Platform Solution in Search of a Business Model

JOHN LORINC

Introduction

In the spring of 2019, Berlin's transit agency, Berliner Verkehrsbetriebe (BVG), announced it would be launching a mobility as a service (MaaS) platform to allow people travelling around the German capital to seamlessly book and pay for trips, taken using a range of services, on a single app called Jelbi (Jelbi, n.d.). The Application Programming Interface (API)-driven platform provided one-stop access to all the various forms of Berlin's extensive transit network -- Deutsche Bahn – as well as taxis, e-scooters, bikes, e-mopeds, and passenger vehicles provided by a car-sharing service called Miles, with pickup locations at hubs around the city. Besides e-ticketing and payments, the app's services included way-finding features, such as those provided on Google Maps, as well as "mixed mode" recommendations, i.e., trips between two locations that involve some combination of mobility modes. Berlin, *Forbes* Magazine declared, was set to become "the Amazon for transportation (with lower fares)" (Walmsley, 2019).

MaaS is a type of "deep integration" app that enables coordination amongst multiple mobility service suppliers and is "important for cities trying to prevent people from clogging up congested roads with their own cars" (Jordans, 2019). Jordans (2019) quotes Andreas Knie, a mobility expert at the Social Science Centre in Berlin, who says that MaaS is a positive step for mobility as "transport services need to be connected, just like cellphone networks, so you can easily roam from one to another."

Developed on a software platform provided by Trafi, a Lithuanian company, Jelbi allows users to create accounts by uploading a selfie, images of their passport and driver's licences, and banking information (Trafi, n.d.). BVG and Jelbi officials suggested that the initial response by mobility providers interested in signing on to the platform had been strong, and they were also encouraged by the number of downloads of the app in the weeks following the launch. Trafi CEO and co-founder Martynas Gudonavicius indicated to Busvine (2019) that "Berlin is becoming the world's largest mobility as a service city." By 2019, the company had raised $14 million from investors and was in talks with other cities, indicating strong interest regarding the launch of Trafi services in additional cities (Busvine, 2019).

Since the mid-2010s, much has been made about the potential for MaaS platforms to revolutionize urban transportation. In most big cities, mobility can be time consuming, expensive, and logistically challenging, not to mention carbon intensive. Urban policymakers try to find ways to shift people out of cars and onto transit while planners promote walkable communities fitted out with cycling infrastructure. Yet the allure and convenience of the private automobile remains powerful. Many people live and work in low-density areas where transit and walking are not especially viable options. Ride-hailing services provide convenience and relatively affordable prices. And automobile manufacturers, in turn, continue to bring out vehicles with battery-powered drive trains that hold out the promise of reduced emissions. The competition among the various modes is more or less explicit, although most city dwellers move around urban regions in a variety of ways, including transferring from one mode to another within a single journey.

While there are numerous definitions of MaaS, a 2021 World Bank evaluation offers some common tenets: they are smartphone-based platforms with online payment features that provide users with an intermodal/one-stop-shopping platform to travelling through urban space. This approach is informed by the hypothesis that the traditionally siloed system for mobility encourages inefficiency, inconvenience, and expense. "By offering multimodal trip planning services, integrating multi-provider single trip fares, and bundling mobility services into monthly subscriptions, the underlying idea of MaaS is to leverage the strength of the growing number of mobility services in each city to facilitate customers' ability to choose the most optimal door-to-door service for each trip" (Bianchi Alves et al., 2021, p. 22).

Early proponents of MaaS argued that journeys could be constructed from some combination of transit, bike-share, car-share, and so on, and the users could even buy monthly subscriptions, not unlike smartphone bundles. They asserted that a MaaS app could act as a broker, putting together optimal routes from the various public and private services that would undercut the intermodal competition that afflicts big cities and drives congestion (Vij & Dühr, 2022).

A 2019 discussion paper expressed the sense of potential and excitement: "It is important to act now to ensure MaaS is the desired tool that our cities can use to build a more sustainable future" (International Association of Public Transport [UITP], 2019, p. 22). Perhaps unsurprisingly, given the hype, MaaS has attracted a significant amount of scholarly attention. The assessment literature to date includes studies on the public policy framework required to implement MaaS platforms, the role of local government in the "governance and diffusion" of MaaS, attitude surveys on its potential, econometric comparisons of business models, ecosystem analyses that seek to examine how best to capture the value inherent in the platform, lessons learned and at least two literature reviews (Arias-Molinares & García-Palomares, 2020; Fenton et al., 2020; Hoveskog et al., 2022; Kanda, 2019; Lajas & Macario, 2020; Matowicki et al., 2022; van der Berg et al., 2022).

In the chapter that follows, I will explore whether MaaS has or can live up to its promise by providing an overview of the context within which this model emerged; itemizing briefly some case studies; offering an analysis of the strengths and weaknesses of this idea; exploring the associated policy and technology issues; and, finally, discussing how MaaS fits into broader conversations about urban transportation, climate, and smart cities.

Context

Mobility has been a service for a very long time, from the hired cabs and carriages of the pre-industrial revolution period to passenger trains, taxis, and public transit. Car rental companies emerged almost simultaneously with private vehicles, in the early twentieth century.

Beginning in the 1990s, car-sharing services, primarily based in big cities, evolved from marginal players in urban mobility markets into large firms with multiple neighbourhood-based pickup locations, online booking, and, eventually, ownership relationships with leading car rental agencies (see Rojas and Miller, chapter 5). A few, like Communauto, a Montreal-based car-sharing service, have remained independent, expanding into multiple markets and relying on on-line reservation services (for a discussion on Communauto in Montreal, see Hashemi, Motaghi, and Tremblay, chapter 9).

In the early 2010s, the advent of ride-hailing services like Uber and Lyft provided a new form of transportation service by massively disrupting the urban taxi sector in large urban markets around the world (Brail, 2017). Ride-hailing firms relied on app-based booking and payment portals to provide transactional convenience and were able to access large driver pools through gig economy relationships. Ride-hailing also set the stage for the use of smartphones as a tool for not just way-finding but also the purchase of mobility services.

The digitalization of transit service proceeds along a different trajectory and begins with the deployment of smart-card-based fare media, such as London's Oyster. The use of tap cards and smartphone apps has steadily replaced tickets, transfers, and tokens, with commuters able to electronically purchase passes of varying durations, which increasingly include options to transition between different transit networks or companies operating within a given region.

That process, ongoing for two decades, has been accompanied by other transit-focused digital innovation, such as online wayfinding or trip-planning tools. Third-party APIs, drawing on municipal open data sources such as transit-vehicle-based GPS sensors, offer scheduling, disruption, and vehicle-arrival information. For example, mapway.com provides all these services via a smartphone app in cities like New York, Boston, and Barcelona. Google Maps, in turn, developed a further enhancement by providing users with optimized trip-planning options, including various transit routes between two points, travel durations, and ads for quotes from ride-hailing firms.

Another related development, in terms of the evolution of MaaS, is the advent of the "as-a-service" business model and brand. The origins can be found in the software industry, and in particular enterprise-based software platforms geared at corporate customers. The concept of "software-as-a-service" was pioneered in the late 1990s by firms like Salesforce, which makes customer relationship management software. The critical innovation turned on replacing the traditional software business model – customers would purchase seats and then routinely install upgrades on their mainframes or servers – with a virtual one, that is, leasing cloud-based software applications on a subscription model. That critical shift freed customers from the problematic business of managing upgrades and maintaining hardware, and it paved the way for the emergence of giant cloud-based software providers such as Google and AWS.

The "as-a-service" business model has become extremely popular in a range of other verticals, from gaming and banking to payments, data, and blockchain. The logical conclusion of this trend has come to be known as "XaaS" or anything-as-a-service, which "describes a general category of services related to cloud computing and remote access," as one cloud services firm, NetApp, describes the acronym. "It recognizes the vast number of products, tools, and technologies that are now delivered to users as a service over the internet" (NetApp, n.d.).

From the perspective of mobility, the significance of this tectonic shift in the computing business model is that it marks the evolution of software from an ownership model to something more akin to leasing or fee-for-service – in other words, from a capital investment to an operating expense. Of course, leasing in the world of heavy equipment is long established – for instance, in the case of fleets, airlines, construction equipment and agriculture. But there's certainly evidence that this change has filtered into other parts of the economy. In the realm of mobility, for example, the shift to the 'aaS' model can be seen in the trend among younger city dwellers away from car ownership, with all its associated encumbrances (insurance, repairs, parking), and towards alternative modes of getting around, such as ride-hailing, car-sharing, and various micromobility options (e.g., e-scooters).

In the pre-pandemic 2010s, MaaS ventures, particularly the first one that launched in Helsinki in 2017, received a lot of media and investor attention. The proliferation of smartphone-based booking/reservation/hailing tools, along with the back-end functionalities associated with mapping, payments, etc., had provided an opportunity to re-conceptualize urban mobility. Whereas a trip was traditionally seen as a journey from A to B using a succession of different mobility modes – typically determined by factors such as time constraints, distance, convenience, cost, and personal preferences – MaaS operators sought to recast this process by providing what is essentially a one-stop-shopping platform that not only provided e-ticketing and reservation functionality but also optimized options, aggregated trips, and incentives for the use of less-carbon-intensive modes. The idea was based on the assumption that commuters, when presented with a choice between an end-to-end private vehicle journey and alternative trips that are less time-consuming or less expensive, will be motivated to select the latter.

As the scope of the platform economy expanded rapidly, the concept underwriting MaaS seemed poised to hit a tipping point. As Goodall et al. (2017), point out, "If Netflix's business model were applied to urban transportation, how might that change the way city-dwellers get around? That's the question at the heart of an ambitious initiative taking shape in Finland's capital, which aims to make it unnecessary for any city resident to own a private car by 2025" (p. 113).

The accelerating climate crisis added even more momentum as MaaS proponents presented their platforms as a way of further expanding incentives for urban dwellers to leave their vehicles at home. The notion of offering travellers a bundled subscription-based service also seemed to be aligned with developments elsewhere in the consumer and business-to-business (B2B) economies, for example, the bundling of cell/data/streaming services, or the shift from ownership to leasing for a range of products, including vehicles, software, and even home appliances such as furnaces.

In 2017, an industry association, MaaS Alliance, was established in Brussels, with a mandate to attract not just mobility companies, but also local and regional governments, tech providers, and academics. The Alliance has since published numerous white papers on various aspects of MaaS and promotes the concept in fora such as COP26.

Yet COVID-19 had a dramatic effect on the mobility sector and served to disrupt the momentum behind MaaS. The widespread shift to work-from-home meant sharp drops in local and regional transit use, *The Economist* observed in 2021: "A survey of American travelling habits by LEK, a consultancy, showed that car journeys declined by just 9% last year, compared with 55-65% for public transport and ride hailing" (New means, 2021). Public health protocols and user caution rocked mobility platforms like Uber and Lyft, while car-sharing services saw a commensurate drop-off in usage. Demand for private vehicles, both new and used, spiked, triggering a domino effect in other mobility verticals like the automobile rental market (Rowe and Bond, 2022).

The pandemic also accelerated the use of a far wider range of personal/micromobility devices, from rented electric scooters to e-bikes, conventional bikes, bike-sharing services, and a range of other devices (e.g., battery powered skateboards and monowheels). While estimates vary greatly, the micromobility sector globally is predicted to see sharp growth for the balance of the decade, and even large automotive manufacturers like Magna International are making investments in the space (Globe and Mail, 2022).

COVID-19's far-reaching impact on all forms and modes of mobility offered both opportunities and challenges to MaaS providers. In the short term, the decline and gradual rebuilding of a customer base posed a significant obstacle for a sector that was in its infancy when COVID-19 struck. The proliferation of mobility modes might seem, on its face, like a huge market opening for MaaS providers. However, this expansion has occurred both among private users (e.g., people who bought and now use e-bikes) and within the sharing/rental economy (e.g., Bird or other e-scooter firms). At the same time, many transit agencies – whose presence within the MaaS ecosystem is crucial – have struggled to rebuild lost ridership and respond to new commuting patterns, including those precipitated by hybrid work arrangements.

As of 2022, the MaaS sector was showing evidence of a rebound, but with some significant changes compared to pre-pandemic (MaaS Alliance, 2022). There have been important technology developments, for example. Trafi, the Lithuanian company that built the back-end for Berlin's Jelbi MaaS app, has expanded to other cities, including Copenhagen and Bogota as well as three cities in a region of Switzerland. The company touts its ability to establish partnerships with mobility firms operating in multiple cities – a strategy that suggests an approach to scaling that wasn't part of the one-MaaS-app/one-city model of the pre-pandemic era.

In fact, some cities have seen and even encouraged the emergence of multiple and competing MaaS providers, marking what Surico (2021) referred to as a "stark departure." Surico (2021) quotes Stijn Vernaillen, a MaaS expert who works for the City of Antwerp, Belgium, who says, "We want to support as many players as possible. But as a city, we are not going to build a MaaS application or solution and put that in the market."

Case Studies

Berlin's Jelbi is an extension of the city's transit authority, but that model is only one of several that together comprise the MaaS sector. Following are some other examples.

Helsinki, Finland: The first MaaS platform, dubbed Whim, launched in 2017 in Helsinki and is operated by MaaS Global. Founded in 2015, MaaS Global provides users with a range of offerings, from single rides to bundles, monthly subscriptions, and various discount offers on taxis, car, and bike rentals (Whim Helsinki, n.d.). The company has raised almost $74 million as of 2022, according to Crunchbase, and spawned imitators in various cities, including Vienna (WienMobil) and Malo, Sweden, as well as pilot projects in Sydney and Pittsburgh (Crunchbase, n.d.; Zipper, 2021). Whim is now available in several cities, including Tokyo, Vienna, Antwerp, Birmingham, and Switzerland.

Paris, France: Unlike other MaaS operators that are anchored to a particular city, BlaBlaCar is a carpooling platform that serves both passengers and drivers and has experienced extremely rapid growth right across Europe. The company, headquartered in Paris, raised US$115 million in 2021 for expansion, and claimed 90 million members in 22 countries (BlaBlaCar, 2021). Unlike urban-based MaaS firms, it functions as a two-sided digital marketplace rather than a broker linking riders with various mobility service providers.

Sydney, Australia: Just prior to the pandemic, the capital city of New South Wales embarked on a plan to test a MaaS system, dubbed

Tripi; the pilot began in late 2019 and included transit, taxis, Uber, car rental agencies, and was offered to employees of IAG, a large Australian insurance firm. The pilot was meant to run for two years but ended when the pandemic lockdowns began in March 2020.

A research team evaluated the foreshortened findings and offered up a set of conclusions and lessons (Hensher et al., 2021). Overall, the participants, who tended to use both cars and public transit, expressed interest in the app, and many said they would have used it more if the trial had not ended so suddenly. In turn, the evaluation concluded that financial incentives, both for customers and private mobility providers as well as other value-added services, would likely be necessary for such a platform to become viable. The authors also offered this piece of advice to other city-regions considering MaaS: "[Pay as you go] by itself is unlikely to make a difference in respect of sustainable outcomes. It is bundle subscribers that decrease their car usage and that are more interested in continuing than PAYG subscribers" (Hensher et al., 2021, p. 6).

Montreal, Canada: In May 2019, Montreal won a federal "smart city challenge" prize of $50 million with a proposal that included another form of MaaS (Infrastructure Canada, 2019). According to the city's pitch, the concept combines many modes of travel already available in the Greater Montreal area and allows users to easily access these various services, thanks to a simplified pricing approach (for more on Montreal's mobility ecosystem, see Hashemi, Motaghi, and Tremblay, chapter 9). It also includes implementing a single mobility account, linked to a shared civic identity. In addition, this innovative platform will provide an intermodal trip-planning tool that gives users greater freedom of choice when determining how to get from point A to point B. This customizable platform is intended to be open, that is, to be able to accept any form of transit that helps reduce car use and make neighbourhoods more pleasant (Montréal, 2019, p. 6).

While two pilot projects preceded Montreal's submission, the service (as of late 2022) has yet to be deployed.

Analysis

An analysis of a novel urban transportation platform like MaaS must consider three interconnected elements: the end user, the service providers and business model, and technology and data.

End Users

In any urban area, the transportation "market" consists of the entire population as well as visitors and business travellers. Within this frame, individual transportation choices are typically driven by a handful of core considerations: cost, convenience, time-savings, and, to a lesser degree, environmental impact. It is also true that most people have predictable travel habits and trips, using established modes and supported by either vehicle ownership, with its attendant costs, or transit passes. For MaaS to catch on, it needs to provide a service that is cheaper, faster, and/or more convenient than the business-as-usual mode, or solves some kind of transportation-related problem, such as last-mile connections, in ways that are superior to status quo options.

According to Maas (2022), "The simple supply of MaaS does not automatically generate user demand. Potential customers must be shown the improvements in the framework conditions, for example, in financial and environmental terms, in order to trigger the consideration of the MaaS offer" (p. 15).

In his review of numerous mobility-as-a-service studies, Maas writes that customer surveys have shown that the users tend to be younger or middle-aged, well educated, use car-sharing services, and want to reduce their own car use for environmental reasons. One group that was less likely to use MaaS are families with children. Maas also cites the findings of a detailed customer evaluation of UbiGo in Sweden. It found that while new MaaS users tended to "over-estimate their transportation needs" when initially subscribing, about two thirds "changed their use of mobility during the project towards more sustainable or shared modes" (Maas, 2022, p. 15).

Other evaluations have come to similar conclusions. In 2022, a Czech/German research team published the results of an online survey of sixty-four hundred regular commuters in England, Germany, the Czech Republic and Poland, all of whom lived in cities with over eighty-thousand residents. The authors found that the greatest interest in MaaS came from commuters who have "environmentally friendly views," value shared mobility, and also sought out economical travel options. The results didn't vary significantly by location, suggesting that the logic of MaaS clicks into place in communities of sufficient size. The authors concluded that cities could market the service to commuters who are both cost-conscious and interested in reducing their carbon footprint (Matowicki et al., 2022).

Providers and Business Model

MaaS companies fall into two categories: private and those owned/operated by municipal/regional transit agencies. In both cases, the platform functions as a marketplace for transportation options, and, as with all such platforms, the more options there are, the better the offering. Typically, MaaS platforms include public transit agencies, bike- or car-share services, car rental, ride-hailing, taxis, and even parking garages/lots.

The range of services varies widely, from bundles/subscriptions to individual tickets, daily/weekly transit passes, and online bookings, all of which are provided through a smart phone app with e-payment and navigation features. Some, like Vienna's WienMobil, also rely on physical hubs, typically at or near transit stations, where users can transfer from one mode to another (e.g., subway to car-share) while booking those arrangements through the app ahead of time. WienMobil's FAQ describes the service this way:

> WienMobil is Wiener Linien's new mobility app, combining the functions of the routing app qando and the Wiener Linien ticket app. You can also use it to find out about additional mobility services (e.g. car sharing, bike sharing or taxis), as well as reserve or book these ... Not only does WienMobil consider public transport when planning a route, but also suggests routes for bicycles, car-sharing vehicles or a combination of different means of transport. Digital tickets can be displayed directly in the app. (Wiener Linien, 2021)

Private MaaS operators have discovered that the model, for all its theoretical potential, bumps up against one significant structural obstacle, and that is the role of municipal transit operators. These agencies have not welcomed private MaaS companies with open arms and typically don't wholesale passes or individual tickets to third parties for re-sale purposes.

As a 2020 Australian study observed, "[A] major challenge for the implementation of full-scale MaaS has often been a lack of a compelling value proposition to existing transport operators that would entice them to integrate their services with other operators and/or join a third-party platform" (Vij & Dühr, 2022, p. 698). The authors go on to note that most MaaS systems "have only been realised with considerable financial and policy support from government" (p. 698).

In higher-density cities where the modal splits reveal an existing preference for transit, bikes, or walking, several existing factors will influence commuter decisions to not use private vehicles, such as parking cost/availability, traffic, and congestion charges. For a MaaS operator, however, the question in such cities is whether a subscription or online ticketing features are, in fact, more useful/convenient than transit passes/smart cards, etc. Indeed, as transit agencies expand their fare structures to include municipally operated bike-share services, there may be even less incentive for commuters to use a private MaaS platform.

What about the motivations of a transit operator? "It can be assumed that for transport operators to be interested in joining a common MaaS platform and integrating with other operators, the MaaS platform will need to offer clear commercial benefits in terms of increased revenues and/or reduced costs," observe Vij and Dühr (2022, p. 699). They itemize six conditions where MaaS might benefit transit operators:

- strengthening "potentially complementary relationships" between services;
- boosting service use through uptake of bundled plans;
- improving revenue with "dynamic and differentiated pricing";
- increasing visibility of existing transit agencies, especially smaller ones;
- increasing utilization rates of underused public and private services;
- reducing operator costs by outsourcing functions to MaaS platforms (Vij & Dühr, 2022).

But, as the authors caution, these benefits only accrue to transit operators if the platform does more than just provide substitute modes of mobility. In other words, if the platform makes it easier, cheaper, or more convenient to make a journey using, for example, a combination of bike-share and ride-hailing instead of a somewhat more time-consuming trip by bus and subway, then there's no upside for the transit agency.

Vij and Dühr (2022) go on to argue that a monthly subscription plan, such as the packages offered by Whim, may also be less attractive or commercially viable than bundled plans offered by telecom companies. By comparison to a full accounting of a household's vehicle costs (including insurance, depreciation, and fuel), a monthly subscription fee may be slightly less costly. But, they note, many people routinely underestimate vehicle carrying costs.

Both MaaS industry consultants and governments have sought to work through the flaws in the business model. The MaaS Alliance has published market playbooks and white papers on how to overcome challenges, such as sales channel restrictions, regulations barring entry to new mobility firms, and establishing public-private partnerships between private MaaS operators and the transit agencies that represent the largest entities on these platforms (MaaS Alliance, n.d.). The market, however, has yet to coalesce around a single model.

Technology and Data

The technology that underpins MaaS service is not cutting edge. Navigation and trip optimization software is well established, as is online ticketing and data/security protections. The platform architecture must reflect the fact that a MaaS service essentially serves as a broker between travellers and transportation providers.

The major technical challenge, according to several evaluations, is interoperability and what Maas (2022) describes as the "standardization of data between transportation companies and data providers." The complexity lies with the high degree of variation in sophistication and versatility between the back-end systems of private firms, like ride-hailing services, and public transit operators. Maas also points out that highly integrated MaaS systems require substantial investment in processing power, likely delivered through cloud-based applications, and may also render established transit smart cards redundant. The technology, therefore, poses important questions about public-sector investment in legacy ticketing systems.

A related technology design issue involves privacy, and specifically the sensitive question of whether such services will assure users that their travel patterns will not be tracked. Maas (2022) notes that a 2018 survey indicated "urgent concerns" about the disclosure of personal data via a MaaS smartphone app.

There's an obvious inherent tension here because MaaS providers might want to use analytics tools/AI and the aggregation of user travel data to develop new pricing plans and services, or as a means of enticing other mobility companies to participate in the platform. In order to collect and analyse the travel data, MaaS companies and government regulators need to employ a privacy-by-design approach and develop robust anonymization standards.

Ultimately, MaaS apps will pose critical policy questions about privacy, access to individual or aggregated customer data, and the integration of artificial intelligence tools as part of the overall suite of services provided by these services (Cottrill, 2020).

Figure 14.1. High-level conceptualization of a MaaS data ecosystem.

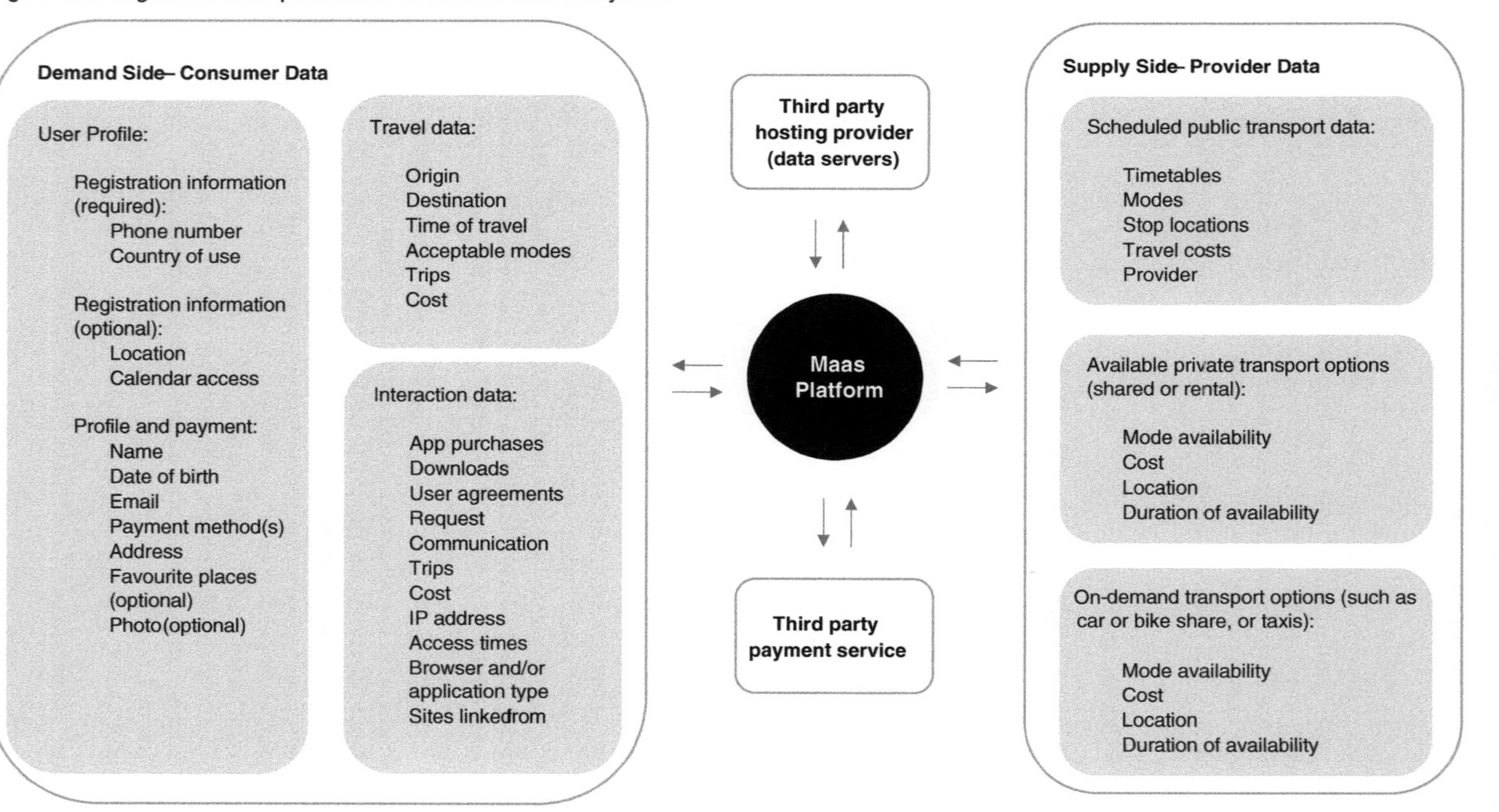

Source: Bianchi Alves et al., 2021, adapted from Cottrill (2020)

Discussion

Several important questions orbit the discussion of the provision of MaaS platforms in urban areas:

- Will this technology offering improve urban mobility, in terms of better connectivity, reduced congestion, increased accessibility, and lower emissions related to transportations?
- Under what conditions will public transit agencies benefit when participating in privately operated MaaS platforms?
- How is success defined? (Zhao et al., 2021)
- Whose interest is served?

The premise behind MaaS is that when looking at an urban mobility ecosystem, the whole should be greater than the sum of the parts. In other words, a MaaS system should, in theory, enable urban dwellers to move more seamlessly around the city, spend less on mobility, and reduce their carbon footprint by limiting reliance on private automobiles.

For private mobility providers, the benefit – again, in theory – is increased access to customers beyond what these firms might expect if operating in their own commercial silo. But the more challenging question is whether MaaS is a zero-sum game – with users shifting between modes, including from public to private – or if it expands that segment of the urban mobility ecosystem in a way that achieves the benefits articulated above (see Gorachinova, Huh, and Wolfe, chapter 11).

One of the main difficulties so far for the MaaS sector has been recruiting municipal transit providers, especially with privately run MaaS platforms. Typically, transit agencies do not sell tickets or passes to wholesalers or resellers, which makes it more difficult to embed financial incentives in, for example, a MaaS subscription or bundle. And while transit agencies that participate in MaaS systems potentially gain access to more customers, they also risk the loss of some share of the travelling public to other modes – a dynamic that may be felt more acutely in the post-pandemic era.

When viewed through a public policy lens, the primary desired outcome is that MaaS platforms shift overall demand for urban mobility away from carbon and space intensive modes, specifically private vehicles, and towards lower carbon modes, for example, shared private electric vehicles, bike-share, or transit. This goal has been one of the core ideas animating private MaaS start-ups like Whim as well as public-sector versions like Jelbi, in Berlin.

Some scholars have proposed key performance indicators (KPIs) that could show whether MaaS services are delivering on these goals.

As one study noted, KPIs include rates of private car usage, vehicle kilometres travelled, fuel consumption, and CO_2 emissions levels (Zhao et al., 2021).

The degree of acceptance should also determine the level of public-sector engagement. If MaaS platforms can be configured and marketed such that they succeed in shifting modal splits towards greener forms (i.e., beyond business-as-usual levels), and do so without financially impairing public transit, then the case for public policy intervention is weak beyond regulation of the uses of data gathered by the MaaS platform. If, on the other hand, these conditions can't be met, local and regional governments may want to invest resources in this emerging sector, assuming they believe that the goal of decarbonizing urban mobility with MaaS is achievable and preferable to other approaches (i.e., opportunity costs).

Besides policy/governance goals relating to improved mobility and sustainability, there are other important considerations that should inform the development of MaaS. The design of a MaaS platform, for example, should take into account goals such as equity and inclusion; after all, a system that merely succeeds in further polarizing the travelling public can't be considered successful (see Palm and Farber, chapter 6, and Kong and Leszczynski, chapter 7).

As the World Bank's study on MaaS points out,

> [f]or MaaS to achieve the ambitious goals of improving mobility, accessibility, equity, and sustainability, it is important that governments prepare for and proactively shape mobility service and technology innovation through mode-agnostic regulations, space allocation, and pricing schemes that reward travel that is greener, more efficient, and inclusive. To achieve these goals, policies that disincentivize private car ownership and use are especially important for the success of MaaS. Financing and funding mechanisms must be made available as a result of these regulations. (Bianchi Alves et al., 2021, p. 16)

It seems clear that municipal transit operators themselves aren't well suited to establish a MaaS service because they have a built-in conflict of interest, even if they are tasked to deliver policy mandates that focus on reduced car use and congestion. A case in point: while most transit authorities now use smart card payment media, fare integration between multiple agencies operating within a given urban region in some regions remains a work in progress.

Arguably, the more appropriate locus for enabling and regulating MaaS service rests with municipal/regional governments that have an explicit mandate to manage transportation across a city. In many places, however, jurisdiction over transportation policy is highly fractured between local and regional governments, and among various municipal departments whose goals may not be well aligned or even consistent. For governance of MaaS to be effective, local or regional governments must begin by deciding what policy goals they wish to achieve, and then work backwards to design the platform so it provides the right incentives.

It is also worth noting that some city-regions may choose, as a matter of policy, to not invest resources and energy into the development of MaaS platforms. An interesting example is Transport for London (TfL), which was established in 2000 and reports directly to the mayor. TfL has overarching responsibility for all aspects of mobility within London, from transit to roads, financing, taxi regulation, cycling, walking, and even river transport on the Thames. TfL's carbon mandate, set out by the mayor, is to ensure that by 2041, 80 per cent of all trips will be made by walking, cycling, or transit, up from two-thirds as of 2022. "We use technology and data to make journeys easier," TfL's website says. "Oyster and contactless payment cards and information in different formats help people move around London" (TfL, n.d.).

Interestingly, London authorities have not rushed to either invest in a MaaS service or encourage a private one to establish a presence. As TfL pointed out, "London has a vision for the city, and it's not MaaS-specific" (Everitt, 2019, p. 11) The authority, though aware of investor and entrepreneurial interest in MaaS platforms, has opted to pursue its 2040 targets by expanding the reach of its contactless fare cards, improving payments, and leveraging system- and city-wide data gathered through TfL's various operations.

Conclusion

This chapter has outlined aspects of the evolution of MaaS platforms, including discussions about the complex questions they pose about how city-regions manage mobility in the era of climate change. These include whether MaaS should exist within a public policy framework, how to assess the success of a MaaS platform, and issues such as data and privacy.

It seems possible that as transit agencies continue to refine their smart cards and online payment/booking services, private mobility services will try to establish partnerships – a kind of organic evolution of MaaS that is very much in evidence in cities like Vienna, with the WienMobil MaaS platform. But as the example of London also suggests, the long-term goal of reducing the use of private automobiles in dense urban areas may also be achieved through land-use planning, infrastructure investment, service integration, governance, and the use of price mechanisms, such as cordon charges or parking levies. In other words, getting to the same destination as MaaS platforms, but using more conventional tools instead of relying on a smart city technology and an untested business model. MaaS if necessary, but not necessarily MaaS.

As of this writing (2022), MaaS remains a very young idea – just seven years old. Indeed, the well-established narrative of other nascent digital technologies suggests that it will require several cycles of growth, retrenchment, refinement, and business model innovation before MaaS matures into something that will become as accepted and useful as the transit smart card is today. Or perhaps it won't, and city-regions will choose instead to pursue other strategies to transform urban mobility in this era of climate change.

A handful of fundamental questions hovers over the future viability of MaaS as conceived: One, is it a solution in search of a problem? Two, why would public transit agencies participate? Three, is there evidence showing that the underlying motivation -- i.e., that such a platform will, in fact, reduce reliance on private automobiles -- bears out?

REFERENCES

Arias-Molinares, D., & García-Palomares, J.C. (2020). The Ws of MaaS: Understanding mobility as a service from a literature review. *International Association of Traffic and Safety Sciences Research*, *44*(3), 253–65. https://doi.org/10.1016/j.iatssr.2020.02.001.

Bianchi Alves, B., Wang, W., Moody, J., Waksberg Guerrini, A., Peralta Quiros, T., Velez, J.P., Ochoa Sepulveda, M.C., & Alonso Gonzalez, M.J. (2021). *Adapting Mobility-as-a-Service for developing cities: A context-sensitive approach.* World Bank Group. http://hdl.handle.net/10986/36787.

BlaBlaCar. (2021, 20 April). *BlaBlaCar announces a $115M funding to boost its growth ambitions.* https://blog.blablacar.com/newsroom/news-list/blablacar-announces-a-115m-funding-to-boost-its-growth-ambitions.

Brail, S. (2017, December). Promoting innovation locally: Municipal regulation as barrier or boost? *Geography Compass*, *11*(12), Article e12349. https://doi.org/10.1111/gec3.12349.

Busvine, D. (2019, 24 September). *From U-Bahn to e-scooters: Berlin mobility app has it all.* Reuters. https://www.reuters.com/article/us-tech-berlin-idUSKBN1W90MG.

Cottrill, C. (2020). MaaS surveillance: Privacy considerations in mobility as a service. *Transportation Research Part A: Policy and Practice, 131*(1), 50–7. https://doi.org/10.1016/j.tra.2019.09.026.

Crunchbase. (n.d.). *MaaS Global – Crunchbase company profile and funding*. https://www.crunchbase.com/organization/maas-finland.

Everitt, V. (2019). *MaaS: The story so far*. Transport for London. https://www.apta.com/wp-content/uploads/transit-ceos-2019-MaaS-The_Theory_So_Far.pdf.

Fenton, P., Chimenti, G., & Kanda, W. (2020). The role of local government in governance and diffusion of Mobility-as-a-Service: Exploring the views of MaaS stakeholders in Stockholm. *Journal of Environmental Planning and Management, 63*(14), 2554–76. https://doi.org/10.1080/09640568.2020.1740655.

Globe and Mail. (2022, 12 September). Magna looks to enter micromobility market with $77-million investment in EV startup Yulu. https://www.theglobeandmail.com/business/article-magna-looks-to-enter-micromobility-market-with-77-million-investment/.

Goodall, W., Fishman, T.D., Bornstein, J., & Bonthron, B. (2017). The rise of mobility as a service: Reshaping how urbanites get around. *Deloitte Review* (20), 113–29. https://www2.deloitte.com/content/dam/Deloitte/nl/Documents/consumer-business/deloitte-nl-cb-ths-rise-of-mobility-as-a-service.pdf.

Hensher, D.A., Ho, C.Q., Reck, D.J., & Smith, G. (2021, March). *The Sydney Mobility as a Service (MaaS) trial: Design, implementation, lessons and the future*. iMOVE Australia. https://imoveaustralia.com/wp-content/uploads/2021/04/iMOVE-Sydney-MaaS-Trial-Final-Report-March-2021.pdf.

Hoveskog, M., Bergquist, M., Esmaeilzadeh, A., & Blanco, H. (2022). Unpacking the complexities of MaaS business models – A relational approach. *Urban, Planning and Transport Research, 10*(1), 433–50. https://doi.org/10.1080/21650020.2022.2107564.

Infrastructure Canada. (2019, 27 February). *Montréal, Quebec*. https://www.infrastructure.gc.ca/cities-villes/videos/montreal-eng.html.

International Association of Public Transport [UITP]. (2019, April). *Mobility as a service*. https://cms.uitp.org/wp/wp-content/uploads/2020/07/Report_MaaS_final.pdf.

Jelbi. (n.d.). *Jelbi – Berlin's public transport and sharing services in one app*. https://www.jelbi.de/en/home/.

Jordans, F. (2019, 18 February). *Berlin to get single travel app for public, private services*. Associated Press. https://apnews.com/article/ba3b94e2fb5347ba91571d5679f3b4ee.

Kanda, T. (2019). *Advancing a mobility-as-a-Service: Lessons learned from leading-edge public agencies*. UCLA Institute of Transportation Studies. https://doi.org/doi:10.17610/T6G59J.

Lajas, R., & Macário, R. (2020). Public policy framework supporting "mobility-as-a-service" implementation. *Research in Transportation Economics, 83.* https://doi.org/10.1016/j.retrec.2020.100905.
MaaS Alliance. (n.d.). *Library.* https://maas-alliance.eu/library/.
MaaS Alliance. (2022, 11 October). *Mobility as a service is still alive.* https://maas-alliance.eu/2022/10/11/mobility-as-a-service-is-still-alive/.
Maas, B. (2022). Literature review of mobility as a service. *Sustainability, 14*(14), Article 8962. https://doi.org/10.3390/su14148962.
Matowicki, M., Amorim, M., Kern, M., Pecherkova, P., Motzer, N., & Pribyl, O. (2022). Understanding the potential of MaaS – A European survey on attitudes. *Travel Behaviour and Society, 27*(4), 204–15. https://doi.org/10.1016/j.tbs.2022.01.009.
Montréal. (2019). *Canada's smart city challenge: Final application by the City of Montréal.* https://laburbain.montreal.ca/sites/default/files/candidature_eng_smart_cities_challenve_vf.pdf.
NetApp. (n.d.). *What is XaaS (anything as a service)?* https://www.netapp.com/cloud-services/what-is-anything-as-a-service-xaas/.
New means of getting from A to B are disrupting carmaking. (2021, 11 April). The Economist. https://www.economist.com/business/2021/04/15/new-means-of-getting-from-a-to-b-are-disrupting-carmaking.
Rowe, B., & Bond, M. (2022, 12 July). Rental car industry struggling to bounce back after COVID-19 pandemic. *CityNews Toronto.* https://toronto.citynews.ca/2022/07/12/rental-car-industry-rebound-covid-pandemic/.
Surico, J. (2021, 5 December). Is an all-encompassing mobility app making a comeback? *New York Times.* https://www.nytimes.com/2021/12/05/business/maas-mobility-app.html.
TfL. (n.d.). *What we do.* https://tfl.gov.uk/corporate/about-tfl/what-we-do?intcmp=2582.
Trafi. (n.d.). *Cities powered by Trafi Mobility as a Service technology.* https://www.trafi.com/cities/.
van der Berg, V.A.C., Meurs, H., & Verhoef, E.T. (2022). Business models for Mobility as a Service (MaaS). *Transportation Research Part B: Methodological, 157*(3), 203–29. https://doi.org/10.1016/j.trb.2022.02.004.
Vij, A., & Dühr, S. (2022). The commercial viability of Mobility-as-a-Service (MaaS): What's in it for existing transport operators, and why should governments intervene? *Transport Review, 42*(5), 695–716. https://doi.org/10.1080/01441647.2022.2028032.
Walmsley, J. (2019, 26 June). *Watch out, Uber. Berlin is the new Amazon for transportation (with lower fares).* Forbes. https://www.forbes.com/sites/juliewalmsley/2019/06/26/watch-out-uber-berlin-is-the-new-amazon-for-transportation-with-lower-fares/?sh=aa5853a269b8.
Wiener Linien. (2021, 1 December). *FAQ WienMobil.* https://www.wienerlinien.at/web/wl-en/faq-wienmobil.

Whim Helsinki. (n.d.). *Plans to unlock Whim Benefits*. https://whimapp.com/helsinki/en/plans/.

Zhao, X., Andruetto, C., Vaddadi, B., & Pernestål, A. (2021). Potential values of maas impacts in future scenarios. *Journal of Urban Mobility, 1*, Article 100005. https://doi.org/10.1016/j.urbmob.2021.100005.

Zipper, D. (2021, 22 July). *Can Pittsburgh make 'mobility as a service' succeed?* Bloomberg. https://www.bloomberg.com/news/articles/2021-07-22/pittsburgh-borrows-a-mobility-concept-from-finland.

SECTION V: MOVING MOBILITY STUDIES FORWARD

15 Conclusion: Mobility in the Twenty-First-Century City

BETSY DONALD AND SHAUNA BRAIL

This book investigates the dramatic changes underway in mobility options and patterns in Canadian cities over the last decade and especially since the COVID-19 pandemic. In the book's central premise, we argue that mobility in the twenty-first-century city has undergone fundamental disruption. Disruption has been sudden, with the impact of COVID-19 and its societal consequences. Lockdowns, remote and hybrid work arrangements, and dramatic downturns in public transit use, especially due to reduced activity in central business districts, have changed mobility. But disruption has also been gradual, with government-led green initiatives to address climate change and firm-led innovations to reimagine mobility. Mobile and ICT technology platform companies are reshaping how we visualize, experience, and move people and goods in the city. While traditional and social media continue to explore the changing nature of mobility, there remains a need for systematic investigations into broader social, economic, and cultural forces and trends shaping mobility patterns in cities. Such research has far-reaching implications for how we plan and govern our cities.

Urban Mobility documents these changes through multiple scholarly perspectives. It brings together a diverse group of researchers to share their latest findings on mobility with particular attention on the Canadian city. Contributors come from a variety of disciplines, including geography, planning, urban studies, business, engineering, labour studies, political science, and journalism. The authors provide a mix of theoretically informed and empirically grounded perspectives on mobility changes in the twenty-first-century city. Together, the researchers make important observations for how we might think about mobility and the future of the truly equitable and sustainable, low-carbon city.

This analysis reveals three core, interrelated themes: (1) the interplay between "shocks" (e.g., COVID) and longer-term structural changes, (2) the power and effect of technological innovation, and (3) the growing climate crisis. After reviewing them separately, we offer some conclusive direction on how these three aspects can help us understand mobility in the twenty-first-century. We then discuss several challenges and implications around (a) questions of how to govern and regulate firm-inspired mobility experiments, (b) the challenge of reaching a more equitable mobility system for a greater number of people, and (c) the prospects for cities of the future. These ideas matter because they all reveal ways to move the agenda on urban mobility forward and demonstrate the power of a multidisciplinary approach as the themes arise throughout the chapters, despite varied approaches. We find some authors present an approach that prioritizes technology-based solutions to mobility studies whereas others focus primarily on government policy and the role of each city to find ways to deepen city engagement. Across all the cases, we see that scholars are bringing fresh solutions to the world's most pressing problems around mobility and its intersections with climate, infectious disease, and the resiliency of urban life.

The Interplay between Shocks and Structural Changes

In a first important theme, several chapters situate urban mobility within exogenous shocks and broader structural forces. Recent shocks such as the COVID-19 pandemic have impacted practices of governance, business, and community relations in profound ways. Cavalli and McGahan (chapter 2) review these in greater detail, including COVID-19's impacts on mobility and proximity, adverse mental-health impacts, and impacts on the structure of work and inflation.

In chapter 3, Hutton argues that many of the latest innovations in urban mobility, such as e-bikes and ride-hailing, represent an evolution of urban economies, especially the age-old link in cities between land use, employment, and housing. Despite recent challenges to commuting patterns exacerbated by the pandemic, Hutton suggests that traditional auto-dependent twentieth-century planning approaches still underlie much transportation planning and investment in Canadian cities. Hutton highlights several major structural changes that have shaped planning and investment in the city, including decades of financialization and gentrification and, more recently, the dire environmental consequences of climate change.

Similarly, in chapter 10, Brail and Donald demonstrate how past structural changes created the right conditions for powerful tech-enabled activities to flourish under recent shocks. For example, rising inequality, the gig economy, and flexible business regulations set the stage for an easy pivot when pandemic-inspired policy forced lockdowns and shutdowns. Hashemi, Motaghi, and Tremblay (chapter 9) also note that longer-term demographic changes in Montreal, with its large and growing immigrant and student populations, facilitate a ready workforce that helps the digital mobile economy to operate.

What is interesting about these chapters is how some authors attach greater emphasis to the broader structural forces (economic, social, political, cultural, and other changes) shaping our mobility patterns in cities, whereas others focus attention on shorter-term solutions tied to new mobility technologies that may or may not help cities adapt to broader disruptions.

The Power and Effect of Technological Innovation

Second, several authors highlight the phenomenon of global platform firms bringing new technological solutions to mobility challenges in cities. In chapter 5, Rojas and Miller survey the state of emerging mobility technologies in Canadian cities, including the rise of ride-hailing, bike-sharing, e-bike-sharing, e-scooters, and demand-responsive transit. McNee and Miller (chapter 8) review bike-sharing ridership trends over the COVID-19 pandemic. Aoyama and Alvarez León (chapter 12) discuss autonomous vehicles while Gorachinova, Huh, and Wolfe (chapter 11) and Lorinc (chapter 14) cover newer mobility on demand (MOD) or mobility as a service (MaaS) digital platforms that allow customers to book and pay for trips from multiple services (e.g., bus, taxi, bike-share) all in one place.

Except for the Montreal examples that Hashemi, Motaghi, and Tremblay (chapter 9) describe, such as Bixi, most of the technological innovations discussed in the book originated outside Canada. Given Canada's history of having a weak business innovation system (Britton, 2002), questions arise as to whether a strengthened policy framework could induce the adoption of made-in-Canada mobile technologies rather than just reacting to technology imported from elsewhere. How beneficial are these foreign mobility technology platforms to Canadian businesses, people, and places? Will Canada simply become a testing ground in the fields of planning and policymaking? Or will this imported technology spur Canadian companies to create new digital opportunities that offer Canada direct economic benefits? In addition to analysing the economic implications of foreign-owned technological innovations, several authors raise questions around other societal effects, such as issues of data surveillance, problems with data ownership (Vinodrai, chapter 4; Aoyama and Alvarez León, chapter 12; Rojas and Miller, chapter 5), and questions of who benefits from these technological experiments (Gorachinova, Huh, and Wolfe, chapter 11). There are also questions around the true environmental costs and prospects for sustainability surfaced by some of these technological solutions. Early entry of private-sector firms, for example, in the e-bike-sharing business left some cities to deal with a lot of e-waste after programs failed. Furthermore, the fact that the private sector has historically treated the environment as a free input can no longer be sustained in the context of the climate crisis.

The Climate Crisis

Cities are both victims of current climate emergencies as well as key contributors to climate change (Bulkeley, 2012; UNpeg.org, 2023). Several authors in the book allude to the role of mobility as a major source of greenhouse gas emissions. In Section II, Hutton, as well as Cavalli and McGahan, set the stage for how crisis can spur innovation, for instance, with respect to climate change. In general, McNee and Miller (chapter 8), Gorachinova, Huh, and Wolfe (chapter 11), Aoyama and Alvarez León (chapter 12), and Lorinc (chapter 14) are quite buoyed by the ability of new technological innovation to move cities closer to a low-carbon, sustainable future. However, other contributors like Hutton (chapter 3) and Kong and Leszczynski (chapter 7) are more cautious about technological fixes and point to the downfalls of cities relying too heavily on firms to lead mobility innovation given the disconnect between new mobility solutions, equity, and climate justice in cities. Regardless of the nuances in focus, most authors in this collection have either implicitly or explicitly stressed the value of clear government policy and action – at multiple levels – for facilitating or enabling a more climate-friendly urban mobility future.

The Technology-Pandemic-Climate Narrative

The heart of this book project was to connect the broad themes of technology, pandemic, and climate more explicitly as a way of understanding what has changed, and is continually changing, about mobility in the twenty-first-century Canadian city. Over the last few years, we have witnessed these trends and their patterns act in both compatible and contradictory ways as Canadian cities work towards a more effective, efficient, equitable, and sustainable approach for moving people and goods around cities. Human settlements have a history of developing within a framework of risk. In the Canadian context under nineteenth-century colonization, this meant locating colonial human settlement in the best geographic and climate locations, next to transportation routes via waterways along the Canada's southern border. By the mid-twentieth century, as cars displaced rail, suburban and exurban populations soared and so did unsustainable commuting and mobility patterns. By the end of the twentieth century, most large Canadian cities were investing heavily in intensification strategies and public transit systems to get more people out of cars and into more sustainable mobility options like transit, light rail, biking, and walking. However, with shifting priorities and budgetary adjustments also came government cutbacks to transit and a lack of new transit and transportation innovation. This left an opening for new firm-led mobility experimentation, including ride-hailing.

Mobility is an essential public good. The objective of contemporary mobility policies is to move as many as people as possible in a city in a non-polluting and safe way. However, the threat of infectious disease threw an unexpected set of obstacles into more traditional goals of sustainably moving large numbers of people. Private technology platforms were able to seize on the pandemic opportunity and create more effective and safe work-from-home options for those in professions that could work from home through online technologies, private transportation ride-hailing, and home goods deliveries. The intensification of these responses had the unintended consequence of cutting into the traditional economic model that depends on transit, physical retail, and the movement of people into geographic workplace regions. Transit systems not only depend on the massive built public infrastructure but also fare box revenues to maintain the public system over time. Retail and the small support businesses (lunch counters, dry-cleaners, business supply shops) in the workplace regions also suffered because fewer people were travelling to those regions to go to work, and they were having many more goods delivered to their home.

What initially seemed like a "pause" from normal pre-pandemic work, home life is increasingly giving way to a new normal around work-life balance and mobility patterns shifting as more people order retail online, decline to use public transit, and no longer frequent as often the central business districts and the accompanying retail and commercial spaces. There is no question that the pandemic disrupted urban life and urban economies. As we ponder both the conclusion to this volume and thoughts of what is ahead, the realities of climate change continue to be ever more present in our daily lives. The intensification of wildfires and extreme weather events reminds us daily of the urgent need to develop and implement sustainable solutions to our mobility needs in Canadian cities.

Challenges to the Technology-Pandemic-Climate trends

The next section reviews some of the challenges and implications related to these new technology-pandemic-climate patterns and trends. This section explores questions around governance and government, data governance, the equity challenges, and options for the future.

Questions of Governance and Government

Before the pandemic, ride-hailing through large-scale private platform companies like Uber and Lyft represented one of the more disruptive mobility options to enter Canadian cities for at least a decade. Prior to their entry, governments had been under financial pressure to cut back on public transportation options and investments. There was also a heavily regulated taxi system, which was understood to be broken for quite some time. When ride-hailing firms entered Canadian cities in 2012, governments reacted in two main ways. Some municipalities embraced the innovative ideas with little regulation or on-the-fly regulation of different transport modes. Other municipalities resisted these innovations to protect existing mobility infrastructure and services, and associated jobs. Under COVID-19, when more pressing urban issues arose and new patterns of mobility emerged, both approaches were tested. In chapter 13, Zwick, Young, and Spicer argue that most debates around ride-hailing regulation ended due to policy convergence, not COVID-19, even though municipalities will adjust regulations in the future. By contrast, Brail and Donald (chapter 10) note that COVID-19 revealed the shortcomings of municipalities moving quickly to regulate ride-hailing without seeing the pervasive role played by ride-hailing firms as technology platforms. This role revealed itself especially in the pivot from ride-hailing to food-hailing during the pandemic, with mobility platforms impacting labour, small business, and physical infrastructure.

Data Governance

Many localities both before and during the pandemic faced the same challenge: they lacked information about the ways platforms were emerging and shifting their focus from moving people to moving goods (Davies, Donald, & Gray, 2023). In chapter 4, Vinodrai writes about the difficulty that governments and researchers encounter when they try to access timely, comparable, and high-quality data for Canadian cities. Large multinational platforms like Google, Uber, and DoorDash retain considerable information, making it increasingly difficult for cities to assess, plan, and prepare. McNee and Miller (chapter 8) also touched on these data challenges, as did Lorinc (chapter 14) and Aoyama and Alvarez León (chapter 12).

In terms of governance, this book documents many experiments at play (see, for example, McNee and Miller, chapter 8, Aoyama and Alvarez León, chapter 12, and Gorachinova, Huh, and Wolfe, chapter 11) but few clear-cut cases of cities at work to improve their mobility options for the largest number of people in their city. Cavalli and McGahan (chapter 2) call on cities to deepen their engagement, recalibrate core systems, and reconstitute innovation priorities to solve the most pressing problems of the day. Their call for city action makes us think about the importance of staying focused, quite literally in your lane, but also working collaboratively – that is, with local and global businesses, governments, citizens – in the best interest of the public in terms of mobility options for a low-carbon and equitable future.

The Equity Challenge

The question of who benefits from mobility experiments leads to another key challenge: equity. COVID-19 has exposed the challenges of reaching more equitable mobility in Canadian cities (Fischer & Winters, 2021). In their study on prospects for public transit equity in the Greater Toronto Area, Palm and Farber (chapter 6) discover that the pandemic altered transit ridership firmly along lines of socio-economic privilege and socio-spatial advantage. Public transit ridership was indeed decimated in the first six months of the pandemic, but for newcomers, people with low incomes, and people without vehicles, transit ridership continued, particularly in those neighbourhoods where walking and cycling were poor alternatives.

In chapter 7, Kong and Leszczynski examine a new pilot of dockless, shared e-scooters in Calgary and find that the initiative benefited those from wealthier parts of the city who had the means to experiment. The authors also argue that the spatial inequity of dockless e-scooter sharing also has implications for climate justice as the two phenomena of climate and inequality are profoundly interlinked.

Building on Hutton's narrative regarding the classic link between employment, housing, and transportation investment, Palm and Farber note that Canadian cities typically built transit lines directing flows of people towards the central business district. In our largest cities, however, investment has lagged in an "everywhere-to-everywhere" network for the growing inner-suburban and non-driving suburban population. This challenge speaks to the urgent need for policymakers to plan cities for the effective, efficient, and equitable movement of people.

The Prospects for Future Cities

Finally, we ask, what is the city of the future? As several authors noted in this book, crises have a way of generating hardships but also stimulating innovation. Experiences in Canadian cities of disease and deaths, dramatic climate events, and growing signs of poverty and dislocation have also sparked conversations about the future of cities and cities of the future. A common theme throughout the book is that cities (and by implication city governments) have a lot to answer for in terms of making our cities vibrant, innovative, sustainable, safe, and prosperous again. Traditionally, city governments have managed transit systems and basic land-use planning, property taxation, and basic mobility regulation, but now in this new mobility era, they seem tasked with a variety of other more complex and world-pressing challenges. As several authors note in this collection, cities take the leadership in experimenting with and regulating new mobility solutions, and they do so in a variety of forms – from city as promoter, to regulator, to arbitrager, to data catalyst (Aoyama and Alvarez León, chapter 12). In some examples, cities are nimble and flexible with innovative ideas whereas in other examples cities have chosen to lock into a particular mobility model because of past investments and practices. What seems clear is that city governments cannot find these solutions on their own as they lack the motivation of efficiency and innovative capacity that the private sector can provide. But for too long the private sector has treated nature and some labour inputs as free. This is no longer appropriate in our double crisis of growing inequality and the climate state of emergency (Donald and Gray, 2019). We need innovative solutions with all players working together.

Moving the Agenda Forward

This book has covered a lot of ground on mobility in the Canadian city that is both empirically grounded and theoretically informed. In this closing section we identify areas for future research. These areas include deeper engagement with global comparisons so Canadian cities can learn from other places. This book brought in several examples from outside Canada to shed light on future directions for more sustainable mobility solutions (Gorachinova, Huh, and Wolfe, chapter 11; Aoyama and Alvarez León, chapter 12; Lorinc, chapter 14). But more work needs to be done. Furthermore, the topic of politics emerged in several papers, but again, more research is needed into the key role that urban, regional/state, and national politics can play in directing and dramatically changing new mobility solutions. Accessibility and proximity are another topic that warrants further study, especially considering changes brought about because of the pandemic (Cavalli and McGahan, chapter 2; Rode et al., 2017). Having access to good publicly available data is another challenge (Vinodrai, chapter 4), especially as many of the changes are driven by private-sector actors who are reluctant to share their data on mobility trends in Canadian cities. We need up-to-date information on commuting trends so researchers and policymakers can know the impacts on the environment, workers, and the public infrastructure such as roads, public transit, bike lanes, and walkways. We also do not yet know the extent to which downtowns might recover, and how our aging and new immigrant populations will navigate our sprawling landscapes without access to safe and affordable transportation and decent housing options. Like mobility, accessibility and proximity are central to the things we want to do in terms of access to jobs, friends, family, and recreation – all experiences wrapped up in the identities of our cities. This collection has sought to start this conversation, and now we must leverage this moment to address long-standing problems and reimagine our cities as places that work for all people while at the same time finding the best way forward to protect against the frequencies of climate emergencies.

This book is one of the first to integrate the iPhone-COVID-climate narrative into thinking about mobility in the twenty-first-century Canadian city. These tripartite trends and shocks – (1) new and innovative disruptive mobility technologies, (2) "shocks" to the system like the pandemic, and (3) the ever-growing and urgent climate crisis – are producing new patterns and trends that have implications for how we currently move around our cities and also how we need to plan for our mobility futures. Recovery, while slow, is evident (Statistics Canada, 2023). However, most of the recovery patterns still mirror traditional auto-dependent travel models, with over eight out of ten commuters across Canada using a car, truck, or van to travel to work in May 2023. In the cities of Toronto and Vancouver, the number of car commuters in the Toronto and Vancouver Census Metropolitan Areas in May 2023 surpasses May 2016 pre-pandemic levels. The share of long commutes (defined as over sixty minutes) rose from 2021 to 2023 but is still down from 2016. Workers returned to the office, but the proportion of people working from home is greater than pre-pandemic. This has implications for downtowns and central business districts and the supporting infrastructure, including transit, in many cities across the country. As discussed throughout the book, the patterns in large cities are also increasingly delineated along socio-economic privilege and socio-spatial advantage. Our cities are becoming more unequal, and our mobility patterns reflect that.

In planning for our cities of the future, we are at a crossroads. We can no longer imprint primarily sprawling, auto-dependent settlement patterns on the Canadian urban landscape. Our social and natural environments are too fragile, and our future survival requires a new resilient vision. We need a clear, multijurisdictional approach to develop communities that are walkable, liveable, inclusive, safe, affordable, and sustainable. This means all players must come to the table and work together – the federal, provincial, and municipal governments, the private sector, and non-profits. Each actor contributes in unique ways: the private sector with innovative solutions, non-profits with their deep knowledge and investments in people and place, and the public sector, which values public goods like equity, justice, climate, health, and safety. This is not an easy task, but our future depends on embracing bold and innovative urban ideas.

REFERENCES

Britton, J. (2002). Does nationality still matter? The new competition and the foreign ownership question revisited. In *The new industrial geography* (pp. 260–86). Routledge.

Bulkeley, H. (2012). *Cities and climate change*. Routledge.

Davies, A., Donald, B., & Gray, M. (2023). The power of platforms – Precarity and place. *Cambridge Journal of Regions, Economy and Society, 16*(2), 245–56, Article rsad017. https://doi.org/10.1093/cjres/rsad017.

Donald, B., & Gray, M. (2019). The double crisis: In what sense a regional problem? *Regional Studies, 53*(2), 297–308. https://doi.org/10.1080/00343404.2018.1490014.

Fischer, J., & Winters, M. (2021). COVID-19 street reallocation in mid-sized Canadian cities: Socio-spatial equity patterns. *Canadian Journal of Public Health, 112*, 376–90. https://doi.org/10.17269/s41997-020-00467-3.

Rode, P., Floater, G., Thomopoulos, N., Docherty, J., Schwinger, P., Mahendra, A., & Fang W. (2017). Accessibility in cities: Transport and urban form. In G. Meyer & S. Shahee (Eds.), *Disrupting mobility: Impacts of sharing economy and innovative transportation on cities* (pp. 239–73). https://doi.org/10.1007/978-3-319-51602-8_15.

Statistics Canada. (2023, 22 August). *Commuting to work by car and public transit grows in 2023.* The Daily. Retrieved from https://www150.statcan.gc.ca/n1/en/daily-quotidien/230822/dq230822b-eng.pdf?st=enOLydX9.

UNep.org. (2023). *Cities and climate change.* Retrieved from www.unep.org/explore-topics/resource-efficiency/what-we-do/cities/cities-and-climate-change.

Contributors

Luis F. Alvarez León (PhD, UCLA 2016) is assistant professor of geography at Dartmouth College. He is a political economic geographer with substantive interests in geospatial data, media, and technologies. His work integrates the geographic, political, and regulatory dimensions of digital economies under capitalism with an emphasis on technologies that manage, represent, navigate, and commodify space. Ongoing research projects examine the geographic transformations surrounding the emergence of autonomous vehicles and the industrial and geopolitical reconfigurations resulting from the proliferation of small satellites.

Yuko Aoyama, PhD, is professor of geography at Clark University. She is an economic/industrial geographer with expertise in globalization, industrial organization, technological innovation, and cultural economy. Her research interest lies in developing geographic understandings of global capitalisms from institutional and comparative perspectives. Her current interests include advancing research on the platform economy, mobility transition (autonomizing and electrifying vehicles), and deglobalization. She has been awarded an Abe Fellowship from the Social Science Research Council, a Bellagio Academic Residency from the Rockefeller Foundation, and research grants from the National Science Foundation, National Geographic Society, and the Association of Asian Studies.

Shauna Brail is an associate professor at the Institute for Management & Innovation, University of Toronto Mississauga, and holds a cross-appointment at the Munk School of Global Affairs and Public Policy, University of Toronto. As an economic geographer and urban planner, her research focuses on the transformation of cities as a result of economic, social, and cultural change. Professor Brail's research encompasses studies of broad urban economic challenges associated with twenty-first-century cities – including the impacts of COVID-19 on cities; the relationship between cities and the digital platform economy, with a particular emphasis on mobility; and shifts in urban governance, policy, and planning in connection to innovation and technological change.

Gabriel Cavalli is a PhD student in the Strategic Management Area at the Rotman School of Management, University of Toronto. His current research investigates how scientific knowledge production is changing organizationally and worldwide. Gabriel holds an MSc degree in economics and a BBA degree, both obtained from Bocconi University (Milan).

Betsy Donald is a professor in the Department of Geography and Planning and associate vice-principal of research at Queen's University. She does research in the field of economic geography with a focus on innovation and regional economic development, urban planning and governance, and sustainable food systems. She is currently an editor of the *Cambridge Journal of Regions, Economy and Society*. Betsy is an active researcher and currently co-principal investigator of a SSHRC Insight Grant along with Professor Shauna Brail (University of Toronto) entitled "Taking Canada for a Ride? Digital Ride-Hailing and Its Impact on Canadian Cities."

Steven Farber is an associate professor in the Department of Human Geography at the University of Toronto Scarborough. He is the director of the Mobilizing Justice partnership and the Suburban Mobilities Cluster of Scholarly Prominence. He is also associate director of the University of Toronto Mobility Network, an Institutional Strategic Initiative.

Elena Gorachinova currently works as a market research analyst assessing Canadian and global socio-economic challenges, trends, and opportunities, and was a post-doctoral fellow at the University of Toronto's Innovation Policy Lab during the course of this research. She holds a PhD from the Department of Political Science and an MA in geography from the University of Toronto.

Moe Hashemi is an assistant professor at the Commerce Department, Mount Allison University. Previously, he was an assistant professor at Concordia University, a part-time lecturer at McGill University, and a post-doc at HEC Montreal. In his research program, Moe is interested in how businesses and non-profit organizations collaborate with various stakeholders to meet society's expectations. Moreover, he is investigating how technological advancements in different areas affect people and businesses and call for changes in public policies and government regulations. Since 2020, Moe Hashemi has been involved in a nationwide project funded by the Social Sciences and Humanities Research Council (SSHRC) on digital mobility platforms, collaborating with five Canadian universities. He is now extending his research program using a newly awarded fund from New Brunswick Innovation Fund (NBIF) in the Atlantic Canada region.

Lisa Huh is a Master of Global Affairs candidate specializing in global markets and innovation policy at the Munk School of Global Affairs and Public Policy at the University of Toronto. Her research interests broadly include sustainability, technology, and entrepreneurship – particularly the role of policy in promoting growth across these areas.

Tom Hutton was educated at the University of British Columbia (modern European history and urban geography) and the University of Oxford (DPhil in geography & anthropology). At the City of Vancouver he worked as a development policy specialist, with an emphasis on Vancouver's engagement with the cities of the Asia-Pacific. At UBC Hutton taught senior undergraduate courses and master's-level courses in the School of Community & Regional Planning. Tom Hutton's principal research outputs include a series of articles and chapters on Vancouver's post-staples/post-industrial development record, with Craig Davis and David Ley; studies of Asian cities (exemplified by *New Economic Spaces in Asian Cities: From Industrial Restructuring to the Cultural Turn*, co-edited with P.W. Daniels and K.C. Ho, 2012); extended study of London's planning and development record (*Millennial Metropolis: Space, Place and Territory in the Remaking of London*, 2022); and continuing work on Metropolitan Vancouver's economy, including collaboration with Trevor Barnes.

Vivian Kong obtained her MSc in geography from Western University. Her thesis explored the spatial equity of micromobility sharing platforms in Canadian cities. Her research interests include digital and urban environments, and the policies surrounding them. As a master's student, Vivian received the Edward G. Pleva Fellowship Award. She is proficient in GIS and statistical analyses using various programs such as ArcGIS and QGIS. Vivian Kong is currently a research associate with

Structure Research, an independent Toronto- and Singapore-based research and consulting firm devoted to the cloud and data centre infrastructure services markets.

Agnieszka Leszczynski is an associate professor in the Department of Geography at Western University. She is one of the editors of *Environment and Planning F: Theory, Philosophy, Models, Methods and Practice*, an article forum editor for *Dialogues in Human Geography*, and a former co-editor of *Big Data & Society*. Her current research examines intensifying intersections of platforms and cities.

John Lorinc is a Toronto journalist and editor. He writes about cities and climate for various publications, including *The Globe and Mail*, *Corporate Knights*, and *Spacing*. Lorinc is the author of four books, including *Dream States: Smart Cities, Technology and the Pursuit of Urban Utopias* (2022), which won the 2022 Writers' Trust Balsillie Prize for Public Policy.

Lisa L. Losada-Rojas is currently an assistant professor at the University of New Mexico. She was a post-doctoral fellow at the University of Toronto – Mobility Network working with Professor Eric Miller. Lisa obtained her PhD and master's degree from Purdue University. Her bachelor's degree in civil engineering was awarded by the National University of Colombia. During her graduate studies, she received several awards and recognitions, such as being named a 2021 MIT CEE Rising Star and 2021 LATinE: Latinx Trailblazers in Engineering Fellow. Her research line aims to advance communities' public health through transportation solutions.

Anita M. McGahan is a Professor and George E. Connell chair in organizations and society at the University of Toronto. Her current research emphasizes private-sector entrepreneurship and innovation in the public interest. McGahan's credits include five books and over 175 articles, case studies, notes, and other published material on competitive advantage, industry evolution, global health, and governance innovation.

Spencer McNee completed his master of applied science at the University of Toronto under the supervision of Dr Eric J. Miller. Spencer's master thesis used microsimulation modelling to examine how Bike Share Toronto network operations could be optimized. Prior to his master's, Spencer worked in infrastructure construction in Toronto. Spencer is interested in all things transportation, housing, and infrastructure.

Eric J. Miller (BASc, MASc, University of Toronto; PhD MIT) is a faculty member in the Department of Civil & Mineral Engineering, University of Toronto, and is Director of U of T's Mobility Network. He is past chair of the US Transportation Research Board (TRB) Committee on Travel Behavior and Values; member emeritus of the TRB Transportation Demand Forecasting Committee; has been a member or chair of over fifty modelling peer review panels; and is recipient of the 2018 IATBR Lifetime Achievement Award. His primary research area is the development of agent-based microsimulation models of urban systems for sustainable policy analysis.

Hamed Motaghi is an associate professor of business technology management at the University of Quebec (Outaouais-UQO). He is managing director of International Entrepreneurship Knowledge Hub (https://ie-knowledgehub.ca/). Prior to his position at UQO, he was a post-doctorate fellow in the marketing department at the Desautels Faculty of Management at McGill University. He holds a PhD with distinction where his dissertation focused on the relationship between creativity and information technology achieved through a joint PhD program administered at University of Quebec of Montreal (UQAM). He also holds a master's degree from University of Paris Dauphine, France, and an engineering degree from University of Sorbonne Université Sciences, Paris. In addition, he was adjunct professor at UQAM, Royal Military College of Canada, HEC Montreal, and Lawrence Technological University, MI, US. His main research interests are innovation/creativity, creative and cultural industries, and the role of technology, ethnic entrepreneurship, and internationalization process of firms.

Matthew Palm is an assistant professor in city and regional planning at the University of North Carolina Chapel Hill. His research covers transportation equity and his prior work spans housing and transportation topics in Australia, Canada, and the United States. This work was supported in part by funding from the Social Sciences and Humanities Research Council's partnership grant: Mobilizing Justice: Towards Evidence-Based Transportation Equity Policy.

Zachary Spicer is an associate professor in the School of Public Policy and Administration and the head of New College at York University in Toronto, Canada.

Diane-Gabrielle Tremblay is professor of labour economics, innovation, and human resources management at Téluq University (www.teluq.

ca/dgtremblay). She was appointed Canada Research Chair on the socio-economic challenges of the knowledge economy from 2002 to 2016 (http://www.teluq.ca/chaireecosavoir/) and director of a CURA on work-life issues in 2009 (www.teluq.ca/aruc-gats). Recently she was admitted as a member of the Royal Society of Canada and of the centre of excellence of University of Québec. She has been invited professor at Université de Paris I, Sorbonne, Université of Lille, of Toulouse, and Lyon 3, Louvain-la-Neuve, HEC, and Liège universities, in Belgium, University of Social Sciences of Hanoi (Vietnam) and the European School of Management. She has done research on the platform economy and multimedia cluster, on telework, coworking, and new forms of work.

Dr Tara Vinodrai is director, Master of Urban Innovation Program, and associate professor, Institute for Management and Innovation at the University of Toronto. She holds a graduate appointment to the Department of Geography and Planning and is a faculty associate of the Innovation Policy Lab at the Munk School of Global Affairs and Public Policy. Dr Vinodrai's research focuses on the dynamics of innovation, economic development, labour markets, and technological change in cities. She has provided advice to both large cities and smaller communities and all levels of government related to innovation, economic development, and regional competitiveness.

David A. Wolfe is professor of political science at the University of Toronto Mississauga and Co-Director of the Innovation Policy Lab at the Munk School of Global Affairs and Public Policy. He has been principal investigator on two SSHRC major collaborative research initiatives and recently completed a six-year SSHRC-funded partnership grant entitled *Creating Digital Opportunity for Canada*. He has been a research associate for the Canadian Institute for Advanced Research (CIFAR) and is the editor or co-editor of ten books and numerous scholarly articles.

Mischa Young is an assistant professor in the Department of Urban Environments at the Université de l'Ontario français.

Austin Zwick is an assistant teaching professor in public administration and international affairs in the Maxwell School of Citizenship and Public Affairs at Syracuse University. He is also a senior research associate at the Autonomous Systems Policy Institute at Syracuse University.

Index

Page numbers in italics represent figures.

www.ingramcontent.com/pod-product-compliance
Ingram Content Group UK Ltd.
Pitfield, Milton Keynes, MK11 3LW, UK
UKHW030706300425
458010UK00001B/92